AF589342

ADVANCES IN NUMERICAL ANALYSIS

VOLUME IV

Theory and Numerics of Ordinary and Partial Differential Equations

Edited by

M. AINSWORTH, J. LEVESLEY,
W. A. LIGHT, and M. MARLETTA

Department of Mathematics and Computer Science,
University of Leicester

CLARENDON PRESS · OXFORD

This book has been printed digitally and produced in a standard specification in order to ensure its continuing availability

Great Clarendon Street, Oxford OX2 6DP

Oxford University Press is a department of the University of Oxford.
It furthers the University's objective of excellence in research, scholarship, and education by publishing worldwide in

Oxford New York

Auckland Cape Town Dar es Salaam Hong Kong Karachi
Kuala Lumpur Madrid Melbourne Mexico City Nairobi
New Delhi Shanghai Taipei Toronto

With offices in

Argentina Austria Brazil Chile Czech Republic France Greece
Guatemala Hungary Italy Japan South Korea Poland Portugal
Singapore Switzerland Thailand Turkey Ukraine Vietnam

Oxford is a registered trade mark of Oxford University Press in the UK and in certain other countries

Published in the United States by Oxford University Press Inc., New York

Reprinted 2011

ISBN 978-0-19-851193-9

Printed and bound in Great Britain by CPI Antony Rowe, Chippenham and Eastbourne

Preface

This volume contains the lecture notes of the six invited speakers at the Sixth SERC Summer School in Numerical Analysis, held at the University of Leicester from the 18th to the 29th of July, 1994. This Summer School was the first in the series to be held in Leicester, the previous five having taken place in Lancaster.

The format and purpose of the Summer School were essentially unchanged from 1992: in each of the two weeks a different subject area was covered, with three invited speakers in each week. Each invited speaker presented five lectures: these were designed to be accessible to beginning graduate students and to progress to a point where, by the final lectures, current research problems could be described. The first week was titled 'Partial Differential Equations', with Claes Johnson, Ian Sloan and Andrew Stuart as the invited speakers; in the second week, 'Ordinary Differential Equations', George Corliss, Linda Petzold and Marino Zennaro presented their lectures. There was also a programme of seminars during each week, which made a substantial contribution to the success of the meeting.

Although the Summer School itself was divided into two very different weeks, we have not perpetuated this dichotomy in the presentation of the lecture notes: they are arranged by alphabetical order of author, rather than in the order in which they were delivered. We are very grateful to the authors for facilitating the publication process by providing their lecture notes in TeX or LaTeX. We are also grateful to Oxford University Press for their assistance in the preparation of the notes for publication.

The successful running of the Summer School depended, as always, on the appointment of 'local experts' for each week: the local experts were in charge of organizing the seminar programme and were also responsible for chairing the sessions, either by acting as chairmen themselves or by delegating this task to some hapless colleague. The local expert for the first week was Ivan Graham, of the University of Bath, to whom we extend our thanks. The local expert for the second week was Marco Marletta: he found organizing nine seminars for the Summer School much easier than persuading the rest of the Organizing Committee to write five paragraphs, for this Preface, upon which we could all agree.

Finally, it is a pleasure to acknowledge the substantial grant we received from the Engineering and Physical Sciences Research Council. This covered the accommodation and subsistence costs of all the UK participants, as well as all the costs of the invited speakers. Without the financial assistance of EPSRC the Summer School would not have been possible.

Leicester *M. Ainsworth, J. Levesley, W. Light, M. Marletta*
December 1994

Contents

Contributors

George F. Corliss Department of Mathematics, Statistics and Computer Science, Marquette University, Milwaukee, Wisconsin 53233, USA.

Kenneth Eriksson Department of Mathematics, Chalmers University of Technology, 412 96 Göteborg, Sweden.

Don Estep School of Mathematics, Georgia Institute of Technology, Atlanta, Georgia 30332, USA.

Peter Hansbo Department of Mathematics, Chalmers University of Technology, 412 96 Göteborg, Sweden.

Claes Johnson Department of Mathematics, Chalmers University of Technology, 412 96 Göteborg, Sweden.

Linda Petzold Department of Computer Science, University of Minnesota, Minneapolis, Minnesota 55455, USA.

Ian Sloan Department of Applied Mathematics, University of New South Wales, P.O. Box 1, Kensington, New South Wales 2033, Australia.

Andrew Stuart Program in Scientific Computing and Computational Mathematics, Division of Applied Mechanics, Durand 252, Stanford University, California 94305-4040, USA.

Marino Zennaro Dipartimento di Scienze Matematiche, Università di Trieste, 34100 Trieste, Italy.

Guaranteed Error Bounds for Ordinary Differential Equations

George F. Corliss

Department of Mathematics, Statistics, and Computer Science,
Marquette University,
Milwaukee,
Wisconsin 53233,
USA

Abstract

Hamming once said, "The purpose of computing is insight, not numbers." If that is so, then the speed of our computers should be measured in insights per year, not operations per second. One key insight we wish from nearly all computing in engineering and scientific applications is, "How accurate is the answer?" Standard numerical analysis has developed techniques of forward and backward error analysis to help provide this insight, but even the best codes for computing approximate answers can be fooled.

In contrast, validated computation

- checks that the hypotheses of appropriate existence and uniqueness theorems are satisfied,
- uses interval arithmetic with directed rounding to capture truncation and rounding errors in computation, and
- organizes the computations to obtain as tight an enclosure of the answer as possible.

These notes for a series of lectures at the Sixth SERC Numerical Analysis Summer School, Leicester University, apply the principles of validated computation to initial value problems in ordinary differential equations (ODEs). We show from concepts familiar to a student of ODEs that validated enclosures are feasible. The concepts of interval analysis are developed from their beginnings, with emphasis on the fundamental problem of computing guaranteed bounds for the range of a function defined on an interval. Much of the discussion of interval techniques is supported by Maple worksheets. We discuss the algorithms for validated solution of ODEs by defect control and by Lohner's AWA program.

A list of open questions invites the reader to join the "interval mafia."

1 Introduction

Consider the initial value problem (IVP)

$$u' = f(t, u), \; u(t_0) = u_0, \text{ with } u \in \mathbb{R}^n. \tag{1.1}$$

We will usually assume the f is analytic in some neighbourhood of the point (t_0, u_0), although weaker conditions could be used. We wish to guarantee the existence of a unique solution to eqn (1.1) and to compute an approximate solution $\widehat{u}(t)$ with guaranteed bounds for the global error $e(t) := u(t) - \widehat{u}(t)$.

This paper is organized for exposition as notes for a sequence of lectures. I assume that the reader is familiar with the standard elementary numerical analysis and numerical methods for ODEs. I assume no previous knowledge of interval analysis. You will learn

- What does it mean to compute validated error bounds?
- How can that be done?
- When are validated bounds worth the cost?

This paper is not organized for reference, nor is it organized in the order you need to attack a problem. It is organized more as a tutorial introduction to validated methods for ODEs. At times, I will tell some half-truths. I will not purposely tell any lies, but I *will* leave many details for the interested reader to follow in the references.

A survey article often attempts to catalogue all known methods for the problem being surveyed. Rihm [56] has written an excellent survey in this style. Since Rihm's survey is available, I have preferred here to emphasize motivation, rationale, and heuristics. I focus primarily on the most widely used software for the validated solution of ODEs, AWA (Anfangswertaufgabe) by Rudolf Lohner [41, 42, 43, 44, 45]. After reading this paper, the reader should be well prepared to read Rihm's survey, or any of several other surveys of interval techniques for ODEs [7, 15, 53, 56, 60, 61]. Extensive bibliographies of literature in the field may be found in the survey papers and in [18, 16], so we do not give an exhaustive bibliography here.

The organization of this paper roughly follows the software lifecycle. In Section 2, we examine the requirements for guaranteed global error bounds or validated solutions. We discuss what that means, when one might be willing to bear the cost, and look at pictures of some examples.

In Section 3, we address the question, "Is it *possible* to compute lower and upper bounds for the solution of an IVP?" Simple exponential functions or chaotic problems from dynamics show that our most optimistic requirements cannot be met, but Picard–Lindelöf iterations and methods based on Taylor's Theorem *are* able to meet more modest, but still useful, requirements.

In Section 4, we introduce the tools of interval arithmetic and automatic differentiation that are required to make validated techniques practical. We also mention some of the more common computing environments for validated computation.

It is easy to get an enclosure of a solution. It is often *much* harder to get *tight* enclosures. In Section 5, we describe several strategies commonly used in self-validating algorithms.

Section 6 forms the central body of the paper, where we describe three different approaches to computing validated enclosures of u. The first approach is based on classical work in the area of differential inequalities as collected by Lakshmikantham and Leela [39] and by Walter [62]. This approach characterizes conditions (including all one-dimensional systems) under which guaranteed global error bounds can be computed by solving two or more point problems.

The second approach considers the defect $d(t) := (\widehat{u})' - f(t, \widehat{u})$. The approximate solution $\widehat{u}$ for the original problem (1.1) is the *exact* solution of the nearby problem

$$u' = f(t, u) + d(t)\,. \tag{1.2}$$

If $||d(t)||$ is "sufficiently small", then the nearby problem (1.2) may be just as satisfactory a model for the physical problem being modeled as was the original problem (1.1).

The third approach is the most general and the most practical. We outline the algorithms used in Lohner's AWA program. Lohner uses a Picard–Lindelöf iteration to guarantee the existence of a unique solution and to provide a coarse a priori bound. Then he uses a Taylor series to compute tighter enclosures and coordinate transformations to reduce the wrapping effect.

Throughout this paper, we denote real, point-valued quantities by Roman letters. Hence, the true solution of Problem (1.1) is denoted by u or $u(t)$. We denote interval-valued quantities by either square brackets or (especially in computer programs) upper case letters. Hence, an enclosure of the true solution is denoted by $[u]$, $[u]\,(t)$, or U. The enclosure of the solution is an interval-valued function with the property that the set of all true solutions lies within the interval bounds, as Section 2 describes in more detail.

Much of the explanations, code segments, and figures were prepared using Maple [11] worksheets and procedures. The LaTeX source file for this paper, the Maple worksheets used to prepare it, and bibliographies are available by anonymous ftp from `boris.mscs.mu.edu` in subdirectory `pub/corliss/ValidODE`, or by sending electronic mail to the author. I especially encourage the interested reader to get and run the Maple worksheets to reproduce the results shown in this paper.

2 Requirements: validated solutions

Purposes of this section:

What does it mean to compute a "validated solution"?
When might one want to do so?

Neumaier [48] characterizes the purposes of validated scientific computation as

- Obtain rigorous verification of existence and uniqueness of solutions, and
- Find guaranteed enclosures with narrow bounds.

He emphasizes that 1) validation, 2) guaranteed enclosures, and 3) tight enclosures are three separate issues. In this section, we address the first: validation. Section 3 addresses issues of validation and guaranteed enclosures. Section 4 describes tools for guaranteed enclosures. Section 5 discusses strategies for making the enclosures tight.

2.1 What is a "validated solution" of an ODE?

Given the problem

$$\begin{aligned}
&u' = f(t, u)\,, \\
&\text{Initial time } t_0 \text{ and final time } t_f \text{ (we assume } t_0 < t_f)\,, \\
&u(t_0) \in [u_0]\,, \\
&\text{a Tolerance} > 0\,, \text{ and} \\
&w([u_0]) < \text{Tolerance} \quad (w(\cdot) \text{ denotes the width of an interval})\,,
\end{aligned}$$

a computer program should either

- Guarantee that there exists a unique solution u in the interval $[t_0, t_f]$,
- Compute an approximate solution $\widehat{u}(t)$ and an interval-valued error function $[e]\,(t)$ such that

  $$\begin{aligned}
  &u(t, u_0) \in \widehat{u}(t) + [e]\,(t)\,, \text{ and} \\
  &w([e]\,(t)) <= \text{Tolerance}\,, \text{ for all } t \in [t_0, t_f], \text{ and all } u_0 \in [u_0]\,,
  \end{aligned}$$

  or equivalently,
- compute an interval-valued function $[u]\,(t) = \widehat{u}(t) + [e]\,(t)$ such that

  $$\begin{aligned}
  &u(t, u_0) \in [u]\,(t)\,, \text{ and} \\
  &w([u]\,(t)) <= \text{Tolerance}\,, \text{ for all } t \in [t_0, t_f], \text{ and all } u_0 \in [u_0]\,,
  \end{aligned}$$

or else

- Notify us that the problem has *no* solution in $[t_0, t_f]$, or
- Notify us that the problem has *more than one* solution in $[t_0, t_f]$,

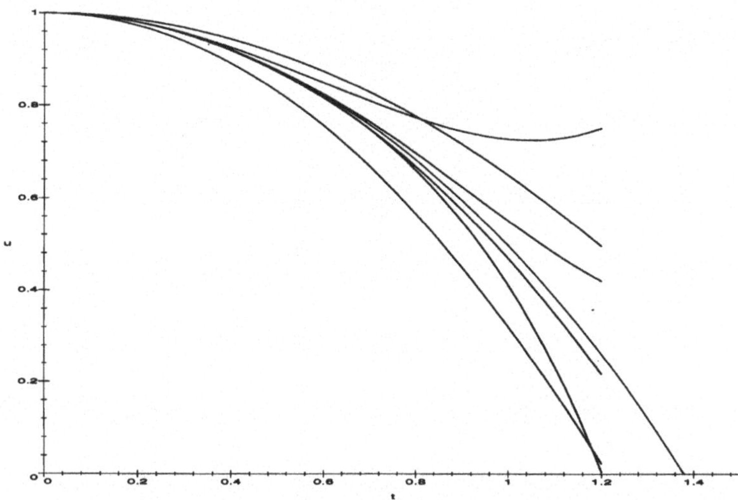

FIG. 1. Enclosures of the solution.

or else

- Notify us that the program cannot do so.

Of course, we can easily meet these requirements by always returning the last alternative, but we would not have a particularly helpful program!

In discussing interval techniques, it is important to distinguish in our notation and in our thinking which quantities are intervals and which are points. We use the notation $[a]$ throughout to denote interval objects, and we will often redundantly describe objects as interval-valued, real-valued, or point-valued to emphasize their natures (their data types, in computer science jargon). The endpoints of the interval $[a] = [\underline{a}, \overline{a}]$ are denoted by $\underline{a} = \text{Inf}\,(a)$ and $\overline{a} = \text{Sup}\,(a)$, respectively.

By "validated solution," we mean an interval-valued function $[u]\,(t)$ such that the real-valued exact solution $u(t, u_0)$, for every point $u_0 \in [u_0]$, lies within $[u]\,(t)$ for each $t \in [t_0, t_f]$, as shown in Fig. 1. In case the right-hand-side f depends on interval-valued parameters $[\alpha]$, the exact point-valued solution for every parameter value $\alpha \in [\alpha]$ must be contained in $[u]\,(t)$. That is, we consider the set of all possible point problems; we consider forming the set of all exact, point-valued solutions; a "validated solution" must enclose them all.

A validated solution encloses all possible point-valued solutions. However, a validated solution also encloses functions that *are not* solutions of any point problem. There may be overestimations in the widths of some

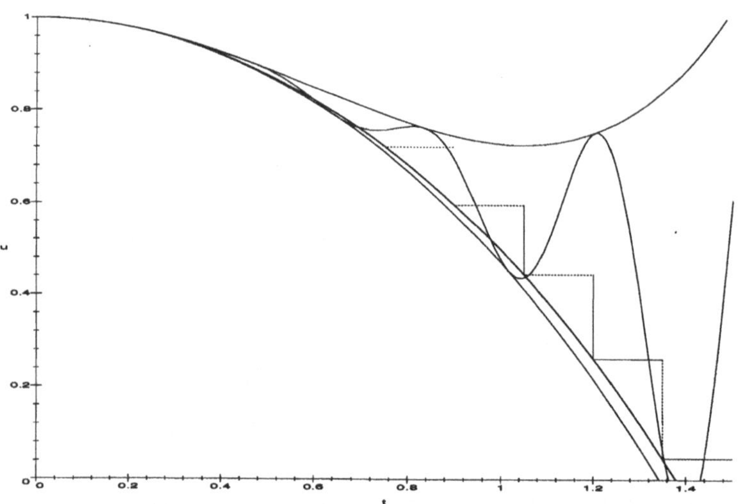

FIG. 2. Validated solutions contain non-solutions, too.

enclosures. More fundamentally, though, the interval-valued function $[u]$ also encloses functions that have no relationship to the ODE or that are nowhere differentiable (see Fig. 2).

Figure 2 shows a lower bound, an upper bound, the solution function, a rapidly oscillating function, and a discontinuous function, all contained within the bounds. The point is that the validated solution contains all solutions, but the validated solution also contains other, non-solution, functions.

2.2 When are validated solutions needed?

It is *not* the case that all numerical solutions should be validated in the sense described above.

- Existing software for computing approximate solutions with error estimation is usually very robust and reliable.
- Existing software for computing approximate solutions is much more varied and flexible than the corresponding software for validated enclosures.

The computational cost of computing validated enclosures is often higher than the cost of computing approximate solutions. Next, we will characterize situations in which the guarantees provided by validated solutions are worth the cost.

2.2.1 *Academic interest*

The reason many academic researchers work in the fields we do is simply because it is fun, and we find the concepts interesting. Interval techniques are intellectually satisfying, use interesting and deep mathematical results for computing useful answers, and are relatively accessible to graduate and undergraduate students.

2.2.2 *Critical computations*

Scientific computing is frequently applied to problems where incorrect results may be responsible for serious loss of life, injury, or major property damage. Even very high quality software may sometimes yield wrong answers. For example, the conditions of use for LANCELOT, one of the leading packages for large-scale, nonlinear optimization, state "It [LANCELOT] should not be relied on as the basis to solve a problem whose incorrect solution could result in injury to persons or property." [12]

For critical computations, we prefer software that provides guaranteed bounds for the errors committed during the computations. Such software has limitations. There may be errors in the algorithms used and bugs in the program. The program may return bounds that are so wide as to be useless, the program may be so slow that the bounds cannot be feasibly computed, or the program may be unable to handle the problem of interest.

Even with these limitations, software for computing validated enclosures eliminates one of the possible modes of failure of the software. In principle, any software claiming to compute an approximation can return a catastrophically incorrect answer. In principle, software claiming to compute an enclosure cannot be fooled in that way.

2.2.3 *Interval or uncertain data*

Some ODEs arise as purely mathematical problems. For example, the first Painlevé transcendent,

$$u'' = 6u^2 + t \tag{2.1}$$

characterizes the first of six classes of second order ODEs having solutions whose only movable singularities are poles (see [30]). The "6" in eqn (2.1) is exact; the solution to

$$u'' = (6 + 10^{-100})u^2 + t$$

does not have the desired property.

In contrast, most ODEs arising from models in biological, chemical, physical, engineering, or other applications lack this precision. ODEs from applications typically contain parameters whose values are at best only approximately known, and we wish to determine the behaviour of the solutions for a range of parameter values. This may suggest hundreds or thousands of simulation runs, especially if there are several such uncertain parameters. Interval techniques are very attractive for ODEs with inexactly known parameters because we can compute an interval-valued solution that is guaranteed to enclose solutions arising from any combination of parameter values in the range of interest. However, the widths of physically motivated input intervals are usually not small enough to allow the execution of Lohner's program for nonlinear ODEs. The interval techniques can be much faster than repeated simulation runs. However, since the validated solution is a worst case analysis, the results are sometimes viewed as being too wide to be of practical use.

2.2.4 *Accuracy is in doubt*

The cost of computing validated solution enclosures may be justified for problems where the accuracy of approximate solutions computed by conventional means is suspect.

2.2.5 *Algorithm development*

Interval techniques may be valuable tools to use in the process of developing conventional algorithms for approximate solutions because they can help distinguish correct answers from *almost* correct answers. For example, if we know that the correct answer is zero, and our approximate algorithm for some problem yields $-1.23456\mathrm{E}{-17}$ in double precision, we might be tempted to accept the result. However, if the "validated solution" for some problem is $[-1.23457\mathrm{E}{-17}, -1.23455\mathrm{E}{-17}]$, then we are guaranteed that our program is wrong, and we must continue debugging. That is, interval techniques can assist the development of reliable algorithms by making our mistakes painfully obvious.

2.2.6 *Prove existence of a periodic solution*

Of particular interest in the context of dynamical systems, interval techniques have been used (see [3, 4, 5]) to prove the existence of both stable and unstable periodic solutions of certain chaotic systems by showing that the Poincairé return map has a fixed point. Another application is to gear drive vibrations where Adams used enclosure algorithms to discover that the standard model in the literature has an infinite number of periodic solutions with the same period T.

3 Feasibility

Purposes of this section:

Demonstrate from familiar mathematical concepts that computing a "validated solution" is in some sense feasible.

In Section 2, we set the goal of computing an approximate solution $\widehat{u}(t)$ and an interval-valued error function $[e]\,(t)$ such that

$$u(t, u_0) \in \widehat{u}(t) + [e]\,(t)\,, \text{ and}$$

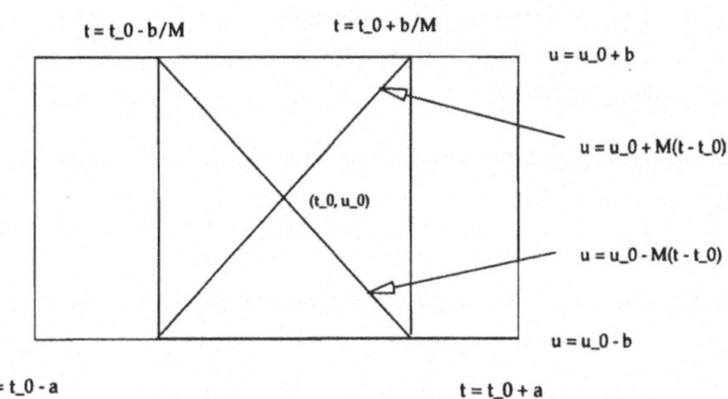

FIG. 3. Domain D for Existence and Uniqueness Theorems.

$$w([e](t)) <= \text{Tolerance}, \quad \text{for all } t \in [t_0, t_f], \text{ and all } u_0 \in [u_0] .$$

In principle, this is impossible for some problems. In practice, it is *very* difficult in others. In this section, we refine the concept of a validated solution initially presented in the previous section. We review the classical theory of existence and uniqueness of solutions. We show how enclosures may be computed by Picard–Lindelöf iterations and by using Taylor series plus a remainder. Then we discuss some of the limitations of methods for computing validated solutions in order that we can appreciate their performance on the wide range of problems for which it is feasible to compute validated solutions.

3.1 Existence and uniqueness

The classical theory of existence and uniqueness of solutions for general, nonlinear ODEs

$$u' = f(t,u), \ u(t_0) = u_0$$

can be found in most standard texts (see eg. Ince [30] or Butcher [10]). Let f be defined and continuous in a rectangular domain $D := \{(t,u) \in \mathbb{R} \times \mathbb{R}^n : |t - t_0| \leq a, ||u - u_0|| \leq b\}$. Then f satisfies a *Lipschitz condition* in u if there exists a Lipschitz constant L such that

$$||f(t,u) - f(t,v)|| \leq L||u - v|| \text{ for all points } (t,u) \text{ and } (t,v) \in D .$$

Figure 3 illustrates the domain D, where in the figure, `t_0` denotes t_0, and u_0 denotes u_0. The outer box is the domain D bounded by intervals centered at the point (t_0, u_0). The diagonal lines and the smaller box illustrate the need to restrict the time interval in the case that M is large enough that $b/M < a$.

Theorem 3.1. (Existence Theorem) *Let f be continuous in D, and let $M := \max_{(t,u)\in D} ||f(t,u)||$. Then there exists a solution $u(t)$ of eqn (1.1) for $|t - t_0| \leq \min(a, b/M)$.*

Theorem 3.2. (Uniqueness Theorem) *If, in addition to the hypotheses of the Existence Theorem, f also satisfies a Lipschitz condition in u, then $u(t)$ is unique.*

As a consequence of these theorems, if our computer programs can find an interval $[t_0, t_1]$ and an interval $[u_0 - b, u_0 + b]$ for which

(1) f is continuous in $D := [t_0, t_1] \times [u_0 - b, u_0 + b]$,
(2) $||f|| \leq M$ in D,
(3) $t_1 = \min(t_1, t_0 + b/M)$, and
(4) all slopes $\dfrac{||f(t,u) - f(t,v)||}{||u - v||} \leq L$ for all points (t, u) and $(t, v) \in D$,

then we can guarantee that eqn (1.1) has a unique solution. After we have introduced interval arithmetic in the next section, we will use the Lorenz system as an example to show exactly how it is possible for our programs to do this validation.

The proofs of both the Existence and the Uniqueness Theorems rely on the operator

$$\Phi(u)(t) := u_0 + \int_{t_0}^{t} f(\tau, u(\tau))\, d\tau\,. \tag{3.1}$$

A more computationally useful formulation uses the Banach Fixed Point Theorem.

Theorem 3.3. (Banach Fixed Point Theorem) *Let $X = (X, d)$ be a nonempty metric space. Suppose that X is complete and let $\Phi : X \to X$ be a contraction on X. Then Φ has precisely one fixed point.*

In the space of continuous functions $u : [t_0, t_1] \to \mathbb{R}^n$, the Picard–Lindelöf iteration map given by eqn (3.1) is a contraction for t_1 sufficiently close to t_0. Hence, if we have an interval-valued function $[u]_0(t)$ with $[u_0] \subset [u]_0(t_0)$ that satisfies

$$[u]_1(t) := \Phi([u]_0)(t) \subset [u]_0(t) \text{ for all } t \in [t_0, t_1]\,,$$

then we have verified that Problem (1.1) has a unique solution for every $u_0 \in [u_0]$. Further, we must have $u(t) \in [u]_0(t)$ (and also $u(t) \in [u]_1(t)$). Rihm [56] discusses the Picard–Lindelöf Theorem [28] in the context of error enclosures.

3.2 Enclosures

Once we have verified from the Existence and Uniqueness Theorems that the eqn (1.1) has a unique solution $u(t)$ in some time interval, we can turn to the issue of finding an enclosure for $u(t)$. Our discussion follows the outlines of Lohner's AWA program. We get a very rough enclosure as a by-product of validating existence and uniqueness. Then the second stage of the algorithm yields tighter enclosures.

Point: There is no magic to getting enclosures for $u(t)$, although there is room for considerable cleverness in computing *tight* enclosures. Some details must wait until we have introduced interval arithmetic in the next section; here we attempt to convey the flavor of two techniques for computing enclosures.

3.2.1 *Picard–Lindelöf iteration*

In the process of showing that f satisfies the hypotheses of the Existence and Uniqueness Theorems, we found intervals $[t_0, t_1]$ and $[u_0 - b, u_0 + b]$ such that in the region $D := [t_0, t_1] \times [u_0 - b, u_0 + b]$, $||f|| \leq M$, and f satisfies a Lipschitz condition with Lipschitz constant L. Then, since $u' = f(t, u)$,

$$\int_{t_0}^{t} -M\, d\tau \leq \int_{t_0}^{t} u'(\tau)\, d\tau = \int_{t_0}^{t} f(\tau, u(\tau))\, d\tau \leq \int_{t_0}^{t} M\, d\tau \text{ for } t \in [t_0, t_1],$$

and it follows that

$$u_0 - M(t - t_0) \leq u(t) \leq u_0 + M(t - t_0)\,. \tag{3.2}$$

Inequalities are to be understood component-wise. Equation (3.2) is our first, crude, enclosure for $u(t)$, enclosing $u(t)$ in a triangular region as shown in Fig. 4. No interval "magic" is required beyond the computation of the bound $||f|| \leq M$. Tighter bounds are possible. For example, if the ODE is a system, the lines $\pm M(t - t_0)$ in eqn (3.2) can be replaced by component-wise lower and upper bounds for the range of f. The point here is that it is easy to get guaranteed bounds for $u(t)$.

3.2.2 *Taylor series plus remainder*

We can do better. We wish to form a truncated Taylor series for $u(t)$ and enclose its remainder.

Theorem 3.4. (Taylor's Theorem) *Let $u : \mathbf{R} \to \mathbf{R}$ have p derivatives on an interval $[t_0, t_1]$. Then there exists some point $\tau \in (t_0, t_1)$ such that*

$$u(t) = u(t_0) + u'(t_0)(t - t_0) + \cdots + \frac{u^{(p-1)}(t_0)}{(p-1)!}(t - t_0)^{p-1} + \frac{u^{(p)}(\tau)}{p!}(t - t_0)^p\,.$$

Does our solution $u(t)$ have derivatives? We know that

- Equation (1.1) has the unique solution $u(t)$,

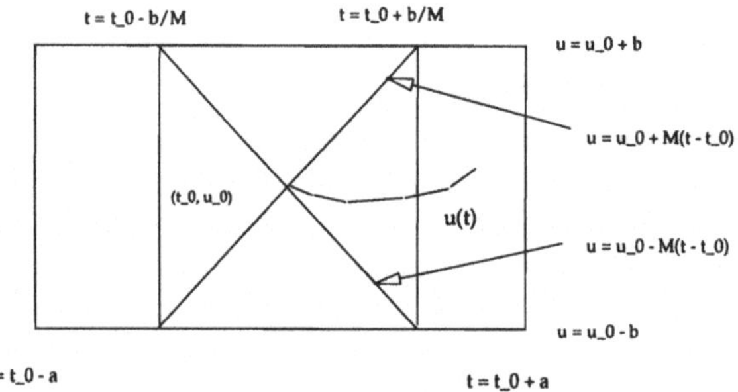

FIG. 4. Enclosure for $u(t)$.

- $u'(t) = f(t, u(t))$, and
- $u''(t) = \dfrac{\partial f(t,u(t))}{\partial t} + u'(t)\dfrac{\partial f(t,u(t))}{\partial u}$.

Do the partial derivatives of f exist? If f is given by a computer program consisting of a composition of $+$, $-$, $*$, $/$, exp, ln, sqrt, sin, cos, etc., with no branches, and if we can evaluate f on D (as we must have done in order to bound it by M), then the only ways f, and hence $u(t)$, can fail to be *analytic* in a neighbourhood are to include in the evaluation of f either

(1) sqrt (0), or
(2) $(\text{non-positive})^{(\text{positive, non-integer})}$.

Of course, the solution may have singularities such as poles, branch points, or essential singularities that restrict the interval of analyticity.

Mathematicians worry about terribly ill-behaved functions. For example, a function may have 1000 derivatives in a neighbourhood, but the 1001-st derivative may have poles, branch points, or essential singularities. Ill-behaved functions can occur in applications. However, if we restrict our programs to what Moore called "the functions most commonly used for computing" [47] with no program branches (no `if` statements, or equivalent), if we can evaluate a function in its domain, and if two easily checked conditions do not occur, then the function may be assumed to be analytic. That is, the functions we can easily compute are *much* better behaved than the full mathematical generality would suggest.

Assuming that neither of the two nasty conditions 1) sqrt (0), or 2) $(\text{non-positive})^{(\text{positive, non-integer})}$ arise in the evaluation of f in D (we can check that assumption), then $u(t)$ is analytic, so it admits a Taylor series. In the next section, we will find the Lipschitz constant L by bounding the Jacobian (all first-order partial derivatives of f), so we can at least say that u'' exists and is bounded by

$$\begin{aligned} \|u''(\tau)\| &= \left\| \frac{\partial f(\tau, u(\tau))}{\partial t} + u'(\tau) \frac{\partial f(\tau, u(\tau))}{\partial u} \right\| \\ &\leq \left\| \frac{\partial f(\tau, u(\tau))}{\partial t} \right\| + \|u'(\tau)\| \cdot \left\| \frac{\partial f(\tau, u(\tau))}{\partial u} \right\| \\ &\leq L + M * L\,. \end{aligned}$$

(The Lipschitz constant L referes to the dependency of f on u, not the dependency of f on t.) Let $u_0 = u(t_0)$, and $u_0' := u'(t_0) = f(t_0, u_0)$. Then

$$u(t) = u(t_0) + u'(t_0)(t - t_0) + \frac{1}{2} u''(\tau)(t - t_0)^2$$

is bounded by

$$\begin{aligned} &u_0 + u_0'(t - t_0) - \frac{1}{2} L(M + 1)(t - t_0)^2 \\ &\quad \leq u(t) \qquad\qquad (3.3) \\ &\quad \leq u_0 + u_0'(t - t_0) + \frac{1}{2} L(M + 1)(t - t_0)^2 \text{ for } t \in [t_0, t_1]\,. \end{aligned}$$

Inequality (3.3) gives guaranteed bounds for u that are parabolic in t in contrast to the linear bounds given by Inequality (3.2). If we prefer to think of an approximation and guaranteed error bounds,

$$\begin{aligned} &\text{Let } \widehat{u}(t) := u_0 + u_0'(t - t_0)\,, \text{ and} \\ &-\frac{1}{2} L(M + 1)(t - t_0)^2 \leq e(t) := u(t) - \widehat{u}(t) \leq \frac{1}{2} L(M + 1)(t - t_0)^2\,. \end{aligned}$$

We can again do better by using component-wise bounds and by using higher order Taylor series, but we have enclosed $u(t)$ with no interval "magic" beyond that required to bound the Jacobian of f.

This section has demonstrated that we can verify in a computer program that a system of ODEs has a unique solution in a neighbourhood of the initial conditions. We have shown how to get enclosures of the solution. In order to get tighter enclosures, we need to know more about the tools of interval arithmetic and automatic differentiation, to which we turn our attention in the next section. We conclude this section by considering some of the limitations of enclosure methods. The point is to gain a better understanding of what we mean by a "validated enclosure" of a solution.

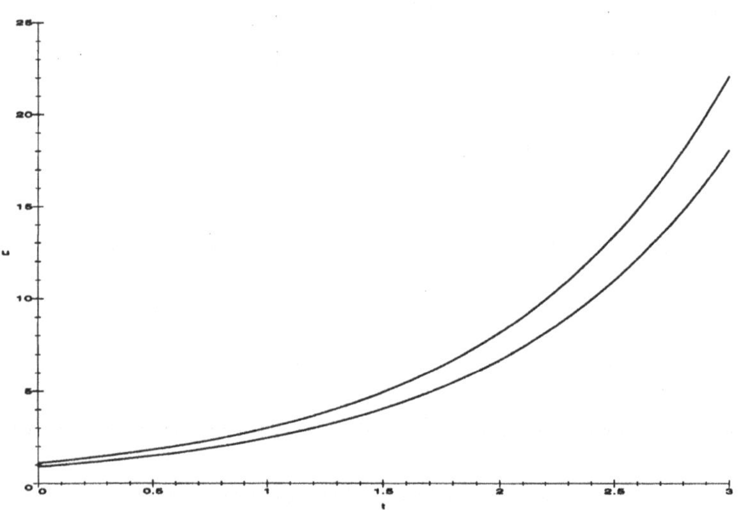

FIG. 5. Exact solutions can diverge: $u' = u$, $u(0) \in [u_0]$.

3.3 When we are doomed to fail

As our goal is stated, we are doomed to fail. True solutions may diverge to the point where the set of solutions cannot be contained in an interval of width $\leq$ Tolerance. Consider the equation

$$u' = u\,, \text{ with } u_0 \in [u_0]\,. \tag{3.4}$$

If $[u_0]$ is thick ($w([u_0]) > 0$), then the two true solutions $u(t, \mathrm{Inf}\,([u_0]))$ and $u(t, \mathrm{Sup}\,([u_0]))$ diverge at the rate of $w([u_0])e^t$ as shown in Figs 5 and 6. Figure 5 shows $0.9e^t$, and $1.1e^t$, the true solutions for the two problems $u' = u$, $u(0) = 0.9$, and $u' = u$, $u(0) = 1.1$. Hence, the set of true solutions for $u' = u$, $u(0) \in [0.9, 1.1]$ must enclose these graphs. Figure 6 shows the error enclosure. If we let $\widehat{u}(t) := e^t$, then the error $e(t) \in [-0.1, 0.1]e^t$. For purposes of illustration, Fig. 6 shows Tolerance $= 1$.

For ODEs whose solutions diverge like those of eqn (3.4), even if we commit no truncation errors in our algorithm and no roundoff errors in our computation, we cannot possibly keep $w([e]\,(t)) \leq$ Tolerance beyond $t = \ln\left(\dfrac{\text{Tolerance}}{w([u_0])}\right)$. Of course, our algorithm does commit truncation errors, and our computations do commit roundoff errors. Our enclosures must enclose these sources of error, as well as the entire set of true solutions. Hence, a more modest goal for problems whose true solutions diverge from one another is to see how large we can make t_f before the enclosed solution

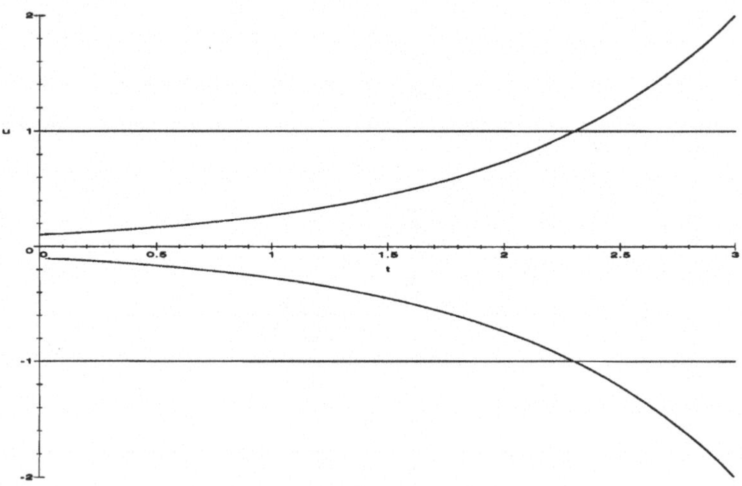

FIG. 6. Error bounds vs Tolerance: $u' = u$, $u(0) \in [u_0]$.

exceeds the allowable width.

The trouble with eqn (3.4) is the thick interval-valued initial condition $[u_0]$ whose width is amplified as the solution propagates. If we restrict ourselves to thin initial conditions (point-valued, or $w([u_0]) = 0$), the situation is theoretically more favourable, but not practically so.

As another example that poses difficulties for interval techniques, consider chaotic dynamical systems such as the Lorenz systems

$$\begin{aligned} x' &= \sigma(y - x)\,, \\ y' &= rx - y - xz\,, \\ z' &= xy - bz\,, \end{aligned} \tag{3.5}$$

whose solution is shown in Fig. 7. One of the definitions of a chaotic system is that it exhibits a sensitive dependence on the initial conditions. Hence even with point-valued initial conditions $x(0) = y(0) = z(0) = 10$, truncation and roundoff errors inevitably force a slight error. If we never committed any further errors, the computed solution would eventually diverge chaotically from the true trajectory. Hence, any enclosure must eventually become wide. In principle, we can reach a prescribed t_f before the error exceeds Tolerance, but arbitrarily high precision may be required to do so.

Further, we cannot know in advance what precision will be required to reach a prescribed t_f. We can begin to compute an enclosure. If the width of the enclosure exceeds Tolerance before we arrive at t_f, we go back and

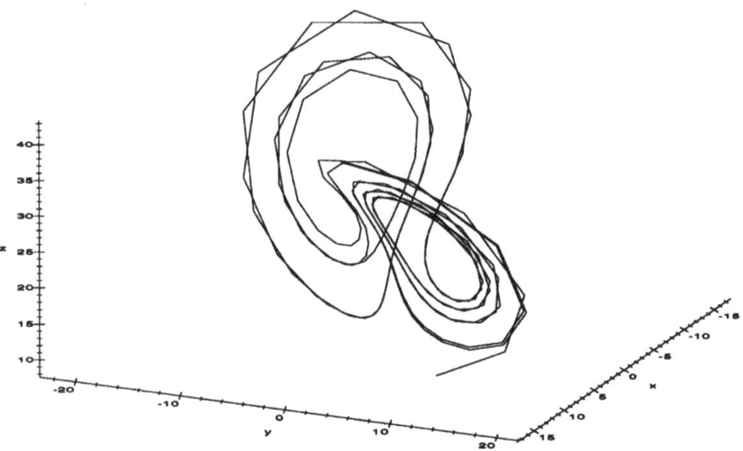

FIG. 7. Lorenz system for $\sigma = 10$, $r = 28$, $b = 8/3$, and $t \in [0, 10]$.

start over with a higher computational precision. We can attempt to do this in a clever, efficient manner, but it is not possible to place an a priori bound on the computational effort that may be required.

In spite of this rather gloomy situation, enclosure methods *have* proven their merit in the study of chaotic systems. For example, Adams and Kühn have used enclosure methods to prove the existence and uniqueness of unstable periodic solutions [3, 4, 5]. Spreuer and Adams [58] have verified and enclosed a homoclinic orbit for the Lorenz system. The game is to see how large a value of t_f we can reach before the enclosure method breaks down.

4 Tools

Purposes of this section:

Interval arithmetic computes guaranteed range bounds.

There are a variety of robust software environments supporting interval arithmetic.

Differentiation arithmetic computes numerical values for derivative objects by recursion, not symbolically, and not by finite differences.

In Section 3, we showed that it is feasible to compute guaranteed error bounds for a solution to a system of ODEs. There, we hinted at the central role of interval arithmetic to compute bounds on the ranges of f and its partial derivatives. In this section, we give a brief introduction to interval arithmetic and to the tools of automatic differentiation that make it painless to compute high-order Taylor coefficients.

4.1 Interval arithmetic

At this point, we begin to introduce into this paper worksheets prepared using Maple V, release 3 [11]. It is more instructive to download the Maple files by ftp from `boris.mscs.mu.edu` (134.48.4.4), subdirectory `pub/corliss/ValidODE/Maple` (or send e-mail to the author), and execute them interactively than simply to read them here. These worksheets will eventually be submitted to the Maple share library. To save space in this paper, I have *omitted some steps.* That means that typing the commands shown here will not give the results shown here. If you wish to reproduce the results shown here, please get the worksheets.

4.1.1 *Maple worksheet intpak_1.ms*

The fundamental capability of interval arithmetic is to compute, with certainty, on a real computer, bounds for the range of values of an expression. We will see that to compute guaranteed bounds is easy; to compute guaranteed bounds that are also tight can be more difficult. Here, we concentrate first on computing bounds; in Section 5, we will discuss how to make them tighter.

Interval arithmetic was widely publicized by Ramon Moore beginning with his book in 1966. The details of the concepts sketched here can be found in many textbooks on interval analysis. For example, see [6, 46, 47, 49, 55].

We denote real, point-valued objects as variables: a, and interval-valued objects either in interval braces: $[a]$, or as upper case variables: A. The endpoints of the interval $[a] = [\underline{a}, \overline{a}]$ are denoted by $\underline{a} = \mathrm{Inf}\,(a)$ and $\overline{a} = \mathrm{Sup}\,(a)$, respectively.

Theoretical interval operations are defined in terms of power sets, for example

$$A + B := \{a + b : a \in A \text{ and } b \in B\},$$
$$\sin(aX + B) := \{\sin(a * x + b) : a \text{ is a point}, x \in X \text{ and } b \in B\}.$$

Maple knows some things about intervals. `Shake` and `evalr` represent intervals as a list of a range.

```
> readlib (evalr):
> A := [4 .. 6];
> B := [-2.1 .. -1.2];     # Integers or floating-point
> 'A + B' = evalr ( A + B );
> 'A * B' = evalr ( A * B );
```

$$A := [\,4..6\,]$$
$$B := [\,-2.1..-1.2\,]$$
$$A + B = [\,1.9..4.8\,]$$
$$AB = [\,-12.6..-4.8\,]$$

```
> 'sin (A)' = evalr (sin (A));     # Takes a bit longer
```

$$\sin(A) = [-1..\sin(6)]$$

There are several situations in which `evalr` is not faithful to the power set model for interval arithmetic. The first is because Maple is fundamentally a SYMBOLIC processor. By default, it simplifies an expression before it evaluates the expression. The correct answer is $[4,6] - [4,6] = [-2,2]$.

```
> 'A - A' = evalr ( A - A );
```

$$0 = 0$$

Another failure comes from `evalr`'s rounding of floating-point results to the nearest floating point number.

```
> A := [4.1 .. 6.2];
> 'sin (A)' = evalr (sin (A));
```

$$A := [4.1..6.2]$$

$$\sin(A) = [-1..-.08308940282]$$

```
> Digits := 20:
> # OUTSIDE the interval given by evalr in the 11th place:
> 'sin (6.2)' = sin (6.2);
> Digits := 10:
```

$$\sin(6.2) = -.083089402817496578001$$

Point: A practical floating-point implementation of interval arithmetic MUST support directed rounding in order to capture all roundoff errors.

INTPAK [13]

One can work around these "lies" told by evalr, but Connell and Corless have contributed a Maple share library package intpak [13]. Intpak uses the more familiar notation [4.0, 6.0] for intervals.

```
> with (share): readshare (intpak, numerics):
> interface (echo = 0): read ('loadquie'): # "Load quietly"
> load ('intpak_2'):        # Corliss extension to share/intpak
> with (intpak):
> A := [4.0, 6.0];
> B := construct (-2.1, -1.2);
> 'A + B' = A &+ B;
> 'A * B' = A &* B;
> 'sin (A)' = &sin (A);
```

$$A := [4.0, 6.0]$$

$$B := [-2.1, -1.2]$$

$$A + B = [1.899999999, 4.800000001]$$

$$AB = [-12.60000001, -4.799999999]$$

$$\sin(A) = [-1., -.2794154980]$$

Intpak does directed rounding so that the power set result is contained in the interval that is displayed. Roundoff errors do not cause intpak to lie:

```
> A := [4.1, 6.2];
> 'sin (A)' = &sin (A);
> Digits := 20:  sin (A[Sup]);  Digits := 10:
```

$$A := [4.1, 6.2]$$

$$\sin(A) = [-1., -.08308940280]$$

$$-.083089402817496578001$$

For a computer implementation, the power set definitions of the operations are replaced by definitions that take advantage of knowledge of monotonicity of the operation, and then round the endpoints outwardly. For example

$$\begin{aligned} A + B &:= \{a + b : a \in A, \text{ and } b \in B\} \\ &= [\mathrm{Inf}(A) + \mathrm{Inf}(B), \mathrm{Sup}(A) + \mathrm{Sup}(B)] \\ &\subset [\text{Round down}(\mathrm{Inf}(A) + \mathrm{Inf}(B)), \\ &\qquad \text{Round up}(\mathrm{Sup}(A) + \mathrm{Sup}(B))]. \end{aligned}$$

COMPUTING RANGES OF FUNCTIONS

Now we know how interval arithmetic works, we are ready to use it to compute guaranteed lower and upper bounds for the range of values of a function. As an example, consider the function

```
> # See [Heck1993a], p. 292:
> f := t -> exp (-t^2) * sin (Pi * t^3);
> T := construct (0.5, 2.0);
```

$$f := t \to e^{(-t^2)} \sin(\pi t^3)$$

$$T := [.5, 2.0]$$

For this function, minima and maxima occur only at critical points. Maple can do that ... It should be easy ... From calculus:

```
> df := D(f);
> solve (df(t) = 0, t);
```

$$df := t \to -2\,t\,e^{(-t^2)} \sin(\pi t^3) + 3\,e^{(-t^2)} \cos(\pi t^3)\,\pi t^2$$

$$0, 0$$

Hmmmm. Perhaps it is not so easy ... $df = 0$ has no easy, closed-form

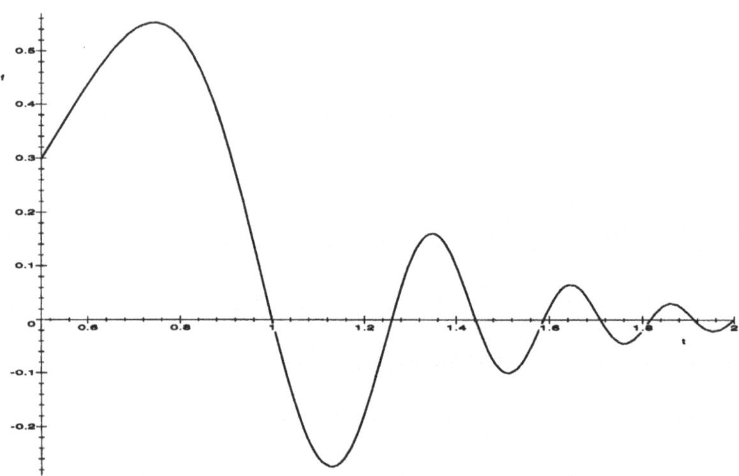

FIG. 8. $f(t) = \exp(-t^2) * \sin(\pi * t^3)$ on $[0, 2]$. Find the range of f.

solution. Try numerical solutions.

```
> root_1 := fsolve ( df(t) = 0, t = T[Inf] .. T[Sup] );
```

$$root\_1 := .7438242226$$

That's nice, but it is only ONE of several roots. Try for more.

```
> root_2 := fsolve ( df(t) = 0, t = 1 .. T[Sup] );
```

$$root\_2 := 1.645957257$$

OK, we have roots of $df = 0$. Evaluate f to find the extrema:

```
> f_extrema := seq ( f (root_.i), i = 1 .. 2 );
> f_range := [ min ( f_extrema ), max ( f_extrema ) ];
> evalf ( f_range );
```

$f\_extrema :=$

$$.5750636929 \sin(.4115389556\,\pi), .06659170279 \sin(4.459186732\,\pi)$$

$f\_range :=$

$$[.06659170279 \sin(4.459186732\,\pi), .5750636929 \sin(.4115389556\,\pi)]$$

$$[.06604506766, .5529992554]$$

Does that range make sense? No! As shown in Fig. 9, `root_2` is for the third highest relative maximum, not for the absolute (or even a relative) minimum. From the graph, we can see that the absolute minimum occurs near

```
> root_3 := fsolve ( df(t) = 0, t = 1 .. 1.3 );
> f_extrema := seq ( evalf ( f (root_.i) ), i = 1 .. 3 );
> f_range := [ min ( f_extrema ), max ( f_extrema ) ];
```

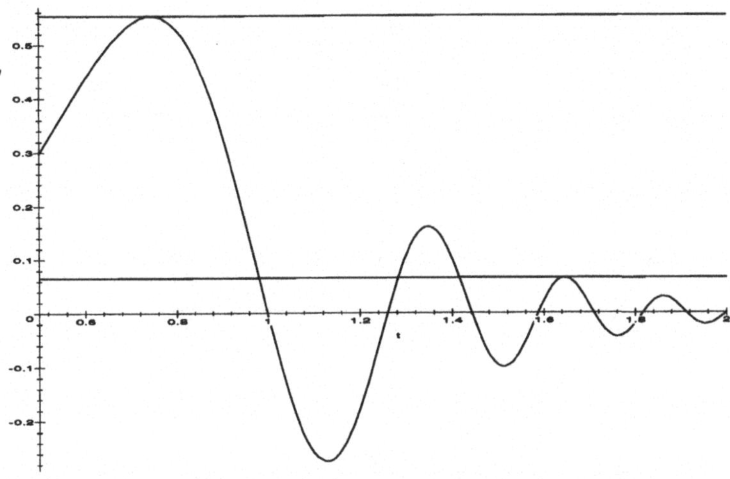

FIG. 9. Not bounds of f!

$$root\_3 := 1.129474479$$
$$f\_extrema := .5529992554, .06604506766, -.2744303367$$
$$f\_range := [-.2744303367, .5529992554]$$

THAT agrees with the graph, and it IS the range of f on T, up to machine roundoff.

Point: This was NOT easy, mechanical, or automatic, even with Maple's powerful features. If f is complicated, the computational cost for forming the derivative and solving may be MUCH larger than the cost of evaluating f.

Point: (Mathematician's threat to the Programmer) It COULD happen that f has a VERY sharp, very narrow peak or valley between adjacent machine numbers. We would miss it.

Point: There still may be values of $t \in T$ for which $f(t)$ is slightly outside f_range:

```
> root_1, root_3, f_range;
> # Checking near the maximum, difference should be positive.
> Digits := 15:
> f_range[2] - evalf ( f (0.74382422255) );
> f_range[2] - evalf ( f (0.7438242226) );
> f_range[2] - evalf ( f (0.74382422265) );
```

$$.7438242226, 1.129474479, [-.2744303367, .5529992554]$$

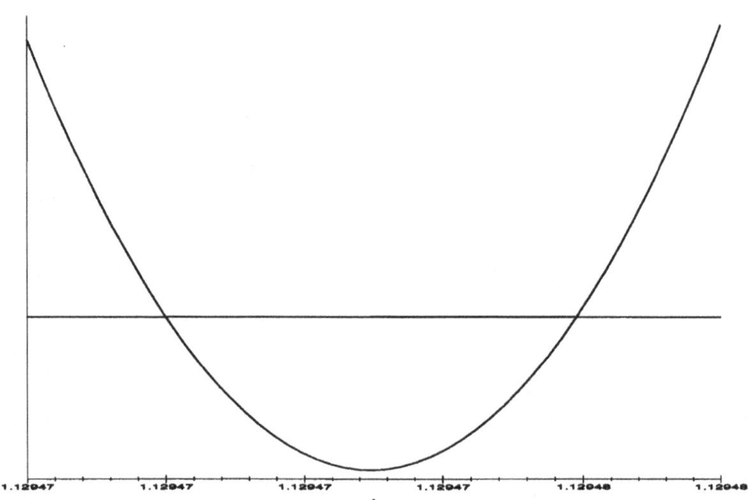

FIG. 10. $f(t) <$ "lower bound" !

$$.62689\,10^{-10}$$

$$.62688\,10^{-10}$$

$$.62689\,10^{-10}$$

OK. Now near the minimum, the difference should be positive:

```
> evalf ( f (1.1294744785) ) - f_range[1];
> evalf ( f (1.129474479) ) - f_range[1];
> evalf ( f (1.1294744795) ) - f_range[1];
```

$$-.44281\,10^{-10}$$

$$-.44282\,10^{-10}$$

$$-.44280\,10^{-10}$$

This is NOT a cooked-up example. Floating-point's round-to-nearest has an equal probability of rounding up or down. Hence, the chances are three out of four that an arbitrary function will show either its minimum overestimated (as in this example) or its maximum underestimated.

We will see that interval arithmetic is much easier than this example turned out to be, and its bounds ARE bounds. To use interval arithmetic to bound the range of f, we replace the point arithmetic in f by calls to interval arithmetic operators and functions:

```
> 'f' = f(t);
> F := inapply (f(t), t);
```

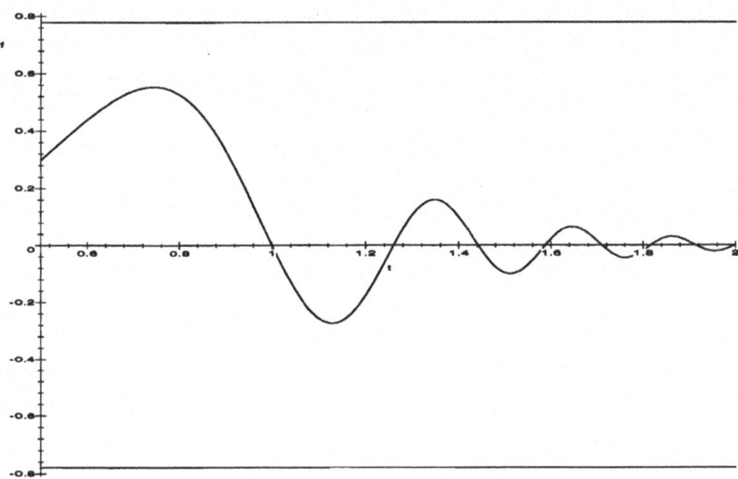

FIG. 11. Function range using interval arithmetic.

$$f = e^{(-t^2)} \sin(\pi t^3)$$

$$F := t \rightarrow \text{`\&exp`}((-1)\ \&*\ (t\ \&\text{intpower}\ 2))\ \&*\ \text{`\&sin`}(\pi\ \&*\ (t\ \&\text{intpower}\ 3))$$

```
> # and evaluate using interval arithmetic:
> f_interval_bounds := F (T);
```

$$f\_interval\_bounds := [-.7788007834, .7788007834]$$

Point: We have computed valid bounds. Very narrow spikes pose no danger of failure. It does not matter that the extrema are assumed at points that are not machine numbers.

Point: Simple algorithm: Evaluate f using interval arithmetic. No derivatives, no solving, no intelligent intervention.

Point: Yes, we may have to get tighter bounds in order to be practical. We will return to that in Section 5.

4.1.2 *Maple worksheet exist_1.ms*

Function range bounds computed using interval arithmetic as described above sometimes appear to give wide enclosures. Here, we show that even wide enclosures may be tight enough to yield useful results. In the next section, we will describe several techniques for computing tighter enclosures.

As an example, we apply the Existence and Uniqueness Theorems to the Lorenz system. As a consequence of the Existence and Uniqueness Theorems, if our computer program can find an interval $[t_0, t_1]$ and an interval $[u_0] = [u_0 - b, u_0 + b]$ for which

(1) f is continuous in $D := [t_0, t_1] \times [u_0 - b, u_0 + b]$,
(2) $||f|| \leq M$ in$D$,
(3) $t_1 = \min(t_1, t_0 + b/M)$, and
(4) all slopes $\dfrac{||f(t,u) - f(t,v)||}{||u - v||} \leq L$ for all points (t, u) and $(t, v) \in D$,

then we can guarantee that the ODE has a unique solution. We use interval arithmetic.

```
> load ('lorenz'):
```

$$\textit{Lorenz system} = \left[\left(\frac{\partial}{\partial t}\mathrm{x}(t)\right) - 10\,\mathrm{y}(t) + 10\,\mathrm{x}(t),\right.$$
$$\left(\frac{\partial}{\partial t}\mathrm{y}(t)\right) - 28\,\mathrm{x}(t) + \mathrm{y}(t) + \mathrm{x}(t)\,\mathrm{z}(t),$$
$$\left.\left(\frac{\partial}{\partial t}\mathrm{z}(t)\right) - \mathrm{x}(t)\,\mathrm{y}(t) + \frac{8}{3}\,\mathrm{z}(t)\right]$$

$$\textit{Initial Time} = 0$$

$$\textit{Initial Value} = [10, 10, 10]$$

$$t\_0 := 0$$
$$x0 := 10$$
$$y0 := 10$$
$$z0 := 10$$
$$t\_1 := 1$$

$$T0 := [0, 1.]$$
$$X0 := [9.899999999, 10.10000001]$$
$$Y0 := [9.899999999, 10.10000001]$$
$$Z0 := [9.899999999, 10.10000001]$$

The region D from the Existence Theorem is $D := T0 \times X0 \times Y0 \times Z0$ in $\mathbb{R}^4$.

```
> f := [ (y - x) * sigma, x * r - y - x * z, x * y - z * b ];
```

$$f := \left[10\,y - 10\,x, 28\,x - y - x\,z, x\,y - \frac{8}{3}\,z\right]$$

Now WE can all see by examination that f is a continuous function of t, x,

y, and z. One way our computer program can conclude that f is continuous is to substitute interval values and evaluate.

```
> F := [ (Y0 &- X0) &* sigma,
>            (X0 &* r) &- Y0 &- (X0 &* Z0),
>            (X0 &* Y0) &- (Z0 &* b) ];
> M := Interval_Sup_Norm (F);
```

$$F := [[-2.000000112, 2.000000112], [165.0899994, 174.8900006], [71.07666659, 75.61000032]]$$

$$M := 174.8900006$$

We conclude that f is defined for all $x \in X0$, $y \in Y0$, and $z \in Z0$. As a by-product, we get the upper bound M required for the range of f. On the other hand, if we attempt a similar evaluation of functions that are NOT defined, we get errors.

```
> F1 := [ 1.0 &/ ((Y0 &- X0) &* sigma),      # Divide by zero
>            (X0 &* r) &- Y0 &- (X0 &* Z0),
>            (X0 &* Y0) &- (Z0 &* b) ];
> F2 := [ (Y0 &- X0) &* sigma,
>            (X0 &* r) &- Y0 &- (X0 &* Z0),
>            &arcsin ((X0 &* Y0) &- (Z0 &* b)) ];
```

$$F1 := [[-\infty, \infty], [165.0899994, 174.8900006], [71.07666659, 75.61000032]]$$

```
Error, (in &arcsin) the arguments must be in the range [-1.,1.]
```

As long as f is a composition of $+$, $-$, $*$, $/$, exp, ln, sqrt, sin, cos, $\cdots$, then f is continuous wherever it is defined. Hence, if we can evaluate f by a computer program with no branches and no discontinuous elementary functions, then f is continuous.

We have an expression for f. To bound the range, we need only evaluate the expression using interval arithmetic operations. We prefer an "automatic" means of converting the expression for f to interval arithmetic, instead of having to convert the expression in our heads and re-type F as above. Let's see how to do that:

```
> f;
```

$$\left[10\,y - 10\,x, 28\,x - y - x\,z, x\,y - \frac{8}{3}\,z\right]$$

```
> # An interval-valued function:
> map (i -> inapply (i, x, y, z)(x, y, z), f);
```

$$\Big[(10 \,\&{*}\, y)\,\&{+}\,((-10)\,\&{*}\, x),$$

$$(28 \,\&{*}\, x)\,\&{+}\,(((-1)\,\&{*}\, y)\,\&{+}\,((-1)\,\&{*}\,(x \,\&{*}\, z))),$$

$$(x \,\&{*}\, y)\,\&{+}\left(\left(\frac{-8}{3}\right)\&{*}\, z\right)\Big]$$

```
> F := map (i -> inapply (i, x, y, z)(X0, Y0, Z0), f);
```

$$F := [[-2.000000221, 2.000000221], [165.0899993, 174.8900006], [71.07666659, 75.61000032]]$$

We have thus given bounds for the range of each component of f.

Another way our program can prove that f is continuous is to form its Jacobian, and then show that the Jacobian is defined for all elements of D.

```
> jacobian_of_f := linalg[jacobian] ( f, [x, y, z] );
```

$$jacobian\_of\_f := \begin{bmatrix} -10 & 10 & 0 \\ 28 - z & -1 & -x \\ y & x & \frac{-8}{3} \end{bmatrix}$$

Now we bound the range of the Jacobian by evaluating it in interval arithmetic:

```
> Jac_F := map ( i -> inapply (i, x, y, z)(X0, Y0, Z0),
>                 jacobian_of_f );
```

$$Jac\_F := \begin{bmatrix} -10, 10, 0 \\ [17.89999997, 18.10000001], -1, [-10.10000002, -9.899999998] \\ [9.899999999, 10.10000001], [9.899999999, 10.10000001], \frac{-8}{3} \end{bmatrix}$$

All of the partial derivatives of f are defined in D. We know that *because we evaluated them.* If we try to do the same with a function that is NOT defined, we get errors.

```
> g := [ 1 / ((y - x) * sigma),
>        x * r - y - x * z,
>        x * y - z * b ];
```

$$g := \left[\frac{1}{10}\frac{1}{y-x}, 28\,x - y - x\,z, x\,y - \frac{8}{3}\,z\right]$$

```
> jac_g := linalg[jacobian] ( g, [x, y, z] );
```

$$jac\_g := \begin{bmatrix} \frac{1}{10}\frac{1}{(y-x)^2} & -\frac{1}{10}\frac{1}{(y-x)^2} & 0 \\ 28 - z & -1 & -x \\ y & x & \frac{-8}{3} \end{bmatrix}$$

```
> Jac_G := map (i -> inapply (i, x, y, z)(X0, Y0, Z0), jac_g);
```

$$Jac\_G := \begin{bmatrix} [-\infty, \infty], [-\infty, \infty], 0 \\ [17.89999997, 18.10000001], -1, [-10.10000002, -9.899999998] \\ [9.899999999, 10.10000001], [9.899999999, 10.10000001], \frac{-8}{3} \end{bmatrix}$$

The function g is NOT defined and continuous in D, as we have discovered by trying unsuccessfully to evaluate it.

As a by-product of computing the Jacobian of f, we have a bound on the Lipschitz constant L,

```
> Abs_F := map ( Interval_Abs, Jac_F );
```

$$Abs\_F := \begin{bmatrix} 10 & 10 & 0 \\ 18.10000001 & 1 & 10.10000002 \\ 10.10000001 & 10.10000001 & \frac{8}{3} \end{bmatrix}$$

```
> L := max ( op ( map ( op, [entries (Abs_F)] ) ) );
```

$$L := 18.10000001$$

In summary, we set out to

- Find an interval $[t_0, t_1]$ and an interval $[u_0] = [u_0 - b, u_0 + b]$ for which
- f is continuous in D,
- $||f|| \leq M$ in$D$,
- $t_1 = \min(t_1, t_0 + b/M)$, and
- all slopes $\dfrac{||f(t,u) - f(t,v)||}{||u - v||} \leq L$ for all points (t, u) and $(t, v) \in D$,

Here is what we have found:
On the time interval

```
> t_1 := min (t_1, t_0 + b_for_u / M):
> T_0 := construct (t_0, t_1);
```

$$T\_0 := [0, .0005717879791]$$

and for [x, y, z] in

```
> [ X0, Y0, Z0 ];
```

$$[[9.899999999, 10.10000001], [9.899999999, 10.10000001], [9.899999999, 10.10000001]]$$

we have $||f||$ bounded by

```
> 'M' = M;
```

$$M = 174.8900006$$

Further, f has been shown to satisfy a Lipschitz condition with

```
> 'L' = L;
```

$$L = 18.10000001$$

Hence, we have proven that the Lorenz system has a unique solution, at least in the interval

```
> '[t_0, t_1]' = T_0;
```

$$[t\_0, t\_1] = [0, .0005717879791]$$

It is a very restrictive time step to only step from $t_0 = 0$ to $t_1 = 0.00057$, but the Lorenz system is a difficult system to solve, and so far we have only used relatively simple interval tools for the interval evaluation of f and its Jacobian.

We construct a picture. First, compute an approximate solution:

```
> ApSol := dsolve (DEs,
>                    convert (map (S -> S(t), SolutionFunctions),
>                             set), series);
```

$$ApSol := \Big\{ \mathrm{x}(t) = 10 + 850\,t^2 - \frac{39050}{9}\,t^3 + \frac{956075}{54}\,t^4 - \frac{17867645}{162}\,t^5 + \mathrm{O}(t^6),$$
$$\mathrm{y}(t) = 10 + 170\,t - \frac{1355}{3}\,t^2 + \frac{74065}{27}\,t^3 - \frac{12131195}{324}\,t^4 - \frac{54827083}{972}\,t^5 + \mathrm{O}(t^6),$$
$$\mathrm{z}(t) = 10 + \frac{220}{3}\,t + \frac{6770}{9}\,t^2 + \frac{53390}{81}\,t^3 + \frac{15404365}{486}\,t^4 - \frac{409237189}{1458}\,t^5 + \mathrm{O}(t^6) \Big\}$$

```
> ApproximateSolution := convert ( op (2, ApSol[3]), polynom);
```

$$ApproximateSolution := 10 + \frac{220}{3}\,t + \frac{6770}{9}\,t^2 + \frac{53390}{81}\,t^3 + \frac{15404365}{486}\,t^4 - \frac{409237189}{1458}\,t^5$$

Zero term Taylor series + enclosure of the remainder.

```
> ConstantBounds := Z0[Inf], Z0[Sup];
```

$$ConstantBounds := 9.899999999, 10.10000001$$

One term Taylor series + enclosure of the remainder.

```
> LinearBounds := z0 - M * (t - t_0), z0 + M * (t - t_0);
```

$$LinearBounds := 10 - 174.8900006\,t, 10 + 174.8900006\,t$$

Two term Taylor series + enclosure of the remainder.

```
> dz0 := subs (x = x0, y = y0, z = z0, f[3]);
> ParabolicBounds := z0 + dz0 * (t - t_0)
>                     - L * (M + 1) * (t - t_0)^2 / 2,
>                   z0 + dz0 * (t - t_0)
>                     + L * (M + 1) * (t - t_0)^2 / 2;
```

$$dz0 := \frac{220}{3}$$

$$ParabolicBounds := 10 + \frac{220}{3}\,t - 1591.804507\,t^2, 10 + \frac{220}{3}\,t + 1591.804507\,t^2$$

Point: The linear bounds remain within the constant bounds in the interval $T0$.

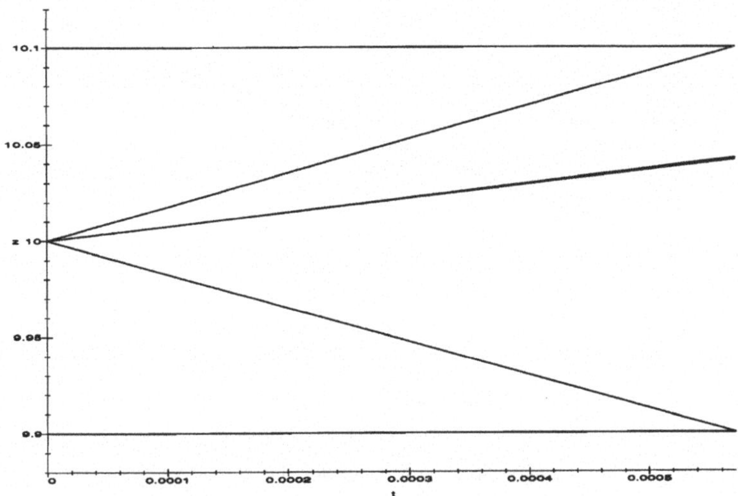

FIG. 12. Solution with constant, linear, and parabolic bounds.

Point: The parabolic bounds enclose the solution and agree with each other to graphical accuracy.

Point: By evaluating f and its Jacobian in interval arithmetic, we have computed range bounds that we have used in turn to enclose the solution of the Lorenz system.

This example is autonomous, so the interval T never figured into any range bounds. From the picture, it appears that the parabolic bounds for z will leave D when the upper parabolic bound meets the upper edge of the box.

```
> ExitPoint := solve (Z0[2] = ParabolicBounds[2], t);
```

$$ExitPoint := -.04739480819, .00132549943$$

Can you prove that in $[t_0, ExitPoint[2]]$, the ODE has a unique solution? We only examined z. We would have to examine x and y similarly. If we can claim that the parabolic bounds staying within D validates the existence of a unique solution, then we can take a step that is over twice as long.

4.1.3 *Implementations*

It used to be the case that interval techniques were not popular because there was no software available. Now there are many high-quality environments available commercially or freely to support interval arithmetic.

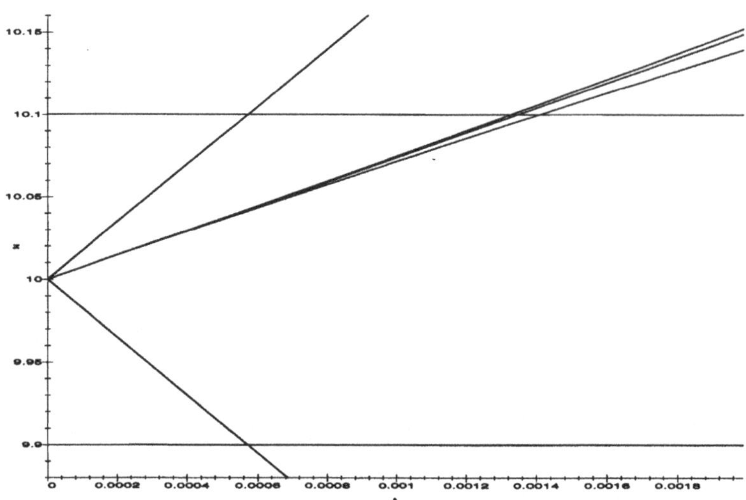

FIG. 13. Parabolic bounds leave D later, in this example.

Corliss [17] surveyed and compared many available packages for interval arithmetic.

Name	Language	Machines	Authors
ABACUS			Rump *et al.*
ACRITH-XSC	Fort 77	IBM S/370	Kulisch *et al.* [29, 63]
AQUARELS	Fort 77		Erhel *et al.* [23]
BIAS	Ada	VAX	Corliss [19]
BIAS	Fort 77	any f77	Neumaier [49]
BIAS	C++		Knüppel [38]
Clemmeson	MS Fort	PC	Wang *et al.* [65]
C-XSC	C++	PC	Kulisch *et al.* [36, 40]
Fortran-XSC	Fort 90	any f90	Walter [64]
Hammer	Pascal-XSC	any P-XSC	Hammer
INTLIB 77	Fort 77	any f77	Kearfott *et al.* [34]
INTLIB 90	Fort 90	any f90	Kearfott [33]
INTPAK	Maple	any Maple	Connell *et al.* [13]
	Mathematica	any Math.	Kieper
Modula-SC			Ullrich *et al.*
Pascal-XSC	Pascal	many	Kulisch *et al.* [37]
PBASIC	Basic	PC	Aberth [1]
Range	C++	some C++	Aberth
TPX	Pascal	PC	Rump *et al.*

The advantages of some of the packages with which I have personal experience include:

ACRITH-XSC: IBM program product. Supported. Full set of high quality elementary functions and utilities. Many problem-solving routines. Contact your IBM representative.

C–XSC: Production quality commercial product. Full set of high quality elementary functions and utilities. Multiple precision. Available for many platforms. Many problem-solving routines. Many literature articles. Contact Michael Neaga (ae18@dkauni2.bitnet).

INTLIB: Portable, fast. Fortran 77 code looks like subroutine calls. Fortran 90 version is more convenient. Free. Contact Baker Kearfott (rbk@usl.edu).

INTPAK: Works within Maple. INTPAK is free from Maple share library. Contact Rob Corless (rmc@watson.ibm.com) or George Corliss (georgec@mscs.mu.edu).

PASCAL–XSC: Production quality commercial product. Full set of high quality elementary functions and utilities. Available for many platforms. Many problem-solving routines. Many literature articles. Contact Michael Neaga (ae18@dkauni2.bitnet).

Neumaier's BIAS: Portable Fortran 77. Fast because it does not do directed rounding. Good for global optimization applications. Contact George Corliss (georgec@mscs.mu.edu).

Wang's Clemmeson: Microsoft assembler and Fortran, IBM PC specific. Uses IEEE directed roundings. Fast with guarantees. Contact George Corliss (georgec@mscs.mu.edu).

I expect that Walter's Fortran 90 module (Wolfgang V. Walter – ae38@dkauni2.bitnet), Ullrich's Modula–SC (Christian Ullrich – Basel), or Knüppel's BIAS (Olaf Knüppel – knueppel@tu-harburg.dbp.de) are at least as good, but I have not used them.

Personally, I prefer INTPAK in Maple for small programs because of Maple's environment for symbolic computation and graphics. I prefer C–XSC for large programs because it is a solid product. The operator overloading in C++ makes it easy to write interval expressions and algorithms. I prefer C–XSC over Pascal–XSC because it is "real" C++, and not a nonstandard language extension. I prefer C-XSC over Fortran 90 INTLIB for no technical reason at all.

4.2 Automatic differentiation

We have seen in previous sections that algorithms for computing guaranteed error bounds for ODEs require us to evaluate Jacobians and univariate Taylor series in interval arithmetic. When the right-hand side f of eqn (1.1) is relatively simple, the symbolic tools in Maple are sufficient, as we have already seen. However, as f becomes more complicated, perhaps given by a procedure rather than by an expression, we need a more powerful tool. Automatic differentiation is the required tool. Automatic differentiation is also called differentiation arithmetic in the literature.

Automatic differentiation (AD) uses well-known recurrence relations derived from the chain rule to propagate derivative values, in contrast to symbolic differentiation, which propagates expressions. Beda *et al.* [8], Wengert [66], and Moore [46] were early proponents of AD. Rall [54] and Griewank [25, 26] give excellent surveys. Griewank and Corliss [27] contain many articles on diverse aspects of AD including an extensive bibliography [16] and a survey of AD software tools [31].

AD can be viewed as a differentiation arithmetic on ordered pairs. It is instructive to compare and contrast differentiation arithmetic with complex arithmetic.

	Complex Numbers	Differentiation Arithmetic
$A = (a_1, a_2)$	$a_1 + ia_2$	$(a(t), a'(t))$
$C := A \pm B$	$(a_1 \pm b_1, a_2 \pm b_2)$	$(a_1 \pm b_1, a_2 \pm b_2)$
$C := A * B$	$(a_1b_1 - a_2b_2, a_1b_2 + a_2b_1)$	$(a_1b_1, a_1b_2 + a_2b_1)$
$C := A/B$	$((a_1b_1 + a_2b_2)/(b_1^2 + b_2^2),$ $(-a_1b_2 + a_2b_1)/(b_1^2 + b_2^2))$	$(a_1/b_1,$ $(a_2 - c_1b_2)/b_1)$
$C := e^A$	$(e^{a_1}\cos(a_2), e^{a_1}\sin(a_2))$	(e^{a_1}, c_1a_2)
$V := \sin(A)$	$(\sin(a_1)\cosh(a_2), \cos(a_2)\sinh(a_2))$	$(\sin(a_1), a_2\cos(a_1))$

Other elementary functions have similar rules for AD.

Point: Differentiation arithmetic is simpler than complex arithmetic.

Point: It is often much less expensive to evaluate the derivative of an elementary function than to evaluate the function itself.

The differentiation arithmetic given above for ordered pairs has a natural generalization to n-tuples for Taylor series using recurrence relations for the Taylor coefficients $(a)_i := \dfrac{a^{(i)}(t_0)}{i!}$. For example,

$$\begin{aligned}(a \cdot b)_k &= \sum_{i=0}^{k} (a)_i (b)_{k-i} \\ (e^a)_k &= \sum_{i=0}^{k-1} \frac{k-i}{k} (e^a)_i (a)_{k-i}\,.\end{aligned}$$

AD can be implemented using operator overloading, using a preprocessor, or by interpreting a code list. AD is not numerical differentiation; it computes derivatives accurately up to roundoff error. AD is not symbolic; it propagates values, not expressions. AD makes heavy use of scalar products. The cost of generating n series terms is $O(n^2)$, and the recurrence relations are easily generated from the ODE, as the following example shows.

Consider the Lorenz system

$$
\begin{aligned}
x' &= \sigma(y - x)\,, \quad x(t_0) = x_0\,,\\
y' &= \rho x - y - xz\,, \quad y(t_0) = y_0\,,\\
z' &= xz - bz\,, \quad z(t_0) = z_0\,.
\end{aligned}
$$

Then the following pseudo-code computes the Taylor series for the solution

$$
\begin{aligned}
&(x)_0 := x_0;\\
&(y)_0 := y_0;\\
&(z)_0 := z_0;\\
&\text{loop for } i = 0 \ldots N-1\\
&\qquad (x)_{i+1} := \sigma((y)_i - (x)_i)/(i+1);\\
&\qquad (y)_{i+1} := \left(\rho(x)_i - (y)_i - \sum_{j=0}^{i} (x)_j (z)_{i-j}\right)/(i+1);\\
&\qquad (z)_{i+1} := \left(\sum_{j=0}^{i} (x)_j (z)_{i-j} - b(z)_i\right)/(i+1);
\end{aligned}
$$

If we execute this program using floating-point arithmetic with $t = t_0$, $u = u_0$, we get approximations for $u(t_0)$, $u'(t_0)$, $u''(t_0)$, etc. If we execute this program using interval arithmetic with $t = t_0$, $u = [u_0]$, we get enclosures for $u(t_0)$, $u'(t_0)$, $u''(t_0)$, etc. If we execute this program using interval arithmetic with $t = [t_0]$, $u = [u_0]_0$, we get enclosures for the remainder terms $u'(\tau)$, $u''(\tau)$, etc.

Maple provides support for AD with the `D` operator and with packages DIFF, GRADIENT, TAYLOR, JACOBIAN, and HESSIAN in the share library. Use of the GRADIENT package is illustrated by a worksheet `gradient.ms` and a document `gradient.tex` distributed with the share library, so we only give a few examples here. We use two examples from [9].

```
> f := proc (t1, t2, t3, t4, t5)
>     t1 * t2 * t3 * t4 * t5
> end;
f := proc(t1,t2,t3,t4,t5) t1*t2*t3*t4*t5 end

> with (share):  readshare (GRADIENT, autodiff):
> grad_f := GRADIENT (f);
grad_f :=
proc(t1,t2,t3,t4,t5)
    RETURN(t2*t3*t4*t5,t1*t3*t4*t5,t1*t2*t4*t5,
           t1*t2*t3*t5,t1*t2*t3*t4)
end
```

As [9] points out, this is correct, but it contains redundant computations. Can we optimize it?

```
> readlib (optimize) (grad_f);
Error, (in optimize/statseq) invalid arguments
```

No, but if we parse the expression for Maple, GRADIENT is able to generate efficient code.

```
> f2 := proc (t1, t2, t3, t4, t5)
>   local s1, s2, s3, s4;
>     s1 := t1 * t2;
>     s2 := t3 * t4;
>     s3 := s1 * s2;
>     s4 := s3 * t5;
>     s4
> end;
```

```
> grad_f2 := GRADIENT (f2);
grad_f2 := proc(t1,t2,t3,t4,t5)
           local ds1,ds2,ds3,s1,s2,s3;
               ds1 := array(1 .. 5);
               ds1[1] := t2;
               ds1[2] := t1;
               s1 := t1*t2;
               ds2[3] := t4;
               ds2[4] := t3;
               s2 := t3*t4;
               ds3[1] := ds1[1]*s2;
               ds3[2] := ds1[2]*s2;
               ds3[3] := s1*ds2[3];
               ds3[4] := s1*ds2[4];
               s3 := s1*s2;
               RETURN(ds3[1]*t5,ds3[2]*t5,
                      ds3[3]*t5,ds3[4]*t5,s3)
           end
```

```
> f := proc (y, z)
>     -y / (z * z * z)
> end;
```

```
> grad_f := GRADIENT (f);
grad_f := proc(y,z) RETURN(-1/z^3,3*y/z^4) end
```

Point: The requirement of interval algorithms for derivatives is no obstacle.

5 Specifications: get tight intervals

Purposes of this section:

Enclosure is easy. Tight enclosure is harder, but feasible.

A good algorithm for computing guaranteed global error bounds should be

Truthful: The claimed bounds should not lie.

Accurate: The enclosures should be tight enough to meet a requested tolerance.

Fast: Of course.

As George Byrne once quipped in the context of reliable approximate solutions, we may choose any two. In the previous section, we showed that truthful algorithms are possible. In this section, we discuss strategies for computing tighter enclosures.

5.1 Maple worksheet tighter.ms

```
>     # Procedures for computing ranges of functions:
> load ('tightr_1'):
>     # See [Heck1993a], p. 292:
> f := t -> exp (-t^2) * sin (Pi * t^3);
> Tf := construct (0.5, 2.0);
```

$$f := t \to e^{(-t^2)} \sin(\pi t^3)$$
$$Tf := [.5, 2.0]$$

The techniques for computing guaranteed range bounds that are tight include

- Naive interval arithmetic (evaluating an expression using interval operators)
- Mean value form
- Slope form
- Intersection of multiple enclosures
- Monotonicity, where present
- Subinterval adaptation
- Approximation + [Error]
- Contractive iterations
- Control of the "wrapping effect".

Neumaier [48] discusses each of these techniques and gives examples. We illustrate the techniques first with a simple polynomial example from Neumaier. Then we return to the example from Heck.

```
> g := t -> t^3 - 3*t^2 + 4*t + 5;
> T := construct (0, 1);
```

$$g := t \to t^3 - 3t^2 + 4t + 5$$
$$T := [0, 1.]$$

```
> # Since g is monotone,
> TrueRange := construct ( g(T[Inf]), g(T[Sup]) );
```

$$TrueRange := [5., 7.]$$

<u>NAIVE INTERVAL ARITHMETIC</u>

We evaluate the expression using interval operators.

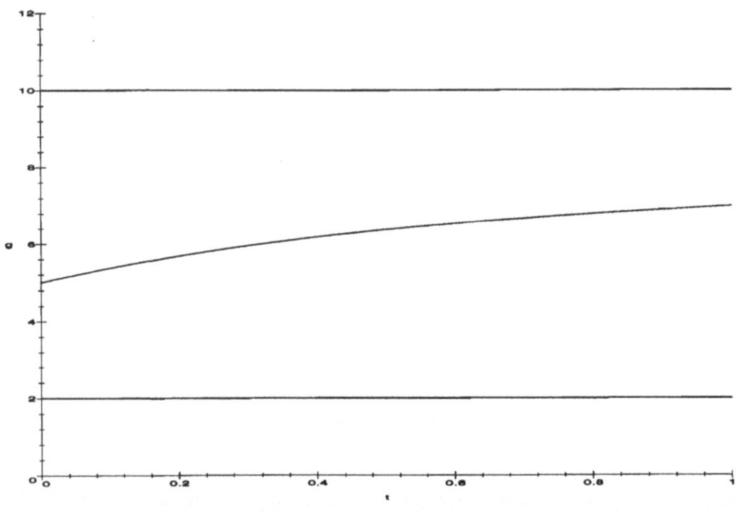

FIG. 14. Naive interval evaluation of $g(t) = t^3 - 3t^2 + 4t + 5$.

```
> inapply ( g(t), t ) (T);           # This is how to do it
>    # Neumaier's example:
> NaiveIntervalBound := ComputeNaiveIntervalRange (g, T);
>    # Heck's example:
> fNaiveBound := ComputeNaiveIntervalRange (f, Tf);
```

$$[\,1.999999993, 10.00000001\,]$$

$$NaiveIntervalBound := [\,1.999999993, 10.00000001\,]$$

$$fNaiveBound := [\,-.7788007834, .7788007834\,]$$

Figure 14 shows that `fNaiveBound` *is* a bound for the range of g. We will compute tighter bounds in the sequel.

MEAN VALUE FORM

According to the Mean Value Theorem,

$$f(t) = f(t_0) + f'(\tau)(t - t_0)\,, \text{ for any } t \text{ and } t_0 \in T.$$

Hence,

$$f(t) \in f(t_0) + f'(T)(t - t_0) \text{ for any } t \text{ and } t_0 \in T.$$

We bound the range of $f'(T)$ using naive interval arithmetic.

```
> dg := D(g);                   # Derivative
> dg_Bound := inapply ( dg(t), t ) (T);
```

$$dg := t \to 3\,t^2 - 6\,t + 4$$

$$dg\_Bound := [-2.000000003, 7.000000006]$$

We can apply the Mean Value Form at any point t in T. We choose the left endpoint, the midpoint, and the right endpoint for illustration.

```
> MeanValueForm := inapply ( g(t), t ) (t_0)
>                  &+ (DBound &* (t &- t_0));
```

$$MeanValueForm := ((t\_0 \ \&\text{intpower}\ 3) \ \&+ \ (((-3) \ \&* \ (t\_0 \ \&\text{intpower}\ 2)) \ \&+ \ ((4 \ \&* \ t\_0) \ \&+ \ 5))) \&+ \ (DBound \ \&* \ (t \ \&- \ t\_0))$$

```
> LeftMVBound
>       := subs ( t = T, t_0 = T[Inf], DBound = dg_Bound,
>                 MeanValueForm );
> 'g(t) is in ', evalf ("), ' for all t in T';
```

$$LeftMVBound := ((0 \ \&\text{intpower}\ 3) \ \&+ \ (((-3) \ \&* \ (0 \ \&\text{intpower}\ 2)) \ \&+ \ ((4 \ \&* \ 0) \ \&+ \ 5))) \ \&+ \ ([-2.000000003, 7.000000006] \ \&* \ ([0, 1.] \ \&- \ 0))$$

$$g(t) \textit{ is in } [2.999999990, 12.00000003] \textit{ for all } t \textit{ in } T$$

```
> LeftMVBound := ComputeMeanValueRange (g, T[Inf], T);
```

$$LeftMVBound := [2.999999990, 12.00000003]$$

Guaranteed bounds use directed rounding.

```
> LeftMVLowBound
>       := s -> subs ( t = s, t_0 = T[Inf], DBound = dg_Bound,
>                      MeanValueForm )[Inf];
> LeftMVHighBound
>       := s -> subs ( t = s, t_0 = T[Inf], DBound = dg_Bound,
>                      MeanValueForm )[Sup];
```

We can also choose the midpoint and the right endpoint as centers for the Mean Value Form.

Mean Value bounds for Heck's example

$$fLeftMVBound := [-48.41491748, 49.01098580]$$

$$fCenterMVBound := [-24.38723220, 24.32571944]$$

$$fRightMVBound := [-48.71295164, 48.71295164]$$

Point: g is contained inside both the constant and the linear bounds in each case.

Point: The midpoint form gives the tightest single enclosure.

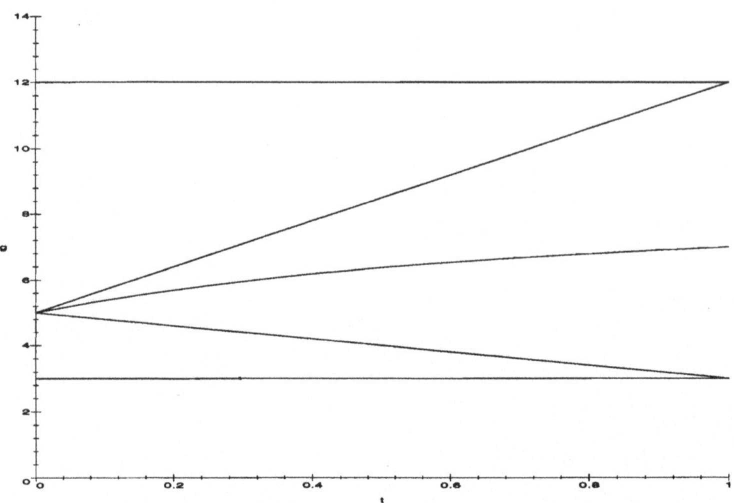

FIG. 15. Mean Value form centered at the left endpoint.

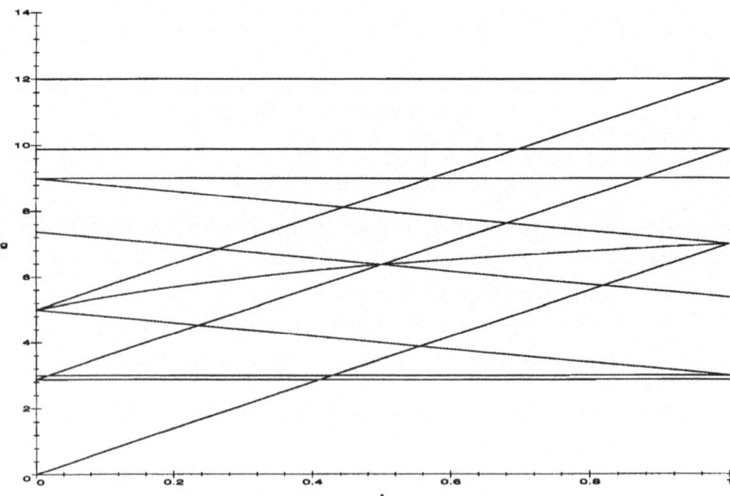

FIG. 16. Mean Value enclosures centered at the left endpoint, the midpoint, and the right endpoint.

Point: The left endpoint gives the best lower bound in this case and the tightest bounds near the left endpoint. The right endpoint gives the best upper bound in this case and the tightest bounds near the right endpoint.

Point: The Mean Value form can be generalized to include more terms of the Taylor series.

SLOPE FORM

The results of the Mean Value form hold if the derivative is replaced by the slope:

```
> ( g(s) - g(t) ) / ( s - t );  # Slope is a divided difference
> slopeg := simplify (( g(s) - g(t) ) / ( s - t ));
```

$$\frac{s^3 - 3\,s^2 + 4\,s - t^3 + 3t^2 - 4\,t}{s - t}$$

$$slopeg := s^2 - 3\,s + s\,t + 4 - 3t + t^2$$

$$f(t) = f(t_0) + \frac{f(t) - f(t_0)}{t - t_0}(t - t_0) \text{ for any } t \text{ and } t_0 \in T.$$

The advantage over the Mean Value form is that if f is a polynomial (or has some polynomial terms), the slope $\frac{f(t)-f(t_0)}{t-t_0}$ can be algebraically simplified BEFORE EVALUATION to reduce the dependency. Hence, the Slope form is

$$f(t) \in f(t_0) + \frac{f(T) - f(t_0)}{T - t_0}(t - t_0) \text{ for any } t \text{ and } t_0 \in T.$$

We bound the range of the slope $\dfrac{f(T) - f(t_0)}{T - t_0}$ using naive interval arithmetic.

```
> SlopeG := inapply (slopeg, t, s);
> SlopeBound := SlopeG (T[Inf], T);
```

$$SlopeG := (\,t, s\,) \to (\,s\,\&\text{intpower}\,2\,)\,\&+(((\,-3\,)\,\&*\,s\,)\,\&+$$
$$((\,s\,\&*\,t\,)\,\&+(\,4\,\&+(((\,-3\,)\,\&*\,t\,)\,\&+(\,t\,\&\text{intpower}\,2\,)))))$$
$$SlopeBound := [\,.9999999968, 5.000000005\,]$$

```
> LeftSlopeBound := ComputeSlopeRange (g, T[Inf], T);
```

$$LeftSlopeBound := [\,4.999999996, 10.00000002\,]$$

Point: The Slope form DOES enclose the range.

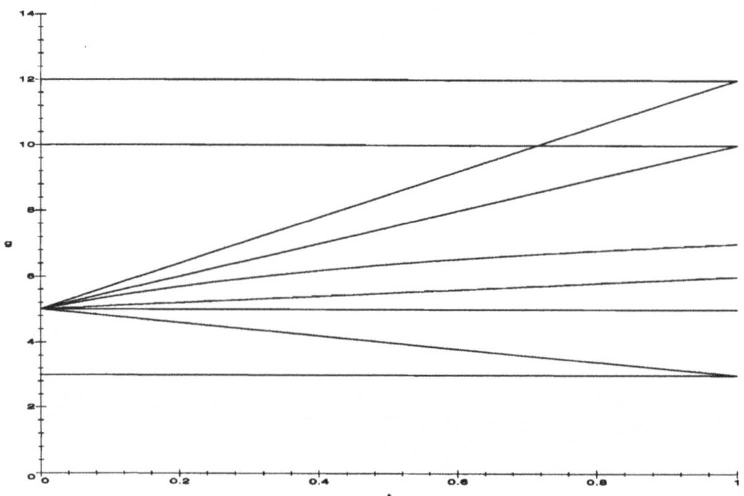

FIG. 17. Slope and Mean Value forms centered at the left endpoint.

Point: The slope form yields a tighter enclosure than the Mean Value Form.

Similarly, we can form the Slope forms centered at the midpoint and at the right endpoint.

For Heck's example, the slope expression contains elementary functions, so it does not simplify. Hence, the derivative is used instead of the slope.

INTERSECTION

We have several different bounds for the range of g.

$$NaiveIntervalBound = [1.999999993, 10.00000001]$$

$$LeftMVBound = [2.999999990, 12.00000003]$$

$$CenterMVBound = [2.874999990, 9.875000010]$$

$$RightMVBound = [-.2300000001\,10^{-7}, 9.000000016]$$

$$LeftSlopeBound = [4.999999996, 10.00000002]$$

$$CenterSlopeBound = [4.249999990, 8.500000010]$$

$$RightSlopeBound = [2.999999976, 8.000000020]$$

Since g(T) is guaranteed to be in each of these intervals, it is also in their

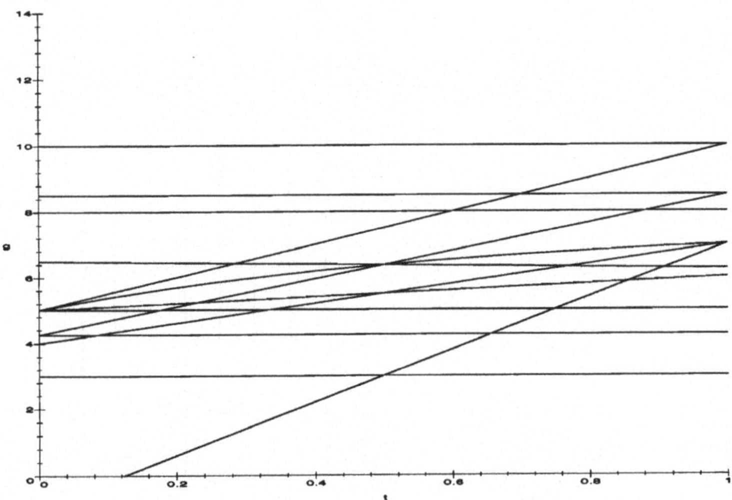

FIG. 18. Slope enclosures centered at the left endpoint, the midpoint, and the right endpoint.

intersection.

```
> IntersectRange := NaiveIntervalBound
>      &intersect LeftMVBound &intersect CenterMVBound
>      &intersect RightMVBound &intersect LeftSlopeBound
>      &intersect CenterSlopeBound &intersect RightSlopeBound;
```

$$IntersectRange := [\,4.999999996, 8.000000020\,]$$

Figure 19 shows the function g. At the far right side are bars representing the seven range bounds computed above. The next bar to the left is their intersection, which is used to determine the two horizontal bounds drawn in the graph. The left-most vertical bar is the true range of g.

Point: The intersection is tighter than any single range bound.

Point: The intersection is determined in this example by the slope form at the endpoints.

MONOTONICITY

If a function f is known to be monotonic, then we get tight bounds for the range by evaluating f at the endpoints using interval arithmetic.

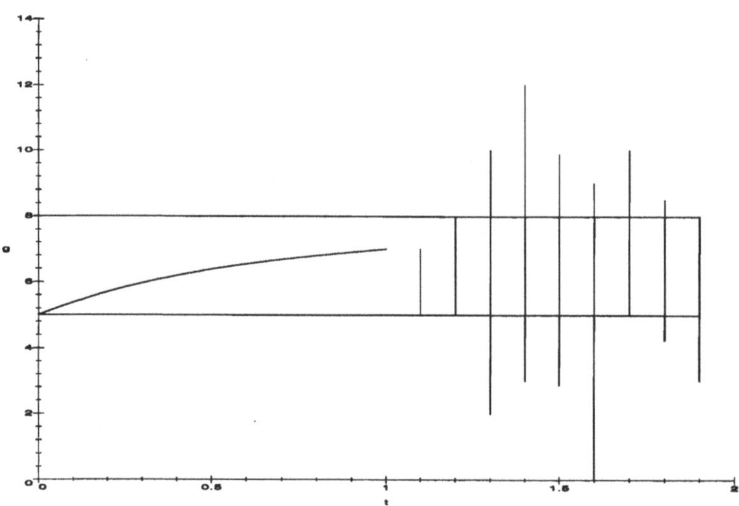

FIG. 19. The intersection of several bounds.

We can often guarantee that f is monotonic by bounding the range of its derivative.

```
> g(t);
> dg(t);
```

$$t^3 - 3t^2 + 4t + 5$$
$$3t^2 - 6t + 4$$

```
> ComputeNaiveIntervalRange (dg, T);
> ComputeMeanValueRange (dg, t_mid, T);
> ComputeSlopeRange (dg, t_mid, T);
```

$$[-2.000000003, 7.000000006]$$
$$[-1.250000005, 4.750000006]$$
$$[-.5000000051, 4.000000007]$$

None of those guarantees that g is monotonic, but

```
> ComputeSlopeRange (dg, T[Inf], T)
>    &intersect ComputeSlopeRange (dg, T[Sup], T) > 0.,
>    ` so g IS monotonic!`;
```

$0 < [.9999999908, 4.000000003]$ *so g IS monotonic!*

```
> LeftG := ComputeNaiveIntervalRange ( g, construct (T[Inf]) );
> RightG := ComputeNaiveIntervalRange ( g, construct (T[Sup]) );
> TrueRange := Interval_Convex_Hull (LeftG, RightG);
```

$$LeftG := [4.999999997, 5.000000003]$$

$$RightG := [\,6.999999991, 7.000000009\,]$$

$$TrueRange := [\,4.999999997, 7.000000009\,]$$

We combine the monotonicity test, naive interval evaluation, and the three Slope forms into a single procedure.

```
> ComputeCombinedRange ( g, T );
```

$$[\,4.999999997, 7.000000009\,]$$

```
> ComputeCombinedRange ( f, Tf );
```

$$[\,-.7788007834, .7788007834\,]$$

SUBINTERVAL ADAPTATION

Our bounds for g are optimally tight, up to roundoff, because g is monotonic. However, Heck's example is not monotonic on the interval Tf.

```
> ComputeMonotonicRange (f, Tf);
```

FAIL

The best bounds we have so far are from the naive interval evaluation.

```
> fIntersectRange = fNaiveBound;
```

$$[\,-.7788007834, .7788007834\,] = [\,-.7788007834, .7788007834\,]$$

We form a list of subintervals of Tf and bounds for the range of f on each subinterval. Initially, the list contains only Tf itself.

$$List\_of\_T := [\,[\,.5, 2.0\,]\,]$$

$$List\_of\_F := [\,[\,-.7788007834, .7788007834\,]\,]$$

Then we identify the victim subinterval on which the width of the range is widest, bisect that subinterval, bound the ranges on each half of the victim subinterval, and recompute the range of f as the convex hull of the range on all the subintervals.

```
> SubdividedRange := SubdivideSubinterval
>         ( f, 'List_of_T', 'List_of_F' );
```

$$SubdividedRange := [\,-.7788007834, .7788007834\,]$$

$$List\_of\_T = [\,[\,.5, 1.250000000\,], [\,1.250000000, 2.0\,]\,]$$

$$List\_of\_F = [\,[\,-.7788007834, .7788007834\,], [\,-.2096113878, .2096113878\,]\,]$$

```
> 'SubdividedRange' = SubdividedRange;
> for i from 1 to 4 do     # Repeat the process a few times ...
>    SubdividedRange := SubdivideSubinterval
>             ( f, 'List_of_T', 'List_of_F' );  od;
```

$$SubdividedRange = [\,-.7788007834, .7788007834\,]$$

$$SubdividedRange := [-.4650431884, .7788007834]$$

$$SubdividedRange := [-.3233867682, .7788007834]$$

$$SubdividedRange := [-.3233867682, .6233443093]$$

$$SubdividedRange := [-.3233867682, .6233443093]$$

$$\begin{aligned} List\_of\_T = [[&1.250000000, 2.0], [1.062500000, 1.250000000], \\ &[.5, .6875000000], [.6875000000, .8750000000], \\ &[.8750000000, .9687500000], [.9687500000, 1.062500000]] \end{aligned}$$

$$\begin{aligned} List\_of\_F = [[&-.2096113878, .2096113878], [-.3233867682, -.03075637724], \\ &[.2707775813, .5586950027], [.3964772214, .6233443093], \\ &[.1101516560, .4003403064], [-.1896402434, .1101516583]] \end{aligned}$$

The bounds given by **SubdividedRange** appear to have converged, but Fig. 20 shows that they are not yet tight. Our bounds will not be improved until we subdivide the subintervals in which the absolute extrema happen to lie. Unfortunately, the range of f on those subintervals is rather smaller than the range on subintervals where f is montonic. Clearly, we can benefit from a more sophisticated strategy for selecting a subinterval to bisect. To explore that farther takes us further into the area of global optimization, so we will be satisfied with a few more iterations of this bisection strategy. The range enclosure shown in Fig. 20 is better, but we still do not have the optimal range bounds. As we said earlier, it is easy to get a guaranteed enclosure for the range, but it may be harder to get TIGHT enclosures.

Other strategies commonly used by interval algorithms include Taylor series plus an enclosure of the remainder term (a generalization of the Mean Value form) and contractive iterations (often Newton's method).

A combination of naive interval evaluation, Mean Value or Slope forms, Taylor series forms, monotonicity tests, intersections of multiple enclosures, and subinterval adaptation can give a tight enclosure for the range of a function. We will apply each of these strategies in the computation of guaranteed error bounds for solutions of ODEs.

5.2 Approximation + [Error]

Interval algorithms tend to use floating-point computations as much as possible, saving the more expensive interval computations for places where they are really necessary [57].

> "In general it is the best in algebraic computations to leave the use of interval arithmetic as late as possible so that it effectively becomes an *a posteriori* weapon." Wilkinson [67]

The algorithms we have already seen for bounding the range of a function use this strategy. The Mean Value, the Slope, and the Taylor forms

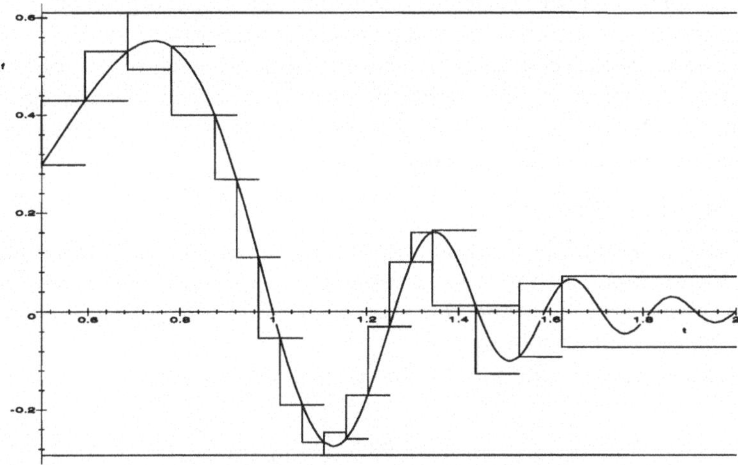

FIG. 20. Range bounds on each subinterval.

each approximate f by points (or a point-valued function, in the case of the Taylor series) plus an enclosure of a remainder (or error) term. The Taylor form illustrates that it is usually advantageous to use an accurate approximation so that the interval-valued remainder term is narrow.

Interval algorithms for ODEs use the Approximation + [Error] strategy, too. We use Taylor series plus an interval-valued remainder, but approximate solutions play other roles in the algorithms as well.

Suppose that $\widehat{u}(t)$ is an approximate solution to $u' = f(t, u)$. Then the error $e(t) := u(t) - \widehat{u}(t)$ satisfies the differential equation

$$\begin{aligned} e' &= u' - \widehat{u}' \\ &= f(t,u) - \widehat{u}' + f(t,\widehat{u}) - f(t,\widehat{u}) \\ &= [f(t,u) - f(t,\widehat{u})] + [f(t,\widehat{u}) - \widehat{u}'] \\ &= \text{defect}(t) + \frac{\partial f}{\partial u}(t, \overline{u}(t))(u - \widehat{u}) \end{aligned}$$

for some function $\overline{u}$ between u and $\widehat{u}$, that is

$$e' = \frac{\partial f}{\partial u}(t, \overline{u}(t))e + \text{defect}(t)\,, \; e(t_0) \in [u_0] - \widehat{u}(t_0)\,. \tag{5.1}$$

Rall [54] gives this formulation. Since $\overline{u}(t)$ is not known, $\dfrac{\partial f}{\partial u}(t, \overline{u}(t))$ is evaluated in interval arithmetic using an interval containing $\overline{u}(t)$. Other, higher-order forms can also be given. This Error ODE (5.1) is *linear* in e.

If $\widehat{u}$ is a good approximation to u, the forcing term defect(t) is small, and eqn (5.1) is smoother and easier to solve than the original problem (1.1).

Another use for point-valued approximate solutions is in [35], where an approximate solution is used to help judge in advance of the validated solution the regions in which neighboring solution trajectories diverge and regions in which they converge. Kerbl proposes that this information be used to inform the step size strategy.

5.3 Wrapping effect

The "wrapping effect" is the name given to the excess width that arises from enclosing regions that are not intervals, mapping them to new non-interval regions, and enclosing them again. The following simple example is due to Moore [46] as told in [56].

The trajectories of individual point-valued solutions of

$$u' = \begin{pmatrix} 0 & 1 \\ -1 & 0 \end{pmatrix} u, \; u_0 \in \begin{pmatrix} [-0.1, 0.1] \\ [0.9, 1.1] \end{pmatrix}$$

are circles in the (u_1, u_2)–phase space. The set of solution values is a rotated rectangle whose interval hull $[u_j]$ is wider than $[u_{j-1}]$, as shown in Fig. 21. Moore shows that the widths of the enclosures grow exponentially, even if the stepsize converges to zero (see [46]), although the actual set of solutions does not increase.

A variety of methods have been proposed to minimize the wrapping effect in the validated solution of ODEs. See surveys in [24, 52, 56]. Moore [46], Eijgenraam [20], Lohner [41], Nickel [51] each used various coordinate transformations. The idea is that instead of propagating the enclosure $[u_j]$ to an interval $u_{j+1} \in [u_{j+1}]$, we compute a *linear transformation* $u_{j+1} \in A_j \, [z_{j+1}]$, for a suitable chosen matrix A_j and interval $[z_{j+1}]$.

6 Design

Purposes of this section:

Outline strategies for validated global error enclosures by methods of differential inequalities, by defect-controlled solutions, and by Lohner's AWA program.

6.1 Differential inequalities

The study of differential inequalities considers theorems of the following forms, in which ____ denotes some hypotheses which vary from one theorem to another.

Theorem 6.1. (Differential Inequalities Template 1) *Assume that $v_0 \leq w_0$ component-wise, and that v and w are the solutions of $u' = f(t, u)$, $u(t_0) = v_0$ and $u' = f(t, u)$, $u(t_0) = w_0$, respectively. If f is ____, then $v(t) \leq w(t)$.*

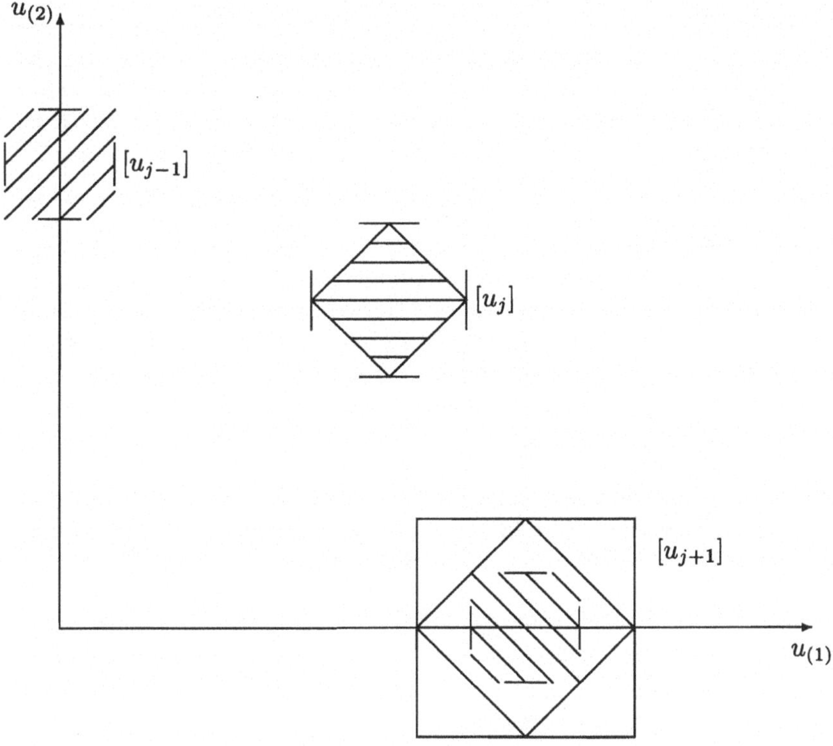

FIG. 21. Wrapping effect.

Theorem 6.2. (Differential Inequalities Template 2) *Assume that v, w are the solutions of $u' = f(t,u)$, $u(t_0) = u_0$ and $u' = g(t,u)$, $u(t_0) = u_0$, respectively. If f and g satisfy ____, then $v(t) \le w(t)$.*

Lakshmikantham and Leela [39] and Walter [62] are excellent texts for differential inequalities. Nickel [50], Adams [2], and others have applied differential inequalities to construct interval algorithms for ODEs. The widths of enclosures computed by algorithms such as Lohner's discussed in Section 6.3 tend to grow from contributions of wrapping, roundoff, and dependency. When conditions ____ hold, the theory of differential inequalities yields optimally tight enclosures. Unfortunately, conditions ____ are quite restrictive.

Definition 6.3. (Quasimonotone [62]) *A function $f(t,u) : \mathbb{R}^{n+1} \to \mathbb{R}^n$ is called quasimonotone if $v \le w$ component-wise and $v_i = w_i$ imply either $(f(t,v))_i \le (f(t,w))_i$ or $(f(t,v))_i \ge (f(t,w))_i$ for each component $i = 1, \ldots, n$. The same inequality must hold for each component and for every t.*

The equation $v_i = w_i$ for $i = 1, \ldots, n$ in Walter's definition does not mean that $v = w$. It means that one component at a time, we let the ith components of v and w be equal, while the other components satisfy $v_j \leq w_j$. Then we check the ith components of $f(t, v)$ and $f(t, w)$ to be sure that $(f(t, v))_i \leq (f(t, w))_i$. We repeat this check for $i = 1, \ldots, n$.

Example 6.1. **($n = 1$).** If $n = 1$, then f is quasimonotone, for $v_i = w_i$ implies that $(f(t, v))_i = (f(t, w))_i$.

Point: A general-purpose suite of programs for computing guaranteed enclosures for solutions of ODEs should treat $n = 1$ as a special case. Even if the interval enclosing the solution is wide (perhaps because of interval-valued initial conditions or parameters in the ODE), we can compute nearly optimal enclosures by following the two endpoints.

Point: A general-purpose suite of programs for computing guaranteed enclosures for solutions of ODEs should check for quasimonotonicity and treat it as a special case.

Theorem 6.4. **(Nickel's Theorem [51])** *Assume that f is quasimonotone. Let v be the point-valued solution to $u' = \underline{f}(t, u)$, $u(t_0) = \underline{u_0}$, and let w be the solution to $u' = \overline{f}(t, u)$, $u(t_0) = \overline{u_0}$. Then $v \leq u \leq w$ componentwise, and these bounds are optimal.*

We can guarantee that f is quasimonotone by bounding the range of the off-diagonal elements of the Jacobian of f. If all of the off-diagonal elements of the Jacobian of f have the same sign for some interval of t and some interval of u, then f is quasimonotone, as long as t and u remain within their assumed ranges, an assumption we can verify.

If f is quasimonotone, Nickel's Theorem says that instead of propagating one relatively wide interval-valued solution $[u_j]$ at $t = t_j$ forward to $t = t_{j+1}$, possibly incurring a growth in the over-estimation, we can propagate 2^n (!!) thin solutions at each of the corners of $[u_j]$. The resulting enclosure is nearly optimal, but it is expensive. As we observed in Section 5, we can achieve two goals from among guaranteed, tight, and fast.

I know of no existing software designed specifically to validate and solve ODEs for quasimonotone right-hand sides.

6.2 Defect-controlled solutions

6.2.1 *Maple worksheet defect_g.ms*

A DEFECT is a function describing the extent to which an approximate solution to an ordinary differential equation fails to satisfy the differential equation. Consider a system of ODEs

$$u' - f(t, u) = 0, u(t_0) = u_0 .$$

We state the ODE in this form so that it will be easier later to verify that a candidate function actually is a solution. A defect-controlled numerical solution must

Loop for each integration step
  Compute an approximate solution $\widehat{u}(t)$
  Compute the defect $d(t) := \widehat{u}'(t) - f(t, \widehat{u}(t))$
  Compute the smallest step for which $||d(t)|| \leq$ Tolerance
  Extend the solution

For discussions of defect-controlled methods, see [14, 21, 22]. This section explores the mathematical concept of using the defect to control the step size of an numerical ODE solver. For example:

```
> DEsystem := [ diff (x(t), t) - y(t) + 3,
>               diff (y(t), t) - x(t) - 1 ];
> InitialTime := 0;
> FinalTime   := 4;
>      # Same order as in SolutionFunctions:
> InitialValue := [ 1, -1];        # [x(t0), y(t0)]
> SolutionFunctions := [x, y];
> Order := 6;
> Tolerance := 1E-6;
```

We give `DEsystem` as a list, although `dsolve` must be passed a set so that we retain control of the order in which the equations are stated. This is important when we form the defect.

DEFINE AN APPROXIMATE SOLUTION

The defect is defined in terms of an `ApproximateSolution`. The "approximate solution" must be

- Defined in a neighborhood of the `InitialTime`
- Differentiable in a neighborhood of the `InitialTime`
- Nearly satisfy the `InitialValue`.

We use a truncated Taylor series as the `ApproximateSolution`, but other forms would work as well.

```
> InitialConditions
>       := convert ( zip ( (fn, val) -> fn(InitialTime) = val,
>                          SolutionFunctions, InitialValue),
>                    set );
```

$$\mathit{InitialConditions} := \{\mathrm{x}(0) = 1, \mathrm{y}(0) = -1\}$$

```
> ApproximateSolution := ComputeApproximateSolution
>          ( DEsystem, InitialConditions,
>            SolutionFunctions, Order );
```

$$\mathit{ApproximateSolution} := \left\{\mathrm{y}(t) = -1 + 2t - 2t^2 + \frac{1}{3}t^3 - \frac{1}{6}t^4 + \frac{1}{60}t^5, \right.$$

$$\left. \mathrm{x}(t) = 1 - 4t + t^2 - \frac{2}{3}t^3 + \frac{1}{12}t^4 - \frac{1}{30}t^5 \right\}$$

KEY IDEA OF DEFECT-CONTROLLED METHODS

Normally, one might explore

$$\text{error}(t) := \text{TrueSolution}(t) - \texttt{ApproximateSolution}(t).$$

Defect methods take a different point of view. We generate a new differential equation

$$u' = f(t, u) + \text{Defect}(t), \tag{6.1}$$

such that

- Defect(t) is "small" so that the new equation is "close to" the original equation, and
- `ApproximateSolution`(t) is the EXACT solution to eqn (6.1).

In many applications, eqn (6.1) is just as good a model of the real world as the specified eqn (1.1), and it has the advantage of admitting an exact solution. Corless and Corliss [14] discuss the significance of defects in modeling.

COMPUTE A DEFECT

The function Defect(t) can be defined in a variety of ways. We use

$$\text{Defect}(t) := \widehat{u}'(t) - f(t, \widehat{u}(t)), \tag{6.2}$$

where $\widehat{u}$ is the `ApproximateSolution`. This Defect has one component for each component of the solution.

```
> Defect := ComputeVectorDefect ( ApproximateSolution,
>                                  DEsystem );
```

$$Defect := \left[-\frac{1}{60}t^5, \frac{1}{30}t^5\right]$$

As you can see, the `Defect` is small near $t = 0$, but it grows rather large. One could also take a norm, or measure the defect relative to some quantity.

COMPUTE AN INTEGRATION STEP SIZE

Define the desired tolerance. We want to take a step as large as possible, while retaining $||\text{Defect}(t)|| \leq$ Tolerance.

Earlier, we said that the `ApproximateSolution` must "nearly" satisfy the `InitialValue`. That was to ensure that ||`Defect(InitialTime)`|| < `Tolerance`.

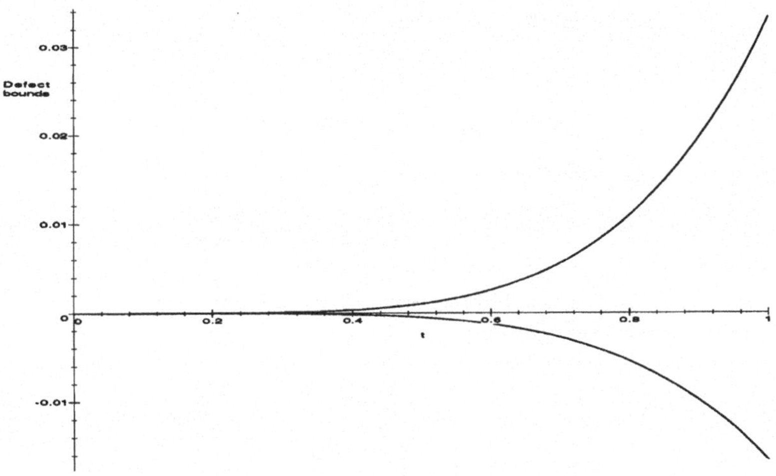

FIG. 22. Defect bounds.

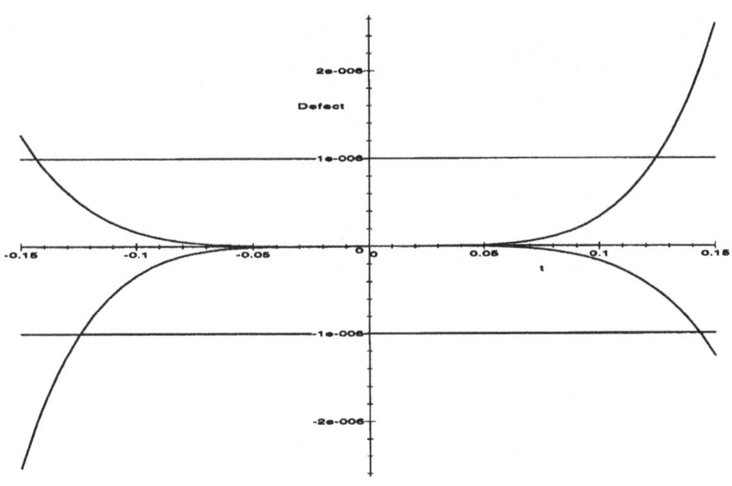

FIG. 23. When does the defect exceed the tolerance?

Earlier, we said that the ApproximateSolution must be differentiable. That was to ensure that the Defect would be continuous. Since

$$||\texttt{Defect(InitialTime)}|| < \texttt{Tolerance},$$

there is some positive step size such that

$$||\texttt{Defect}(t)|| \leq \texttt{Tolerance},$$

for $t \in [\texttt{InitialTime}, \texttt{NextTime} := \texttt{InitialTime} + \text{step size}]$.

```
> NextTime := ComputeStepSize (Defect, InitialTime,
>                               FinalTime, Tolerance);
```

$$NextTime := .1245730940$$

On the time interval [InitialTime, NextTime], the function

```
> Y := ApproximateSolution;
```

$$Y := \left\{ y(t) = -1 + 2t - 2t^2 + \frac{1}{3}t^3 - \frac{1}{6}t^4 + \frac{1}{60}t^5, \right.$$
$$\left. x(t) = 1 - 4t + t^2 - \frac{2}{3}t^3 + \frac{1}{12}t^4 - \frac{1}{30}t^5 \right\}$$

is the EXACT SOLUTION to

$$u' = f(t, u) + \texttt{Defect}(t)$$

```
> DEsolvedExactly := zip ( (deq, def) -> deq - def,
>                          DEsystem, Defect );
```

$$DEsolvedExactly := \left[\left(\frac{\partial}{\partial t} x(t) \right) - y(t) + 3 + \frac{1}{60}t^5, \right.$$
$$\left. \left(\frac{\partial}{\partial t} y(t) \right) - x(t) - 1 - \frac{1}{30}t^5 \right]$$

```
> subs ( Y, DEsolvedExactly );
```

$$\left[\left(\frac{\partial}{\partial t} \left(1 - 4t + t^2 - \frac{2}{3}t^3 + \frac{1}{12}t^4 - \frac{1}{30}t^5 \right) \right) \right.$$
$$+4 - 2t + 2t^2 - \frac{1}{3}t^3 + \frac{1}{6}t^4,$$
$$\left(\frac{\partial}{\partial t} \left(-1 + 2t - 2t^2 + \frac{1}{3}t^3 - \frac{1}{6}t^4 + \frac{1}{60}t^5 \right) \right)$$
$$\left. -2 + 4t - t^2 + \frac{2}{3}t^3 - \frac{1}{12}t^4 \right]$$

```
> expand ( " );
```

$$[0, 0]$$

That is, ApproximateSolution IS the exact solution of eqn (6.1). Figure

24 shows that ||**Defect**(t)|| < **Tolerance**, for $t \in$ [**InitialTime**, **NextTime**].

DEFECT-CONTROLLED ALGORITHM

We have now walked through the steps of a defect-controlled algorithm:

Loop for each integration step
    Compute an approximate solution $\widehat{u}(t)$
    Compute the defect $d(t) = \widehat{u}'(t) - f(t, \widehat{u}(t))$
    Compute the smallest step for which $||d(t)|| \leq$ Tolerance
    Extend the solution

This algorithm computes an **ApproximateSolution** for the original system of ODEs. The **Approximate Solution** and the **Defect** are defined piecewise. We have defined a special data type, **PiecewiseType**, to hold the solutions and the defects on each subinterval. We are now ready to solve the DEsystem on the entire interval of integration. Initialize structures to hold the solution and the defect:

```
> PW_Solution := ConstructPiecewise ();
> PW_Defect   := ConstructPiecewise ();
```

We may take one defect-controlled step. This single procedure call does what we stepped through above.

```
> TakeDefectControlledStep
>       (InitialTime, InitialValue, FinalTime,Tolerance,
>        DEsystem, SolutionFunctions, Order,
>        't1', 'y1', 'PW_Solution', 'PW_Defect');
> print (t1, y1);
```

$$.1245730940, [\,.5159563603, -.7812859666\,]$$

```
> DefectControlledSolution
>       (t0, y0, FinalTime, Tolerance, Steps_to_Take,
>        DEsystem, SolutionFunctions,
>        't0', 'y0', PW_Solution, PW_Defect, 'Statistics'):
> print (PW_Defect);
```

table([
$domain = [\,0, 4.\,]$
$current = 5$
$count = 6$
$0 =$ table([
$domain = [\,0, .76593025837896383629\,]$
$$function = \left[t \to -\frac{1}{181440}\,t^9, t \to \frac{1}{90720}\,t^9\right]$$
])
$1 =$ table([
$domain = [\,.76593025837896383629, 1.5421895446686786659\,]$

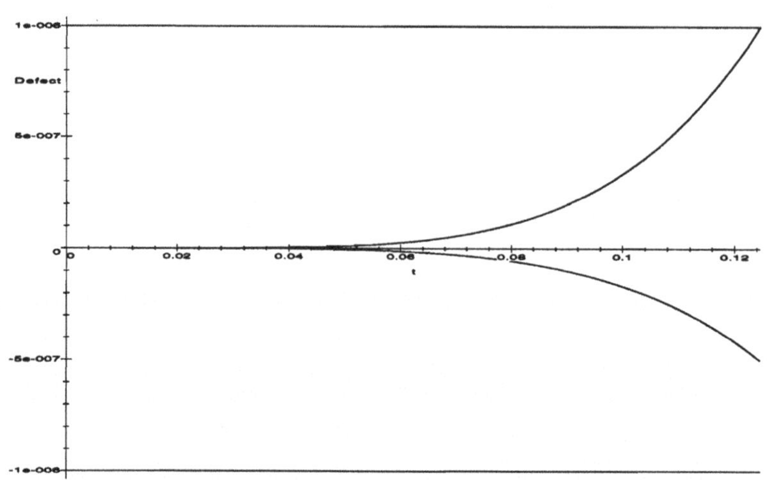

FIG. 24. $||$`Defect`$(t)|| <$ `Tolerance`, for $t \in$ [`InitialTime`, `NextTime`].

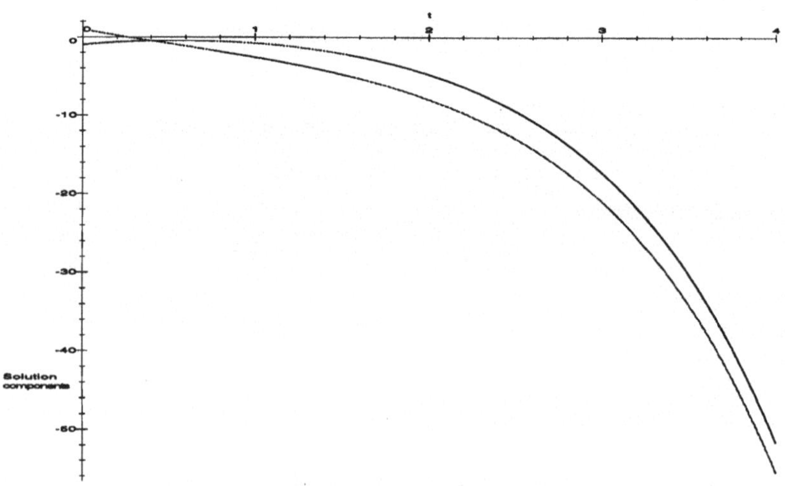

FIG. 25. Solution components from the defect-controlled algorithm.

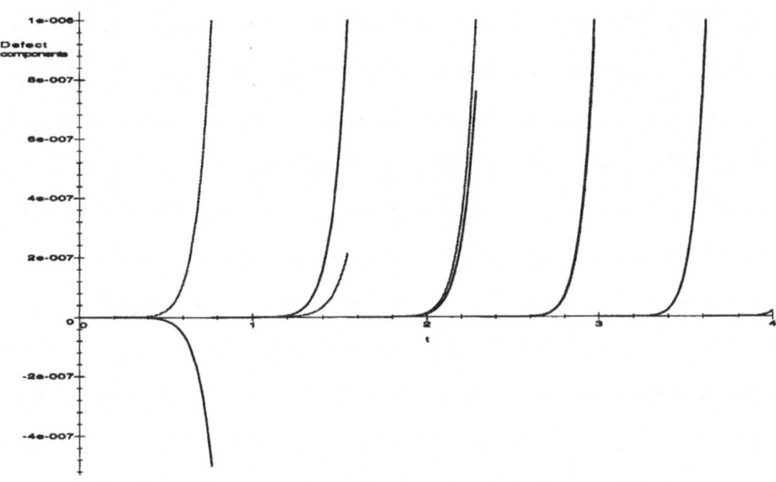

FIG. 26. Defect components.

$$
\begin{aligned}
&\mathit{function} = [t \rightarrow .22216828024627210231\,10^{-5}\,t\\
&+ .000044015605800319514 62\,t^7 - .000078663396754472471 99\,t^6\\
&- .000011602533145326757 39\,t^2 + .000035346008390362963 5\,t^3\\
&- .000069221723525788586 768\,t^4 + .000090376013701680068 2\,t^5\\
&+ .20841344343110900099\,10^{-5}\,t^9 - .000014366714631915499236\,t^8\\
&- .18907267588066961071\,10^{-6}, t \rightarrow .000010415857904682489840\,t\\
&+ .000206357223945951888 53\,t^7 - .000368795564329006810 9\,t^6\\
&- .000054395855448913021 0\,t^2 + .000165711775093929966 54\,t^3\\
&- .000324530412425500008 42\,t^4 + .000423707522813297947 88\,t^5\\
&+ .97709934550413794669\,10^{-5}\,t^9 - .000067355095874751075059\,t^8\\
&- .88642452624124474218\,10^{-6}]
\end{aligned}
$$

])

...

])

The defect-controlled method we have just described has given guarantees on how far from the original Problem (1.1) is a problem for which $\widehat{u}$ is the exact solution. It does not consider how far $\widehat{u}$ might be from the true solution of Problem (1.1). Stetter [59, 60] has given an interval version of a defect correction algorithm. After the computations outlined above, he goes on to solve eqn (6.1), whose exact solution is the point-valued func-

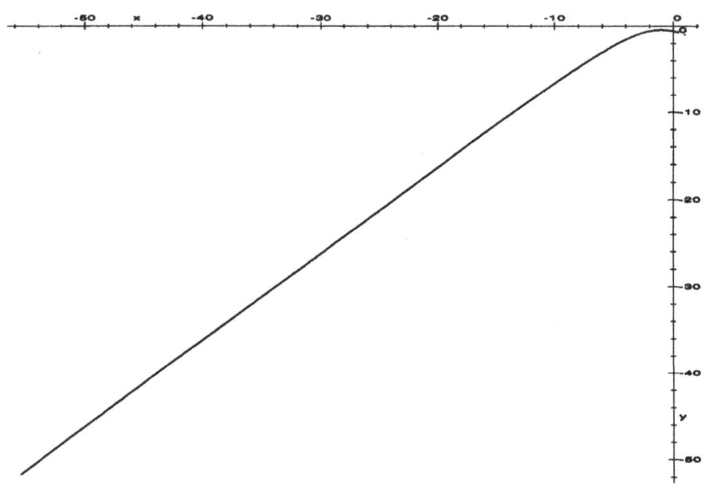

FIG. 27. Phase portrait of the solution.

tion $\widehat{u}$. By comparing the difference between the approximate and the true solutions to eqn (6.1), he bounds the error in $\widehat{u}$.

6.3 Lohner's method

The most widely used software for the validated solution of ODEs is Rudolf Lohner's AWA (Anfangswertaufgabe) [41, 42, 43, 44, 45]. Most of the pieces we need to understand AWA are already in place. Lohner uses a one-step local expansion as suggested by Moore [46]. Much of the material in the rest of this section is drawn from Lohner's thesis [43].

As we discussed in Section 5.3, Lohner attempts to minimize the wrapping effects by representing the enclosure of the solution at each time step as a linear transformation of an interval: $u(t_j) \in A_0 \cdots A_{j-1} [x_j]$ at time $t = t_j$. He advances the solution to $t = t_{j+1} := t_j + h$ using two algorithms:

Algorithm I. Validate existence and uniqueness

Input:

- Problem: $u' = f(t, u)$
- Initial conditions: t_j, $u(t_j) \in [u_j]$
- Final time: t_f
- Tolerance: Tolerance

Input assertions:

Width $w([u_j]) <$ Tolerance

$f(t_j, [u_j])$ can be evaluated with no $\sqrt{0}$ or fractional powers of non-posit numbers.

Output:

Either

Coarse enclosure: $[u_j]_0$

Step size: h

or else

Failure

Output assertions:

$u' = f(t, u)$, $u(t_j) = u_j \in [u_j]$ has a unique solution in $[t_j, t_j + h] \times [u_j]_0$

$u(t) \in [u_j]_0$ for all $t \in [t_j, t_j + h]$

Basic idea:

Picard–Lindelöf iteration

Algorithm II. Compute tighter enclosure

Input:

Problem: $u' = f(t, u)$

Initial conditions: t_j, $u(t_j) \in [u_j]$

Current solution: matrices $A_0, A_1, \ldots, A_{j-1}$ and interval $[x_j]$

Coarse enclosure: $[u_j]_0$

Step size: h

Tolerance: Tol

Input assertion:

$u(t) \in [u_j]_0$ for all $t \in [t_j, t_j + h]$

Output:

Either

Matrix: A_j

Interval: $[x_{j+1}]$

Next node: $t_{j+1} \leq t_j + h$

Solution: $[u_{j+1}] :=$ Convex hull $(A_0 A_1 \cdots A_j \, [x_{j+1}])$ with $w([u_{j+1}]) \leq$ Tol

or else

Failure

Also available:

Approximation solution: $\widehat{u}(t)$

Continuous enclosure on each subinterval: $[u_j]_1 \, (t)$

Output assertion:

$u(t_{j+1}) \in A_0 A_1 \cdots A_j \, [x_{j+1}]$

Basic idea:

Truncated Taylor series plus the remainder term

Local coordinate transformations

Next, we discuss Algorithms I and II in greater detail. Lohner gives a complete discussion in his thesis [43].

6.3.1 *Algorithm I. Validate existence and uniqueness*

Algorithm I attempts to validate the existence of a unique solution to the problem $u' = f(t, u)$, $u(t_j) \in [u_j]$ by applying the Existence and Uniqueness Theorems. It applies the Banach Fixed Point Theorem to the Picard–Lindelöf operator

$$\Phi(u)(t) := u_j + \int_{t_j}^{t} f(\tau, u(\tau))\, d\tau$$

mapping the set of continuous functions on the interval $T_j := [t_j, t_j + h]$ into itself, as we illustrated in Section 4.1.2 for the Lorenz equations. We discuss the steps of Algorithm I in turn.

Algorithm I:

1. Guess an a priori enclosure $[u_j]_0$; guess a step size h;
2. Compute $[u_j]_1 := \Phi([u_j]_0)$;
3. If $[u_j]_1 \subseteq [u_j]_0$
   then Algorithm I is successful;
4. else try again.

Step I.1. Guess an a priori enclosure $[u_j]_0$ and a step size h.
The initial guesses for $[u_j]_0$ and h are tightly coupled, and there are difficult trade-offs. As long as $w([u_j]) <$ Tolerance (as we have assumed), there is *some* step that can be taken. We want to take a step h as large as possible for efficiency. In general, the larger we choose h, the larger we must choose $[u_j]_0$, because the true solution trajectories will move farther from their initial conditions $(t_j, [u_j])$. However, as the widths of the intervals T_j and $[u_j]_0$ grow, so does the width of $\Phi([u_j]_0)$, making it less likely that $[u_j]_1 \subseteq [u_j]_0$. For an initial guessed h, if $[u_j]_1$ is not $\subseteq [u_j]_0$, we can always find a new, smaller h_1 for which $[u_j]_1 \subseteq [u_j]_0$. However, that h_1 is usually smaller than would have been necessary if we had chosen a smaller h to start with. At any rate, we can always reduce the step size, but it is harder to increase it.

Lohner uses a strategy developed by Eijgenraam [20] based on the current solution value and its derivative. The current version of AWA uses a fixed step size h. Most of the current activity in the area of validated techniques for ODEs is directed at a better step size selection strategy. Given a guess for h, define the interval

$$[u_j]_0 := [u_j] + [0, h]\,[f(t_j, [u_j])] + h \cdot \beta \cdot [-1, 1]\,,$$

for some algorithm tuning parameter β. The solution starts the step at the point $(t_j, [u_j])$ with slope $f(t_j, [u_j])$. If that is as large as the slope gets, the solution value could change by $h \cdot f(t_j, [u_j])$. Of course, the slope in the interval $T_j := [t_j, t_j + h]$ can differ from $f(t_j, [u_j])$, so we inflate with the term $h \cdot \beta \cdot [-1, 1]$, as shown in Fig. 28.

Another strategy for discovering an initial guess for a suitable $[u_j]_0$ is to use an approximate ODE solver. Take as $[u_j]_0$ the range of the approximate solution $\widehat{u}$, plus a small inflation to enclose (hopefully) the truncation and roundoff errors in the approximate method. A Taylor series is a natural choice since we need it in Algorithm II. The step size chosen by the approximate solver might be a reasonable initial guess for h, too. Alternatively, $\widehat{u}$ can be used to form the error equation to which the methods being described here are applied.

Step I.2. $[u_j]_1 := \Phi([u_j]_0)$.
We wish to bound the range of the function

$$\Phi(u)(t) := u_j + \int_{t_j}^{t} f(\tau, u(\tau))\, d\tau$$

for $u \in [u_j]_0$ and $t \in T_j = [t_j, t_j + h]$. Any of the strategies discussed in Section 5 for computing tight range bounds may be applied.

Naive interval evaluation:

$$[u_j]_1 := \Phi([u_j]_0)(T_j) \subseteq [u_j] + [0, h] \cdot \left[f(T_j, [u_j]_0)\right] .$$

Instead of a constant enclosure, we get a linear inclusion if we replace $[0, h]$ by $(t - t_j)$

$$[u_j]_1\,(t) := [u_j] + \left[f(T_j, [u_j]_0)\right] \cdot (t - t_j) .$$

In this form, if $[u_j]_1\,(t_j + h)$ is not $\subseteq [u_j]_0$, we CAN find a smaller $0 < h_1 < h$ for which we DO have $[u_j]_1\,(t_j + h_1) \subseteq [u_j]_0$. We use $h = h_1$.

As an example, consider the Lorenz system:

$$\textit{Lorenz system} = \left[\left(\frac{\partial}{\partial t}\mathrm{x}(t)\right) - 10\,\mathrm{y}(t) + 10\,\mathrm{x}(t),\right.$$
$$\left(\frac{\partial}{\partial t}\mathrm{y}(t)\right) - 28\,\mathrm{x}(t) + \mathrm{y}(t) + \mathrm{x}(t)\,\mathrm{z}(t),$$
$$\left.\left(\frac{\partial}{\partial t}\mathrm{z}(t)\right) - \mathrm{x}(t)\,\mathrm{y}(t) + \frac{8}{3}\mathrm{z}(t)\right]$$
$$t\_j := 0$$

$$\textit{Initial Value} = [10, 10, 10]$$

AWA uses $[u_j]_0 := [u_j] + [0, h][f(t_j, [u_j])] + h\beta[-1, 1]$. In order to make the graphs look more instructive, we inflate to interval-valued initial conditions.

$$U\_j := [9.899999999, 10.10000001], [9.899999999, 10.10000001],$$
$$[9.899999999, 10.10000001]$$

```
> EstimatedSlopeBnd := map (i -> inapply (i, x, y, z)(U_j), f);
```

$$EstimatedSlopeBnd := [[-2.000000221, 2.000000221], [165.0899993, 174.8900006], [71.07666659, 75.61000032]]$$

$$InitialGuessFor\_h := .1$$
$$H := [0, .1]$$
$$T\_j := 0...1$$
$$BETA := [-3.000000001, 3.000000001]$$

```
> U_j_0 := NULL:
> for i from 1 to n do
>    U_j_0 := U_j_0, ( U_j[i] &+ (H &* EstimatedSlopeBnd[i])
>                      &+ (H &* BETA) ):
> od:
```

At $t = 0$,

$$\textit{Initial guess for } h = .1$$

$$\textit{with initial conditions} = ([9.899999999, 10.10000001], [9.899999999, 10.10000001], [9.899999999, 10.10000001])$$

Our guess for *a priori* enclosures of each component of the solution to the Lorenz equation:

```
> U_j_0;
```

$$[9.399999975, 10.60000005], [9.599999997, 27.88900010], [9.599999997, 17.96100006]$$

Next, we draw lines whose slopes are the estimated and the inflated bounds for u':

```
> component := 3;
> EstimatedSlopeLine
>       := U_j[component][Inf]
>            + EstimatedSlopeBnd[component][Inf] * (t - t_j),
>          U_j[component][Sup]
>            + EstimatedSlopeBnd[component][Sup] * (t - t_j);
> InflatedSlopeLine
>        := EstimatedSlopeLine[1] - beta * (t - t_j),
>           EstimatedSlopeLine[2] + beta * (t - t_j);
> APrioriBound := U_j_0[component][Inf], U_j_0[component][Sup];
```

If we replace $[0, h]$ by $(t - t_j)$ we get a linear enclosure

$$[u_j]_1(t) := [u_j] + [f(T_j, [u_j]_0)](t - t_j)$$

Constant guess for a priori bounds:

$$U\_j\_0 = [9.399999975, 10.60000005], [9.599999997, 27.88900010], [9.599999997, 17.96100006]$$

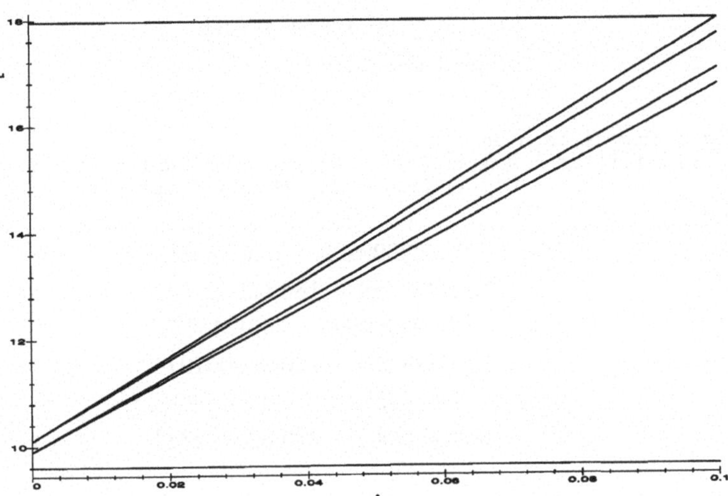

FIG. 28. Guess an a priori enclosure $[u_j]_0$ and a step size h.

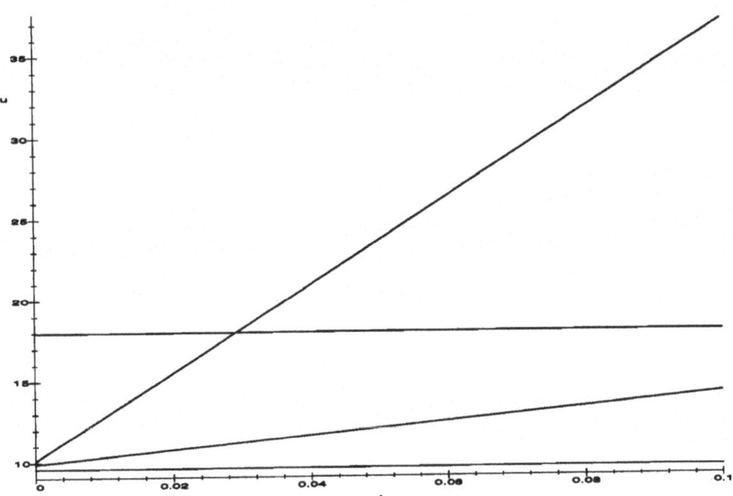

FIG. 29. Constant and linear a priori enclosures.

```
> SlopeBnd := map (i -> inapply (i, x, y, z)(U_j_0), f);
```

$$SlopeBnd := [[-10.00000065, 184.8900015], [44.92439729, 196.9600019], [42.34399953, 270.0234027]]$$

```
> for i from 1 to n do
>    U_j_1[i] := [ U_j[i][Inf] + SlopeBnd[i][Inf] * (t - t_j),
>                  U_j[i][Sup] + SlopeBnd[i][Sup] * (t - t_j) ]:
> od;
```

$$U\_j\_1_1 := [9.899999999 - 10.00000065\,t, 10.10000001 + 184.8900015\,t]$$
$$U\_j\_1_2 := [9.899999999 + 44.92439729\,t, 10.10000001 + 196.9600019\,t]$$
$$U\_j\_1_3 := [9.899999999 + 42.34399953\,t, 10.10000001 + 270.0234027\,t]$$

When do the linear bounds exit the region?

```
> for i from 1 to n do
>    ExitPoint[i]
>       := [ solve ( U_j_1[i][Inf] = U_j_0[i][Inf] ),
>            solve ( U_j_1[i][Sup] = U_j_0[i][Sup] ) ];
> od;
```

$$ExitPoint_1 := [.04999999915, .002704310866]$$
$$ExitPoint_2 := [-.006677885962, .09031783062]$$
$$ExitPoint_3 := [-.007084829145, .02911229164]$$

```
> h := InitialGuessFor_h;
> for i from 1 to n do
>    temp := ExitPoint[i][Inf];
>    if ( 0 < temp ) and ( temp < h) then h := temp fi;
>    temp := ExitPoint[i][Sup];
>    if ( 0 < temp ) and ( temp < h) then h := temp fi;
> od;
> `Step size` = h;
```

$$\textit{Step size} = .002704310866$$

Notice that this is a larger step than the 0.00057 we validated in Section 4.1.2. We have validated that for t in

$$T\_j := [0, .002704310866]$$

the components of the solution for the Lorenz system

1) exist,
2) are unique, and
3) lie in the constant bound enclosure

$$U\_j\_0 \;=\; [\,9.399999975, 10.60000005\,], [\,9.599999997, 27.88900010\,], [\,9.599999997, 17.96100006\,]$$

and in the linear enclosure

```
> print ( U_j_1 );
```

table([
$1 = [\,9.899999999 - 10.00000065\,t, 10.10000001 + 184.8900015\,t\,]$
$2 = [\,9.899999999 + 44.92439729\,t, 10.10000001 + 196.9600019\,t\,]$
$3 = [\,9.899999999 + 42.34399953\,t, 10.10000001 + 270.0234027\,t\,]$
])

Mean Value Form (see [32, 56])
If we use the Mean Value Form

$$f(\tau, [u]\,(\tau)) = f(\tau, \tilde{u}(\tau)) + \frac{\partial f}{\partial u}(\tau, [u]\,(\tau)) \cdot ([u]\,(\tau) - \tilde{u}(\tau)),$$

then

$$[u_j]_1\,(t) := [u_j] + \int_{t_j}^{t} f(\tau, \tilde{u}(\tau))\, d\tau + \int_{t_j}^{t} \frac{\partial f}{\partial u}(\tau, [u]\,(\tau)) \cdot ([u]\,(\tau) - \tilde{u}(\tau))\, d\tau\,.$$

The disadvantage of the validation procedures described above is that they all restrict the step size that may be taken by the algorithm to roughly the step size for Euler's method, even if high order Taylor series are used for the tightening in Algorithm II.

Lohner has proposed a method for validating an a priori enclosure using Taylor expansions of higher order. Guess an interval $[u_p] = [\underline{u_p}, \overline{u_p}]$, where $\underline{u_p}$, $\overline{u_p} \in \mathbb{R}^n$, which we hope encloses $u^{(p)}$. We might guess $[u_p]$ by extrapolating from $u^{(p)}$ on the previous interval or by solving the ODE with an approximate high-order Taylor series method. Let $[u_p]\,(t)$ denote

$$[u_p]\,(t) := \{x \in C(T_j) : \underline{u_p} \le x(t) \le \overline{u_p},\; t \in T_j\},$$

the set of functions continuous on the interval $T_j = [t_j, t_j + h]$ whose range is contained in $[u_p]$. Let $V(t)$ denote

$$V(t) := \Big\{v(t) \;=\; u(t_j) + u'(t_j)(t - t_j) + \cdots + \frac{u^{(p-1)}(t_j)}{(p-1)!}(t - t_j)^{p-1} + [u_p]\,(t)(t - t_j)^p\Big\}\,. \quad (6.3)$$

$V(t)$ is a set of continuous functions. The derivative values in eqn (6.3) are exact (although unknown) values. Now, view Φ as a map from $V(t)$ to

$V(t)$. If $V_0(t)$ is a function strip of the form (6.3) with $[u_p] = [u_p]_0$, let

$$V_1(t) := \Phi(V_0(t))(t) = u(t_j) + \int_{t_j}^{t} f(\tau, V_0(\tau))\, d\tau\,. \tag{6.4}$$

The beauty of eqn (6.4) is that we do not need to compute with the low order terms. Since the low order terms of functions in $V_0(t)$ are assumed to be the exact values of the solution to the ODE, they MUST be the same on both sides of eqn (6.4). Hence, we need only compute an enclosure of the remainder term. Lohner has written automatic differentiation operators following the work of Eiermann to do the required computations. If $V_1(t) \subseteq V_0(t)$ for all $t \in T_j$, then the ODE has a unique solution in $V_1(t)$.

The effect of Lohner's idea is to attempt to enclose u in a time-dependent strip, rather than in a constant band. The resulting computations in the integration of $f(\tau, V_0(\tau))$ are more subtle, but the intervals are narrower, and the resulting step sizes may be taken roughly as large as would be appropriate for a method of order p.

For example, consider the ODE

$$u' = \frac{2(u - t - 1/2)}{(u - t + 1)(u - t - 1)}\,.$$

Trajectories are vertical on the lines $u = t \pm 1$; there are horizontal tangents on the line $u = t + 1/2$; and trajectories converge toward the line $u = t$. If we wish to follow a trajectory near the line $u = t$, any constant a priori bounds have corners in regions with relatively high slopes, while a band of the form (6.3) can enclose the desired solution while staying in a region with slopes near 1.

Another possible technique to validate the existence of a unique solution is the Taylor series enclosures illustrated in Fig. 1. By Taylor's Theorem,

$$\begin{aligned} u(t) &= u(t_j) + u'(t_j)(t - t_j) + \ldots + \frac{u^{(p-1)}(t_j)}{(p-1)!}(t - t_j)^{p-1} \\ &\quad + \frac{u^{(p)}(\tau)}{p!}(t - t_j)^p \text{ for } t, \tau \in T_j \\ &\in u(t_j) + u'(t_j)(t - t_j) + \ldots + \frac{u^{(p-1)}(t_j)}{(p-1)!}(t - t_j)^{p-1} \\ &\quad + \frac{u^{(p)}(T_j)}{p!}(t - t_j)^p\,. \end{aligned}$$

Let

$$[u_j]_1(t) := u(t_j) + u'(t_j)(t - t_j) + \ldots + \frac{u^{(p-1)}(t_j)}{(p-1)!}(t - t_j)^{p-1}$$

$$+\frac{u^{(p)}(T_j)}{p!}(t - t_j)^p .$$

The Taylor coefficients $u^{(i)}(t_j)/i!$ are readily computed from the ODE using automatic differentiation. In practice, we must evaluate the recurrence relations given by AD in interval arithmetic beginning with the interval initial conditions $u(t_j) \in [u_j]$. The interval-valued remainder $u^{(p)}(T_j)/p!$ is computed by evaluating exactly the same recurrence relations in interval arithmetic using the intervals $t = T_j$ and the assumed a priori bound $[u_j]_0$.

Conjecture 6.5 *If* $[u_j]_1(t) \subseteq [u_j]_0$ *for* $t \in T_j$, *then* $u' = f(t,u)$, $u(t_j) \in [u_j]$ *has a unique solution* u *in* T_j *with* $u(t) \in [u_j]_1(t)$.

A proof of this conjecture would exhibit some integral equation that is a contractive map in some space of continuous functions. The hard part is to show existence and uniqueness; the enclosure is Taylor's Theorem. A counter-example would be VERY interesting, especially if its right-hand side involved only arithmetic operations and elementary functions.

Step I.4. If Algorithm I is NOT successful, try again.
We have discussed various strategies for validating that the ODE has a unique solution in $[u_j]_0$ for $t \in T_j := [t_j, t_j + h]$. If we are successful, then Algorithm I has done its job, and we proceed to use Algorithm II to tighten the bounds. When $[u_j]_1 \subseteq [u_j]_0$, we can often tighten the enclosure by repeating the Picard–Lindelöf iteration. AWA does not do this, but leaves the task of tightening the validated enclosure to the more powerful Algorithm II.

If Algorithm I is NOT successful, we must try again.

If $w([u_j]) <$ Tolerance (as we have assumed), then we can always use the linear enclosure to find a step size h that works. However, that step is often too small, so we prefer to explore some strategies for trying again. In the repeated execution of the construction, the alternatives are:

1. Try a smaller initial guess for h.
2. Try a narrower initial guess for $[u_j]_0$, in the hope of reducing over-estimations from wide intervals.
3. Try a wider initial guess for $[u_j]_0$, in the hope of making it wide enough to enclose the true solution.

AWA uses the third strategy. If the initial guesses for h and $[u_j]_0$ do not yield a validated coarse enclosure, $[u_j]_0$ is widened, and the validation is repeated. If repeated inflations of $[u_j]_0$ still do not yield a validated enclosure, the step size is reduced.

The careful reader may have noticed that AWA has not verified that f satisfies a Lipschitz condition in u. In his thesis [43], Lohner applies the following theorem to show it suffices to verify the enclosure $[u_0]_1 \subseteq [u_0]_0$.

Theorem 6.6. (**[43, 62]**) *If we can find intervals* $[t_0, t_0 + h]$ *and* $[u_0]_0$ *such that*

$$[u_0]_1 := u_0 + [0, h] \cdot f([t_0, t_0 + h], [u_0]_0) \subseteq [u_0]_0 \,,$$

then the ODE $u' = f(t, u)$, $u(t_0) = u_0$ *has a unique solution.*

6.3.2 *Algorithm II. Compute tighter enclosure*

AWA enters Algorithm II with a guarantee that the ODE has a unique solution u, and that $u(t) \in [u_j]_0$ for all $t \in T_j = [t_j, t_j + h]$. Hence, a local Taylor expansion encloses u:

$$\begin{aligned} u(t) \quad \in \quad & [u_j]_1(t) := [u_j] + [u'](t_j)(t - t_j) + \ldots \\ & + \frac{[u^{(p-1)}](t_j)}{(p-1)!}(t - t_j)^{p-1} + \frac{[u^{(p)}](T_j)}{p!}(t - t_j)^p \,. \end{aligned}$$

If we simply let $[u_{j+1}] := [u_j]_1(t_j + h)$, we have $w([u_{j+1}]) > w([u_j])$ because of the widths contributed by the higher-order series terms. Enclosures computed by this algorithm do not contract, even when the true trajectories are strongly contracting. Hence, we need to be more clever. AWA uses the essential idea of Eijgenraam [20] and Lohner [41, 42] to express the enclosure as

$$\begin{aligned} u(t) \quad \in \quad & \begin{pmatrix} \text{point-valued truncated Taylor} \\ \text{series for approximate solution} \end{pmatrix} + \begin{pmatrix} \text{old} \\ \text{error} \end{pmatrix} \\ & + \begin{pmatrix} \text{enclosure of local} \\ \text{truncation error} \end{pmatrix} . \end{aligned}$$

AWA computes the enclosure of the local truncation error by solving a variational equation. This allows AWA to follow contractive flows and to control the wrapping effect.

Algorithm II:

1. Compute point-valued approximate solution $\widehat{u}_j(t)$;
2. Compute enclosure $[z_{j+1}]$ of the local error of $\widehat{u}_j(t)$;
3. Compute enclosure for the transformation matrix $[A_j]$;
4. Prepare for next solution step;

Initially, let $\tilde{u}_0 :=$ midpoint $([u_0])$, $s_0 := 0$, and $[z_0] := [u_0]$.

Step II.1. Point-valued approximate solution $\widehat{u}_j(t)$.

We need to compute a point-valued approximate solution value at time $t_{j+1} := t_j + h$. Any method is satisfactory, provided that we can compute the necessary enclosures of the local truncation errors. AWA uses a local Taylor series expansion because

subsequent error analysis is easy,

Taylor series approximations are fast and accurate,

the order can be adjusted easily,

a local Taylor expansion may have already been computed by Algorithm I,

an estimate for $u^{(p)}(t_{j+1})$ required by Algorithm I at the next step is readily available, and

the AD machinery for Taylor expansions is necessary anyway.

The point $\tilde{u}_j$ is in the interval $[u_j]$. Usually, $\tilde{u}_j$ is taken as an approximate midpoint of $[u_j]$. Let $\widehat{u}_j(t)$ be the truncated point-valued Taylor series expanded at the point $(t_j, \tilde{u}_j)$ computed by automatic differentiation.

$$\widehat{u}_j(t) := \tilde{u}_j + \tilde{u}_j'(t - t_j) + \ldots + \frac{\tilde{u}_j^{(p-1)}}{(p-1)!}(t - t_j)^{p-1}.$$

Step II.2. Enclosure $[z_{j+1}]$ of the local error of $\widehat{u}_j(t)$ at t_{j+1}.
The true solution to $u' = f(t, u)$, $u(t_j) = \tilde{u}_j$, has the local truncation error

$$\frac{1}{p!}u^{(p)}(\tau)(t_{j+1} - t_j)^p \subseteq \frac{1}{p!}\left[u^{(p)}\right](T_j)h^p.$$

AWA computes $[z_{j+1}]$ as an enclosure of $\left[u^{(p)}\right](T_j)h^p/p!$ by evaluating the recurrence relation from the ODE in interval arithmetic. The resulting bound for $\left[u^{(p)}\right](T_j)h^p/p!$ may be an overestimate; [19] describes adaptations of the strategies discussed in Section 4.1.1 to the problem of computing bounds on the range of derivatives.

Let $\tilde{u}_{j+1} :=$ midpoint $([z_{j+1}])$.

Step II.3. Enclosure for the transformation matrix $[A_j]$.
In Section 5.3, we described the wrapping effect. AWA attempts to reduce the effect by propagating not interval enclosures of the solution, but *linear transformations* of intervals. The necessary linear transformation matrices are computed in Step II.3. AWA views the solution at t_{j+1} as a function depending on all of the previously committed local errors

$$u_{j+1} = u_{j+1}(z_0, z_1, \ldots, z_{j+1}).$$

The values z_0, z_1, ..., z_{j+1} are the exact (but unknown) local errors committed at each step. We expand u_{j+1} by the Mean Value Theorem

$$u_{j+1} = \widehat{u}_j(t_{j+1}) + \sum_{k=0}^{j+1} \frac{\partial u_{j+1}}{\partial z_k}(\widehat{z}) \cdot (z_k - s_k), \tag{6.5}$$

where $\widehat{u}_j(t_{j+1}) = u_{j+1}(s_0, s_1, \ldots, s_{j+1})$. That is, $s_{j+1} := \tilde{u}_{j+1} - \widehat{u}_j(t_{j+1})$. The derivatives appearing in eqn (6.5) can be computed recursively. Let

$$\begin{aligned}
A_{j+1,k} &:= \frac{\partial u_{j+1}}{\partial z_k}, \text{ for } k \leq j+1 \\
A_{j+1,j+1} &:= I \\
A_j &:= \frac{\partial u_{j+1}}{\partial u_j} \\
A_{j+1,k} &:= \frac{\partial u_{j+1}}{\partial z_k} = \frac{\partial u_{j+1}}{\partial u_j} \cdot \frac{\partial u_j}{\partial z_k} = A_j \cdot A_{j,k} = A_j A_{j-1} \cdots A_{j-k} .
\end{aligned}$$

Then we have

$$\begin{aligned}
u_{j+1} &= \widehat{u}_j(t_{j+1}) + \sum_{k=0}^{j} A_{j+1,k}(z_k - s_k) + z_{j+1} \\
&= \widehat{u}_j(t_{j+1}) + A_j \sum_{k=0}^{j} A_{j,k}(z_k - s_k) + z_{j+1} \\
&= \widehat{u}_j(t_{j+1}) + A_j A_{j-1} \cdots A_0(z_0 - s_0) \\
&\qquad + A_j A_{j-1} \cdots A_1(z_1 - s_1) \\
&\qquad \vdots \\
&\qquad + A_j(z_j - s_j) \\
&\qquad + I(z_{j+1} - s_{j+1}) . \qquad (6.6)
\end{aligned}$$

Equation (6.6) is in terms of exact, but unknown values. In order to compute, we apply enclosures. Let $y(t) \in \mathbb{R}^{n \times n}$ be the solution to the variational equation

$$y' = \frac{\partial f}{\partial u} \cdot y, \; y(t_j) = I, \qquad (6.7)$$

where the partial derivative $\frac{\partial f}{\partial u}$ is evaluated along the true solution of $u' = f(t, u)$. Then

$$A_j = y(t_j) + y'(t_j)h + \cdots + \frac{y^{(p-1)}(t_j)}{(p-1)!}h^{p-1},$$

a truncated Taylor series with no remainder. Now we want an enclosure of A_j, so it suffices to solve the $n + n \times n$-dimensional ODE system

$$\begin{aligned}
u' &= f(t, u), \; u(t_j) = [u_j] \\
y' &= \frac{\partial f}{\partial u}(t, u) \cdot y, \; y(t_j) = I . \qquad (6.8)
\end{aligned}$$

The solution of system (6.8) gives the desired enclosure $[A_j]$. We solve the variational equation (6.7) by an interval Taylor series method.

Step II.4. Prepare for next solution step.
In Step II.2, AWA computed $[z_{j+1}]$ and $\tilde{u}_{j+1} :=$ midpoint $([z_{j+1}])$. In Step II.3, AWA computed $s_{j+1} := \tilde{u}_{j+1} - \widehat{u}_j(t_{j+1})$ and $[A_j]$. Now we have

$$\begin{aligned} u(t_{j+1}) \in [u_{j+1}] \quad := \quad & \widehat{u}_j(t_{j+1}) + [A_j]\,[A_{j-1}]\cdots[A_0]\,([z_0] - s_0) \\ & + [A_j]\,[A_{j-1}]\cdots[A_1]\,([z_1] - s_1) \\ & \vdots \\ & + [A_j]\,([z_j] - s_j) \\ & + I([z_{j+1}] - s_{j+1}) \,. \end{aligned}$$

This completes the information required by Algorithm II to go on to the next integration step.

AWA includes several variants of the representation of the matrices A_j given above. For details, see [43]. In many problems, a QR factorization is an effective representation.

7 Conclusions and further research directions

These lectures have attempted to convey an appreciation for the tools of interval arithmetic as they support the guaranteed computation of global error bounds for ordinary differential equations. We have seen how one can compute validated enclosures of solutions, and we have discussed AWA, Lohner's program for computing enclosures.

This is an exciting area of research. The concepts are accessible even to undergraduate students. New results or new software products promise to significantly improve the reliability of software used for scientific computation in critical applications. Current software is slow, so there is much room for improvements. And the workers in the field are few, so there are more than enough interesting problems for all who care to participate. Here are a few of the open problems

Quasimonotone problems: There is no readily available software for quasimonotone problems, especially for the important special case of 1-dimensional problems.

Step size control: AWA is currently a fixed step size code. Several experimental step size control strategies have been attempted. Among the most promising strategies are Eijgenraam's original strategy of choosing the step to maintain the local error below the tolerance and the strategy proposed by Beermann in his thesis.

Variable order: AWA is currently a fixed-order method. A successful program will adapt its order, as well as its step size.

Global error "prediction": Kerbl explored a step size control based on using an approximate solution to predict regions where the trajectories are converging and where they are diverging.

Coarse enclosure: Algorithm I of AWA currently allows steps only of the order of those for Euler's method, no matter how high an order is used by Algorithm II. A successful program must be able to take longer steps.

Representing the solution: Lohner has explored several representations of A_j, but there is no clearly best solution.

Stiff problems: Very little is known about validated techniques for stiff problems.

DAEs: Very little is known about validated techniques for DAEs. We can generate Taylor series for solutions for many DAEs, so there is hope for validation.

Production-quality software: AWA has been used successfully to solve many applied problems, but there is no widely available, production-quality software for the validated solution of ODEs.

Applications: AWA has been used to solve real-world applications problems, but there are PLENTY more to do.

Acknowledgments

Robert Rihm has offered much assistance, especially in his survey article [56]. Ernst Adams offered many helpful suggestions from a careful reading of an early draft of this paper. I have engaged in many discussions of interval techniques for ODEs with Hans Stetter, Louis Rall, Ernst Adams, Siegfried Rump, Rudolf Lohner, Gabriella Schranz-Kirlinger, Robert Corless, and Gary Krenz. Professor Kulisch has generously assisted me during many visits to Karlsruhe.

I especially wish to thank Professor Light for the invitation to participate in the Sixth SERC Numerical Analysis Summer School at Leicester University.

Bibliography

1. Aberth, O. (1988). *Precise Numerical Analysis.* William Brown, Dubuque, IA.
2. Adams, E. (1980). On methods for the construction of the boundaries of sets of solutions for differential equations or finite-dimensional approximations with input sets. In *Fundamentals of Numerical Computation (Computer Oriented Numerical Analysis)*, edited by G. Alefeld and R. D. Grigorieff. Computing Supplement 2, Springer-Verlag, Vienna.
3. Adams, E. (1990). Periodic solutions: Enclosure, verification, and ap-

plications. In *Computer Arithmetic and Self–Validating Numerical Methods*, edited by C. Ullrich. Academic Press, San Diego.

4. Adams, E. (1993). The reliability question for discretizations of evolution problems, part I: Theoretical consideration on failures, part II: Practical failures. In *Scientific Computing with Automatic Result Verification*, edited by E. Adams and U. Kulisch. Academic Press, San Diego.
5. Adams, E., Ames, W., Kühn, W., Rufeger, W. and Spreuer, W. (1993). Computational chaos may be due to a single local error. *J. Comput. Phys.*, **104**, 241–250.
6. Alefeld, G. and Herzberger, J. (1983). *Introduction to Interval Computations.* Academic Press, New York.
7. Bauch, H. and Kimmel, W. (1989). Solving ordinary initial value problems with guaranteed bounds. *Z. Angew. Math. Mech.*, **69**, 110–112.
8. Beda, L.M., Korolev, L.N., Sukkikh, N.V. and Frolova, T.S. (1959). *Programs for automatic differentiation for the machine BESM.* Technical Report, Institute for Precise Mechanics and Computation Techniques, Academy of Science, Moscow, USSR. (In Russian).
9. Bischof, C., Carle, A., Corliss, G., Griewank, A. and Hovland, P. (1992). ADIFOR – Generating derivative codes from Fortran programs. *Scientific Programming*, **1**, 11–29.
10. Butcher, J.C. (1989). *The Numerical Analysis of Ordinary Differential Equations : Runge-Kutta and General Linear Methods.* Wiley, New York.
11. Char, B.W., Geddes, K.O., Gonnet, G.H., Leong, B.L., Monagan, M.B. and Watt, S.M. (1992). *First Leaves: A Tutorial Introduction to Maple V.* Springer-Verlag, Berlin.
12. Conn, A.R., Gould, N.I.M. and Toint, P.L. (1992). *LANCELOT : A Fortran Package for Large-Scale Nonlinear Optimization (Release A).* Springer Series in Computational Mathematics No. 17, Springer-Verlag, Berlin.
13. Connell, A. and Corless, R.M. (1993). An experimental interval arithmetic package in Maple. *Interval Computations*, **2**, 120–134.
14. Corless, R.M. and Corliss, G.F. (1992). Rationale for guaranteed ODE defect control. In *Computer Arithmetic and Enclosure Methods*, edited by L. Atanassova and J. Herzberger. North-Holland, Amsterdam.
15. Corliss, G.F. (1989). Survey of interval algorithms for ordinary differential equations. *Appl. Math. Comput.*, **31**, 112–120.
16. Corliss, G.F. (1991). Automatic differentiation bibliography. In *Automatic Differentiation of Algorithms: Theory, Implementation, and Application*, edited by A. Griewank and G. F. Corliss. SIAM, Philadel-

phia, PA.

17. Corliss, G.F. (1993). *Comparing software packages for interval arithmetic.* Presented at SCAN '93, September 1993, Vienna.
18. Corliss, G.F., Davis, P.H. and Krenz, G.S. (1988). Bibliography on interval methods for the solution of ordinary differential equations. Technical Report No. 289, Department of Mathematics, Statistics and Computer Science, Marquette University, Milwaukee, WI.
19. Corliss, G.F. and Rall, L.B. (1991). Computing the range of derivatives. In *Computer Arithmetic, Scientific Computation and Mathematical Modelling*, edited by E. Kaucher, S. M. Markov and G. Mayer. Vol. 12 of IMACS Annals on Computing and Applied Mathematics. J. C. Baltzer AG, Basel.
20. Eijgenraam, P. (1981). The Solution of Initial Value Problems Using Interval Arithmetic. Mathematical Centre Tracts No. 144, Stichting Mathematisch Centrum, Amsterdam.
21. Enright, W.H. (1989). Analysis of error control strategies for continuous Runge-Kutta methods. *SIAM J. Numer. Anal.*, **26**, 588–599.
22. Enright, W.H. (1989). A new error-control for initial value solvers. *Applied Mathematics and Computation*, **31**, 288–301.
23. Erhel, J. and Philippe, B. (1992). Design of a toolbox to control arithmetic reliability. In *Computer Arithmetic and Enclosure Methods*, edited by L. Atanassova and J. Herzberger. North-Holland, Amsterdam.
24. Gambill, T. and Skeel, R. (1988). Logarithmic reduction of the wrapping effect with application to ordinary differential equations. *SIAM J. Numer. Anal.*, **25**, 153–162.
25. Griewank, A. (1989). On automatic differentiation. In *Mathematical Programming: Recent Developments and Applications*, edited by M. Iri and K. Tanabe. Kluwer Academic Publishers, Dordrecht.
26. Griewank, A. (1991). The chain rule revisited in scientific computing. *SIAM News*, **24**, no. 3, p. 20 and no. 4, p. 8.
27. Griewank, A. and Corliss, G.F. (eds.) (1991). *Automatic Differentiation of Algorithms: Theory, Implementation, and Application.* SIAM, Philadelphia, PA.
28. Heuser, H. (1991). *Gewöhnliche Differentialgleichungen* (2nd edn). B.G. Teubner, Stuttgart.
29. IBM (1990). *IBM High Accuracy Arithmetic - Extended Scientific Computation.* IBM Publication No. SC33–6462–00, IBM, Mechanicsburg, PA.
30. Ince, I.L. (1956). *Ordinary Differential Equations.* Dover, New York.
31. Juedes, D. (1991). A taxonomy of automatic differentiation tools.

In *Automatic Differentiation of Algorithms: Theory, Implementation, and Application*, edited by A. Griewank and G. F. Corliss. SIAM, Philadelphia, PA.

32. Kaucher, E.W. and Miranker, W.L. (1984). *Self-Validating Numerics for Function Space Problems — Computation with Guarantees for Differential and Integral Equations.* Academic Press, New York.
33. Kearfott, R.B. (1994). Fortran 90 environment for research and prototyping of enclosure algorithms for constrained and unconstrained nonlinear equations. To appear in *ACM Trans. Math. Software.*
34. Kearfott, R.B., Dawande, M., Du, K.S. and Hu, C.-Y. (1994). INTLIB: A portable FORTRAN 77 interval standard function library. To appear in *ACM Trans. Math. Software.*
35. Kerbl, M. (1991). Stepsize strategies for inclusion algorithms for ode's. In *Computer Arithmetic, Scientific Computation and Mathematical Modelling,* edited by E. Kaucher, S. Markov and G. Mayer. IMACS Annals on Computing and Appl. Math. 12, J.C. Baltzer, Basel.
36. Klatte, R., Kulisch, U., Lawo, C., Rauch, M. and Wiethoff, A. (1993). *C–XSC: A C++ Class Library for Extended Scientific Computation.* Springer-Verlag, Berlin.
37. Klatte, R., Kulisch, U., Neaga, M., Ratz, D. and Ullrich, C. (1991). *PASCAL-XSC: A PASCAL Extension for Scientific Computation.* Springer-Verlag, Berlin.
38. Knüppel, O. (1993). BIAS – basic interval arithmetic subroutines. Technical Report 99.3, Berichte des Forschungsschwerpunktes Informations- und Kommunikationstechnik, Technische Universität Hamburg-Harburg.
39. Lakshmikantham, V. and Leela, S. (1969). *Differential and Integral Inequalities: Theory and Applications.* Academic Press, New York.
40. Lawo, C. (1992). C–XSC, A programming environment for verified scientific computing and numerical data processing. In *Scientific Computing with Automatic Result Verification*, edited by E. Adams and U. Kulisch. Academic Press, Orlando, FA.
41. Lohner, R.J. (1978). *Anfangswertaufgaben im* $\mathbb{R}^n$ *mit kompakten Mengen für Anfangswerte und Parameter.* Diplomarbeit, Inst. f. Angew. Math., Universität Karlsruhe.
42. Lohner, R.J. (1987). Enclosing the solutions of ordinary initial and boundary value problems. In *Computer Arithmetic: Scientific Computation and Programming Languages*, edited by E. W. Kaucher, U. W. Kulisch and C. Ullrich. Wiley-Teubner Series in Computer Science, Stuttgart.
43. Lohner, R.J. (1988). *Einschließung der Lösung gewöhnlicher Anfangs– und Randwertaufgaben und Anwendungen.* PhD thesis, Universität

Karlsruhe.

44. Lohner, R.J. (1993). Interval arithmetic in staggered correction format. In *Scientific Computing with Automatic Result Verification*, edited by E. Adams and U. Kulisch. Academic Press, San Diego.
45. Lohner, R.J. (1993). *On step size and order control in the verified solution of ordinary initial value problems.* Presented at SCAN '93, September 1993, Vienna.
46. Moore, R.E. (1966). *Interval Analysis.* Prentice-Hall, Englewood Cliffs, N.J.
47. Moore, R.E. (1979). *Methods and Applications of Interval Analysis.* SIAM, Philadelphia, PA.
48. Neumaier, A. (1990). *Interval Methods for Systems of Equations.* Cambridge University Press, Cambridge.
49. Neumaier, A. (1991). *BIAS Fortran interval library.*
50. Nickel, K. (1978). Ein Zusammenhang zwischen Aufgaben monotoner Art und Intervall–Mathematik. In *Numerical Treatment of Differential Equations*, Proc. of a Conf. Held at Oberwolfach, July 4–10, 1976, edited by R. Bulirsch, R. Grigorieff and J. Schröder, eds. Lecture Notes in Mathematics No. 631, Springer, Berlin.
51. Nickel, K. (1977). Bounds for the set of solutions of functional–differential equations. Freiburger Intervall–Berichte, **9** (1979). Also appeared as MRC Techn. Summary Report No. 1782, Univ. of Wisconsin, Madison (1977).
52. Nickel, K. (1985). How to fight the wrapping effect. In *Interval Analysis 1985*, edited by K. Nickel. Lecture Notes in Computer Science No. 212, Springer, Berlin.
53. Nickel, K. (1986). Using interval methods for the numerical solution of ODE's. *Z. angew. Math. Mech.*, **66**, 513–523.
54. Rall, L.B. (1981). *Automatic Differentiation: Techniques and Applications.* Vol. 120 of Lecture Notes in Computer Science, Springer-Verlag, Berlin.
55. Ratschek, H. and Rokne, J. (1984). *Computer Methods for the Range of Functions.* Halsted Press, Wiley, New York.
56. Rihm, R. (1994). Interval methods for initial value problems in ODEs. In *Topics in Validated Computations: Proceedings of the IMACS-GAMM International Workshop on Validated Computations, University of Oldenburg*, edited by J. Herzberger. Elsevier Studies in Computational Mathematics, Elsevier.
57. Rump, S.M. (1994). Verification methods for dense and sparse systems of equations. In *Topics in Validated Computations: Proceedings of the IMACS-GAMM International Workshop on Validated Computations,*

University of Oldenburg, edited by J. Herzberger. Elsevier Studies in Computational Mathematics, Elsevier.

58. Spreuer, H. and Adams, E. (1993). On the existence and the verified determination of homoclinic and heteroclinic orbits of the origin for the Lorenz system. *Computing Suppl.*, **9**, 233–246.
59. Stetter, H.J. (1984). Sequential defect correction for high–accuracy floating–point algorithms. In *Numerical Analysis: Proceedings of the 10th Biennial Conference (Dundee, 1983)*, edited by D. Griffiths. Lecture Notes in Mathematics No. 1066, Springer-Verlag, Berlin.
60. Stetter, H.J. (1986). Algorithms for the inclusion of solutions of ordinary initial value problems. In *Equadiff 6: Proceedings of the International Conference on Differential Equations and Their Applications (Brno, 1985)*, edited by J. Vosmanský and M. Zlámal. Lecture Notes in Mathematics No. 1192, Springer-Verlag, Berlin.
61. Stetter, H.J. (1990). Validated solution of initial value problems for ODEs. In *Computer Arithmetic and Self-Validating Numerical Methods*, edited by C. Ullrich. Academic Press, New York.
62. Walter, W. (1970). *Differential and Integral Inequalities.* Springer, Heidelberg.
63. Walter, W.V. (1992). ACRITH–XSC: A FORTRAN-like language for verified scientific computing. In *Scientific Computing with Automatic Result Verification*, edited by E. Adams and U. Kulisch. Academic Press, Orlando, FA.
64. Walter, W.V. (1993). *FORTRAN–XSC: A portable Fortran 90 module library for accurate and reliable scientific computation.* To appear in Kulisch 60th birthday volume.
65. Wang, M.C. and Kennedy, W.J. (1993). *A self-validating numerical method for computation of central and non-central F probabilities and percentiles.* To appear in *Statistics and Computing.*
66. Wengert, R.E. (1964). A simple automatic derivative evaluation program. *Comm. ACM*, **7**, 463–464.
67. Wilkinson, J.H. (1971). Modern error analysis. *SIAM Rev.*, **13**, 548–568.

Introduction to Computational Methods for Differential Equations

Kenneth Eriksson, Don Estep, Peter Hansbo and Claes Johnson

Don Estep is at
School of Mathematics,
Georgia Institute of Technology,
Atlanta,
Georgia 30332,
USA;
the other authors are at
Department of Mathematics,
Chalmers University of Technology,
412 96 Göteborg,
Sweden.

Abstract

We give an introduction to a unified approach to computational methods for differential equations based on Galerkin methods with piecewise polynomial approximation. We present adaptive methods with reliable and efficient control of computational errors and also discuss estimation and control of data and modeling errors. As model problems we consider two-point boundary value problems, initial value problems for ordinary differential equations, the Poisson equation and the heat equation. We present sample results using the Femlab software.

Knowing thus the Algorithm of this calculus, which I call Differential Calculus, all differential equations can be solved by a common method. (Gottfried Wilhelm von Leibniz, 1646-1719)

When, several years ago, I saw for the first time an instrument which, when carried, automatically records the number of steps taken by a pedestrian, it occurred to me at once that the entire arithmetic could be subjected to a similar kind of machinery so that not only addition and subtraction, but also multiplication and division could be accomplished by a suitable arranged machine easily, promptly and with sure results.... For it is unworthy of excellent men to lose hours like slaves in the labour of calculations, which could safely be left to anyone else if the machine was used.... And now that we may give final praise to the machine, we may say that it will be desirable to all who are engaged in computations which, as is well known, are the managers of financial affairs, the administrators of others' estates, merchants, surveyors, navigators, astronomers, and those connected with any of the crafts that use mathematics. (Leibniz)

1 Leibniz's vision

Newton and Leibniz invented Calculus in the late 17th century and laid the foundation for the revolutionary development of science and technology leading to the present day. A second revolution started in the 1940's with the invention of the modern computer. Today, we are experiencing a process that we may describe as the marriage of Calculus and Computation, or as the final realization of the original Leibniz vision. Already three hundred years ago, Leibniz sought to realize such an idea, but failed because the Calculator he invented was far from being powerful enough. A concrete evidence of the "marriage" is the rapid development and spread of mathematical software such as Mathematica, Matlab, Maple, and the large number of finite element codes.

The basic mathematical models of science and engineering very often take the form of differential equations that relate derivatives of undetermined functions and given data. A differential equation typically expresses a law of Physics such as conservation of mass or momentum. If the differential equation has derivatives with regard to only one independent variable, then it is called an ordinary differential equation (ODE), otherwise it is a partial differential equation (PDE). Information about the physical process being modeled is gained by finding a function that satisfies the differential equation, or *solving* the differential equation.

Exact solutions of differential equations may sometimes be determined through symbolic computation by hand or by means of software such as Maple or Mathematica. In most cases, however, this not feasible and the alternative is to determine an approximate solution for given data through numerical computations using a computer. The basic idea is to *discretize* a given differential equation to obtain a finite-dimensional system of equations that is solved using a computer to produce an approximate solution. The finite-dimensional problem is referred to as a *discrete problem* and the corresponding differential equation as a *continuous problem*. Discrete problems derived from physical models are usually computationally intensive, and hence, the rapid increase of computer power is opening fascinating new possibilities for this approach. Today, realistic problems in three dimensions may be solved numerically on desktop workstations.

Weather prediction is a familiar example of mathematical modeling and computer simulation. Weather forecasts are sometimes based on numerically solving a system of partial differential equations that model the evolution of the atmosphere starting from initial data obtained from measuring the current weather, i.e. temperature, wind speed, etc., at given locations. Of course, weather forecasts are not always reliable, but they do sometimes give correct predictions. There are three sources of errors affecting the reliability: (i) data, (ii) modeling and (iii) computation. The initial data at the start of the computer simulation are known only approximately; the set of differential equations in the computer model only approximately describe the evolution of the atmosphere; and finally the numerical solution of the differential equations is only an approximation of the true solution. All this adds up to form the total prediction error, which may be large. It is essential to be able to estimate the total error by estimating individually the contribution from the sources (i)-(iii) and improve the precision where possible. This example contains the issues in mathematical modeling and computer simulation that are basic in all applications.

In these notes, we give an introduction to a general method for approximating differential equations that is further discussed in [8], [9] and [10]. Taken as a whole, this material leads the reader from the elementary right to the frontiers of research in numerical methods for differential equations. At every step, we emphasize a close interplay between computation and mathematical analysis, and the freely-available software package Femlab is meant to serve as a laboratory to test the algorithms.

The basic classification of differential equations is into problems of *elliptic*, *parabolic* or *hyperbolic* type. One motivation for this classification is that Galerkin methods are fully optimal for elliptic and parabolic problems, but in general not fully optimal for hyperbolic problems.

We begin by considering one-dimensional model problems of each type. We then discuss the general nonlinear initial value problem for ordinary differential equations. Finally, we consider as multidimensional model problem, the Poisson equation and the heat equation. Before going into these different model problems, we give an introductory section on interpolation using piecewise polynomial functions.

2 The discretization method

$(Fe)^m$: Finite elements For everything For everyone "For ever" (m=4)

As a general method for numerical solution of differential equations, we consider *the finite element method* Fem. This method is based on

- *Galerkin's method*
- *piecewise polynomial approximation.*

In a Galerkin method, the approximate solution is determined as the element from a certain finite-dimensional vector space of *trial functions* for which the *residual error* is orthogonal to a certain set of *test functions.*

In the simplest case, the trial and test space are the same. The residual error, or simply the residual, is obtained by inserting the approximate solution into the given differential equation and computing. The residual of the exact solution is zero, while the residual of approximate solutions deviates from zero because they do not solve the equation exactly.

In Fem, the trial and test functions are piecewise polynomials. A *piecewise polynomial* is a function that is equal to a polynomial, e.g a linear function, on each subdomain of a partition (or division) of a given domain in space or space-time. The subdivision is referred to as a *mesh* and the subdomains as *elements.*

Thus, the basic idea of Fem is to seek an approximate solution in a trial space of piecewise polynomial functions on a mesh, such that the residual of the approximate solution is orthogonal to a test space of piecewise polynomial functions on the same mesh. Applying this method for a given set of data leads to a system of equations that is solved using a computer.

3 An introductory example

It is necessary to solve differential equations. (Newton)

As a concrete example, we consider the following model problem: find $u = u(x)$ such that

$$\begin{aligned} -u'' &= f, \quad \text{in } I = (0,1), \\ u(0) &= 0, \quad u(1) = 0, \end{aligned} \tag{3.1}$$

where $f = f(x)$ is a given source term, $v' = \frac{dv}{dx}$, and $v'' = \frac{d^2v}{dx^2}$. This problem models time-independent heat flow in a heat conducting rod where f is the heat production and u is the unknown temperature. This problem may be solved in analytic form by symbolic computation in the rare situation when f and its anti-derivative can be integrated exactly. Otherwise, we have to rely on a numerical method. For example, if f is given by measurements, then we cannot integrate f exactly.

We formulate the simplest finite element method for (3.1) which is based on continuous piecewise linear approximation. We first need to define the trial and test space V_h of continuous piecewise linear functions and then apply Galerkin's method. To this end, let $T_h : 0 = x_0 < x_1 < ... < x_{M+1} = 1$ be a partition of $I = (0,1)$ into intervals or elements $I_j = (x_{j-1}, x_j)$ of size $h_j = x_j - x_{j-1}$, and let V_h be the vector space of continuous piecewise linear functions on T_h that vanish for $x = 0$ and $x = 1$. Each function v in V_h is determined by its values $v(x_j)$ at the interior nodes x_j, $j = 1, ..., M$.

To apply Galerkin's method, we give 3.1 a *variational formulation* as follows: find a function u with $u(0) = u(1) = 0$ such that

$$\int_I u'v'\,dx - \int_I fv\,dx = 0, \tag{3.2}$$

for all functions v such that $v(0) = v(1) = 0$. This relation results from multiplying (3.1) by the test function v, integrating over I , and integrating by parts over I moving one derivative from u'' to v:

$$-\int_I u''v\,dx = -u'(1)v(1) + u'(0)v(0) + \int_I u'v'\,dx = \int_I u'v'\,dx, \tag{3.3}$$

where we used $v(0) = v(1) = 0$.

The finite element method for (3.1) is the following finite-dimensional analog of (3.2): find U in V_h such that

$$\int_I U'v'\,dx - \int_I fv\,dx = 0, \qquad \text{for all } v \text{ in } V_h. \tag{3.4}$$

Note that since U is linear on each element and also continuous, U' is constant on each element, see Fig. 1b below. The relation (3.2) expresses the fact that the residual of the exact solution is orthogonal to **all** test functions v, and the relation (3.4) the fact that the residual of the finite element solution U is orthogonal to the **restricted** set of test functions v in the discrete test space V_h.

We shall see that (3.4) is equivalent to a linear system of equations of the form $A\xi - b = 0$, or

$$A\xi = b, \tag{3.5}$$

where $\xi = (\xi_j)_{j=1}^M$ is the vector of nodal values $\xi_j = U(x_j)$ of the Fem-solution U at the interior nodes x_j, $A = (a_{ij})$ is an $M \times M$ matrix called the *stiffness matrix* and $b = (b_j)$ is an M vector called the *load vector*. We observe that the stiffness matrix A is a square matrix because the trial and test space are the same. The stiffness matrix A is computable once the nodal points x_j are determined and the load vector b when, in addition, f is specified. It turns out that the stiffness matrix A is positive definite and tridiagonal and may be solved by Gaussian elimination to give an approximate solution U in V_h for a given f. We note that the dimension or number of degrees of freedom of V_h is equal to M, and we expect the precision and the cost of the approximation to increase with M.

Example 1 For a first example, we choose

$$f(x) = 10(1 - 10(x - 0.5)^2)\,e^{-5(x-0.5)^2}.$$

This is an exceptional case in which the solution can be computed exactly;

$$u(x) = e^{-5(x-0.5)^2} - e^{-5/4};$$

We chose it so that we can compare the approximation with the solution.

Straightforward computation shows that (3.5) is equivalent to the following system of equations:

$$-\frac{1}{h_i}\xi_{i-1} + \left(\frac{1}{h_i} + \frac{1}{h_{i+1}}\right)\xi_i + \frac{1}{h_{i+1}}\xi_{i+1} =$$

$$\int_{x_{i-1}}^{x_i} f(x)\frac{(x - x_{i-1})}{h_i}dx + \int_{x_i}^{x_{i+1}} f(x)\frac{(x - x_{i+1})}{h_{i+1}}dx, \qquad i = 1, \cdots, M.$$

It is important that we allow the mesh to be *non-uniform*, or to have unequal steps. The data f varies a good deal throughout the interval, and obtaining an accurate approximation of the solution depends of course on resolving the behavior of the data. We will see that a non-uniform mesh gives an optimal approximation in terms of the computational work needed to resolve the behavior of the data.

Specifically, we compute using the following mesh sizes

$$\begin{array}{lcl} h_1 = h_{16} & \approx & .201 \\ h_2 = h_{15} & \approx & .0673 \\ h_3 = h_{14} & \approx & .0493 \\ h_4 = h_{13} & \approx & .0417 \\ h_5 = h_{12} & \approx & .0376 \\ h_6 = h_{11} & \approx & .0353 \\ h_7 = h_{10} & \approx & .0341 \\ h_8 = h_9 & \approx & .0335 \end{array}$$

that are chosen according to a procedure described below which guarantees an optimal distribution of error for this solution. This *adaptive* procedure does **not** require any knowledge about the solution, which is unknown in general. Instead, it uses information obtained only from the data and the computed approximation.

In Fig. 1a, we plot the exact solution u and the piecewise linear Fem approximation U, and in Fig. 1b, we plot u' and U'. Note that U' is a step-function because U is linear on each element. The adaptive procedure has clustered points towards the middle of the interval in order to keep the root mean square area under $u' - U'$ the same on each interval, yielding a total error of 0.25. The error of $u - U$ measured by the root mean square area is less than 0.01. If we perform the computation while insisting that the mesh have uniform spacing, we find that $h = h_i = 1/24$ for all i must be used in order to obtain the same accuracy given by the non-uniform mesh. In other words, we must use 50% more elements and solve a correspondingly bigger system to obtain the same accuracy. This disparity increases as the number of elements is increased to make the error smaller.

This example motivates the following basic questions

(i) How is U computed?

(ii) How big is the *discretization error* $u - U$ or $u' - U'$?

(iii) How can the the mesh be chosen to control the discretization error

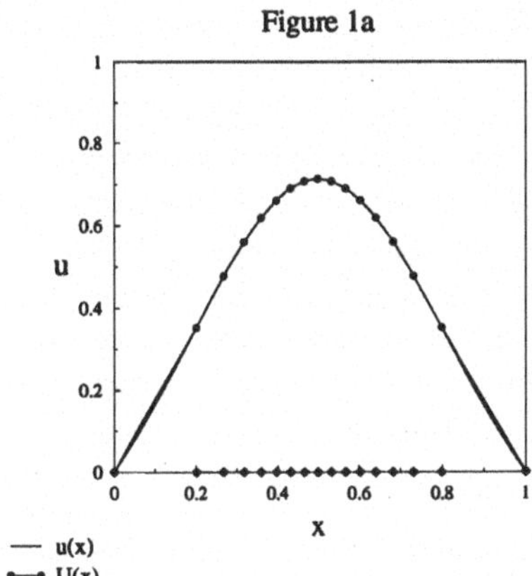

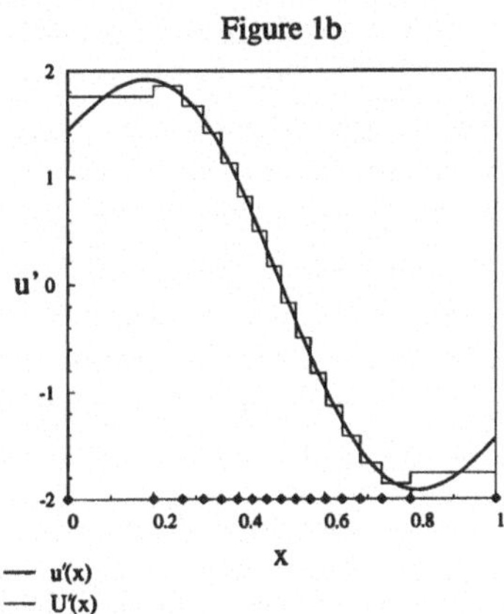

FIG. 1. Solution and Fem approximation.

$u - U$ or $u' - U'$ to a given tolerance?

4 Numerical methods

We now introduce the basic concepts that are fundamental to all numerical methods for differential equations. The student will find it useful to return to this section many times, each time gaining a more concrete understanding. A first reading can only give a vague idea of these issues.

4.1 Reliability and efficiency

The goal of the design of a numerical method is to achieve

- *reliability*
- *efficiency.*

Reliability means that the discretization error is controlled on a given tolerance level; for instance, the numerical solution is guaranteed to be within 1 percent of the exact solution at every point. Efficiency means that the computational work to compute a solution within the given tolerance is essentially as small as possible. As illustrated in Example 1, realizing these goals in general requires an adaptive numerical method.

An *adaptive method* consists of a discretization method together with an *adaptive algorithm*. An adaptive algorithm consists of

- a *stopping criterion* guaranteeing error control to a given tolerance level
- a *mesh selection strategy* in case the stopping criterion is not satisfied.

The goal of the adaptive algorithm is to determine through computation the "right" mesh realizing both reliability and efficiency. In practice, this usually results from an iterative process, where at each step an approximate solution is computed on a given mesh. If the stopping criterion is satisfied, then the mesh and the corresponding approximate solution are accepted. If the stopping criterion is not satisfied, then a new mesh is determined through the mesh selection strategy and the process is continued. To start the procedure, a first coarse mesh is required. Central in this process is the idea of *feedback* in the computational process. Information concerning the nature of the underlying problem and exact solution is revealed through computation, and this information is used to determine the right mesh.

4.2 A priori and a posteriori error estimates

The adaptive algorithm is built on *error estimates*, which we now discuss. These come in two forms

- *a priori error estimates*
- *a posteriori error estimates.*

An a priori estimate predicts the error between the exact and the approximate solution in terms of properties of both the exact (unknown) solution and the computed approximation. In an a posteriori estimate, the error is evaluated in terms of the computed approximate solution. The stopping criterion is based solely on the a posteriori error estimate. The mesh modification strategy in addition may build on a priori error estimates.

4.3 Accuracy and stability

The basic properties of a numerical method are

- *accuracy*
- *stability.*

These properties enter in different forms in the a posteriori and the a priori error estimates. The a posteriori version may be expressed conceptually as

$$\text{small residual} + \text{stability of the continuous problem} \rightarrow \text{error small.} \quad (4.1)$$

The a priori version takes the conceptual form

$$\text{small interpolation error} + \text{stability of the discrete problem} \rightarrow \text{error small,} \quad (4.2)$$

where the interpolation error is the difference between the exact solution and a piecewise polynomial that interpolates the exact solution at given points, where it takes the same values as the exact solution. Note that the a posteriori error estimate involves the stability of the continuous problem and the a priori estimate the stability of the discrete problem.

Accuracy is connected to the size of the residual in the a posteriori case and to the interpolation error in the a priori case.

Both the residual and the interpolation error accumulate to make up the total error. The concept of stability measures this accumulation and therefore is fundamental. The stability is typically measured in a stability estimate involving a multiplicative ***stability factor***. The size of this factor measures the computational difficulty of the problem. If the stability factor is large, then the problem is sensitive to perturbations from the numerical solution process, and more computational work is needed to reach a certain error tolerance.

A successful numerical method has to be both stable and accurate, and the basic difficulty is to design methods with both qualities. For a large class of problems Galerkin's method satisfies this requirement. In some problems, Galerkin's method is applied in modified form to enhance stability.

4.4 The total error

There are three sources of error in mathematical modeling of physical phenomena

- *modeling*
- *data*
- *numerical computation.*

Modeling errors result from simplifications in the mathematical description of real phenomena, for example in the derivatives and the coefficients in differential equations related to material properties. Data errors come about because physical conditions can never be measured or set exactly in a given situation. The computational error is our main concern in these notes, and it has three sources

- *Galerkin discretization*
- *quadrature*
- *solution of the discrete problem.*

The Galerkin discretization error arises because the solution is approximated by a piecewise polynomial, assuming that the integrals arising in the Galerkin variational formulation are evaluated exactly and that the resulting discrete equations are solved exactly. The quadrature error comes from evaluating the integrals arising in the Galerkin formulation using numerical quadrature, and the discrete solution error is the error resulting from solving the discrete system of equations only approximately.

Different perturbations propagate and accumulate differently, or in other words are connected to different stability factors. In some problems, Galerkin perturbations propagate in a more favorable way as compared to the other perturbations.

Though we focus on the Galerkin and quadrature error in these notes, we remark that a complete adaptive algorithm controls not only these errors, but also, in principle, the discrete solution error and some kinds of modeling error. It is natural to seek to control all these errors on the same level, that is typically the level of the error in the given data. Thus, it is important not only to adaptively control the numerical error, but in addition, estimate the effects of modeling and data error.

4.5 Femlab

A software package Femlab, containing a basic implementation of an adaptive Fem, complements the notes and allows the reader to test the algorithms and solve a large variety of problems easily. Directions for obtaining Femlab are given below.

5 Piecewise polynomial approximation in one dimension

We focus on the simple cases of discontinuous piecewise constant approximation and continuous piecewise linear approximation on an interval $I = (a, b)$ in space or time.

Assuming first that $I = (0, 1)$ is a space interval, let $0 = x_0 < x_1 < x_2 < ... < x_{M+1} = 1$ be a subdivision of I into subintervals $I_j = (x_{j-1}, x_j)$ of length $h_j = x_j - x_{j-1}$. We use the notation $T_h = \{I_j\}$ for the corresponding mesh and define its mesh function $h(x)$ indicating the local mesh size of the mesh T_h, by $h(x) = h_j$, for x in I_j. A mesh T_h is depicted in Fig. 2.

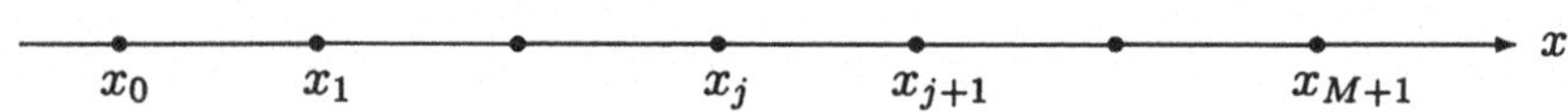

FIG. 2. The mesh T_h.

We also consider the situation in which I is a time interval, for example $I = (0, \infty)$. In this case, the mesh is given by a sequence of discrete time levels $0 = t_0 < t_1 < .... < t_n < ..$, with corresponding time intervals $I_n = (t_{n-1}, t_n)$, time steps $k_n = t_n - t_{n-1}$ and mesh function $k(t)$ defined by $k(t) = k_n$, for t in I_n. We denote the corresponding mesh by T_k.

We consider first piecewise constant discontinuous and then piecewise linear continuous approximations on the space mesh T_h or the analogous time mesh T_k.

5.1 Discontinuous piecewise constant approximation

A piecewise constant function v on T_h is defined by its constant value v_j on each subinterval I_j:

Let W_h denote the space of discontinuous piecewise constant functions $v = v(x)$ on I such that v is constant on each subinterval I_j, $j = 1, .., M+1$. A basis $\{\varphi_j\}_{j=1}^{M+1}$ for W_h is given by the functions φ_j in W_h defined by (see

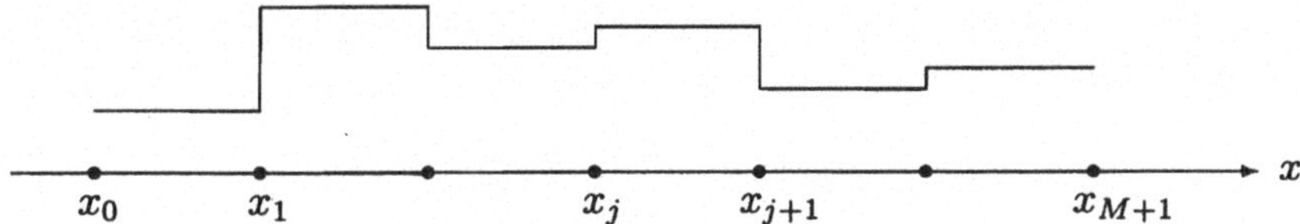

FIG. 3. A piecewise constant function v on T_h.

Fig. 4)

$$\varphi_j(x) = 1 \text{ if } x \text{ in } I_j, \qquad \varphi_j(x) = 0 \text{ otherwise.} \tag{5.1}$$

In other words, an arbitrary function v in W_h can be represented as

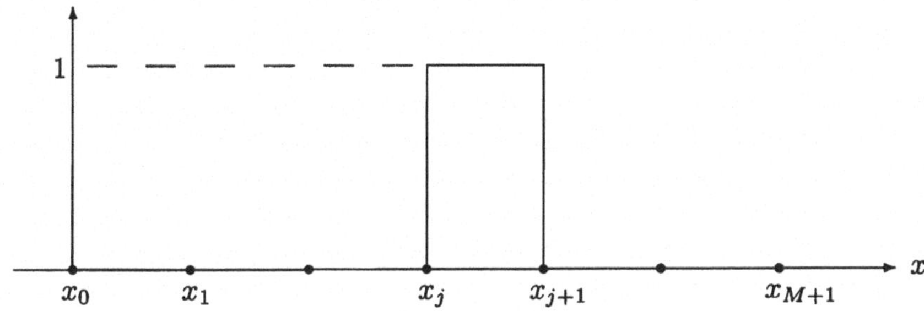

FIG. 4. The basis function φ_{j+1}.

$$v(x) = \sum_{j=1}^{M+1} v_j \varphi_j(x), \tag{5.2}$$

where v_j is the value of v on interval I_j. We note that W_h thus has dimension $M+1$.

In order to understand the approximation properties of functions in W_h, we let v be a given function on I and let $\pi_h v$ in W_h denote an *interpolant* of v, that is, a function $\pi_h v$ in W_h that approximates v. It is natural to define the interpolant $\pi_h v$, in the present case of piecewise constant approximation, by

$$\int_{I_j} (v - \pi_h v)\, dx = 0, \tag{5.3}$$

so that

$$\pi_h v = \frac{1}{h_j} \int_{I_j} v(x)\, dx \quad \text{on} \quad I_j, \qquad j = 1, ..., M+1, \tag{5.4}$$

is the average of v on each element. Using the equation $v(x) = v(z) +$

$\int_z^x v'(y)dy$, it is easy to show that the interpolation error $v - \pi_h v$ satisfies the following estimate:

$$\|v - \pi_h v\|_{I_j} \leq \int_{I_j} |v'|dx \leq \|hv'\|_{I_j}, \tag{5.5}$$

where

$$\|v\|_{I_j} = \max_{x \in I_j} |v(x)|. \tag{5.6}$$

The estimate (5.5) shows that the interpolation error $v - \pi_h v$ for a given function v decreases with decreasing mesh size depending on the slope v'; see Fig. 5.

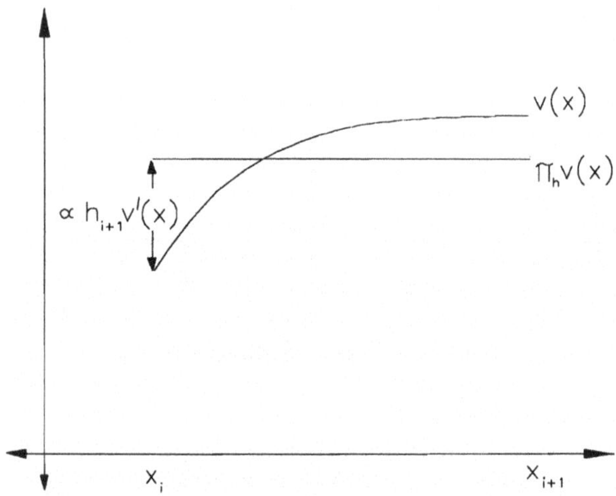

FIG. 5. Piecewise constant interpolant.

5.2 Continuous piecewise linear approximation

A linear function is defined by its values at the ends of an interval, hence a continuous piecewise linear function v on a mesh T_h is defined by its value at each mesh point x_j and has the general appearance displayed in Fig. 6.

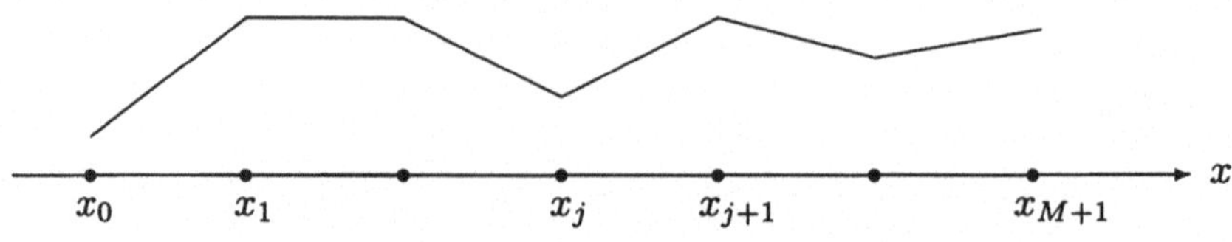

FIG. 6. A continuous piecewise linear function on T_h.

We use V_h to denote the vector space of continuous piecewise linear functions v on I such that v is linear on each subinterval I_j, $j = 1, ..., M+1$. A basis $\{\varphi_j\}_{j=0}^{M+1}$ for V_h is given by the basis functions φ_j in V_h defined by

$$\varphi_j(x_i) = 1 \text{ if } i = j, \qquad \varphi_j(x_i) = 0 \text{ otherwise.} \tag{5.7}$$

The basis functions φ_j are also referred to as "hat" functions or nodal basis functions. An example is plotted in Fig. 7.

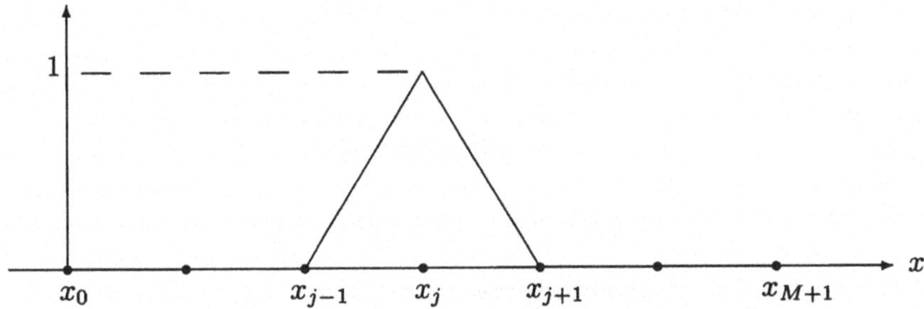

FIG. 7. The "hat" basis function φ_j.

Note that each basis function φ_j is defined on the whole interval I, but vanishes outside the intervals meeting at the node x_j. An arbitrary function v in V_h can be written

$$v(x) = \sum_{j=0}^{M+1} v_j \varphi_j(x), \tag{5.8}$$

where $v_j = v(x_j)$ is the value of v at the mesh point x_j. Hence V_h has dimension $M+2$. We also use the subspace V_h^0 of V_h consisting of the functions v such that $v(0) = v(1) = 0$. A basis for V_h^0 is given by $\{\varphi_j\}_{j=1}^{M}$ and the dimension of V_h^0 is M.

As a representative approximation in V_h of a given continuous function v, we consider the nodal interpolant $\pi_h v$ in V_h defined by

$$\pi_h v(x_j) = v(x_j), \ j = 0, .., M+1. \tag{5.9}$$

We now state a basic estimate for the interpolation error $v - \pi_h v$. The estimate shows that the interpolation errors $v - \pi_h v$ and $(v - \pi_h v)'$ are quadratic and linear, respectively, in the mesh size h with a multiplicative factor depending on v'': see Fig. 8.

Theorem 5.1 *We have, for $j = 1, ...M+1$,*

$$\|v - \pi_h v\|_{I_j} \leq \min(\|h_j v'\|_{I_j}, \frac{1}{8}\|h_j^2 v''\|_{I_j}), \tag{5.10}$$

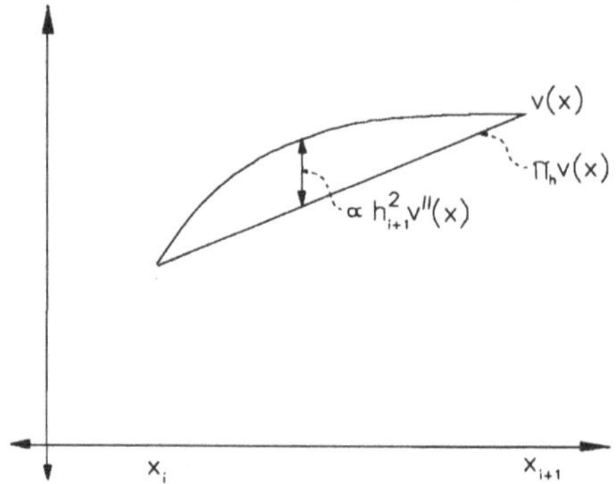

FIG. 8. Piecewise linear interpolant.

$$\|(v - \pi_h v)'\|_{I_j} \leq \frac{1}{2}\|h_j v''\|_{I_j}. \tag{5.11}$$

5.3 Galerkin approximations

We consider Galerkin methods based on the spaces W_h, V_h and V_h^0 of piecewise polynomials just defined. To compute the Galerkin discretization error, we compare it with the interpolation error. If the Galerkin discretization error is of the same size as the interpolation error, then we say that the Galerkin method is *optimal*. Note that an interpolant requires complete knowledge of the function, so it is not useful for approximating the unknown solution of a differential equation. The Galerkin approximation is computed from the same data that determines the solution through the differential equation.

We refer to Galerkin's method based on discontinuous piecewise constant functions as *discontinuous Galerkin* dG(0) and based on continuous piecewise linear functions as *continuous Galerkin* cG(1). These methods extend directly to higher-order piecewise polynomials and are then referred to as dG(q) or cG(q) with q being the degree of the polynomials involved. We shall see that dG-methods may be used for differential equations with derivatives of order at most one, and cG-methods if the order is at most two. Problems with higher-order derivatives may be reduced by reformulation into a system of first-order differential equations.

6 An elliptic model problem

We consider the following elliptic two-point boundary value problem, a generalization of (3.1)

$$\begin{aligned} -(a(x)u'(x))' &= f(x), \quad \text{in } I = (0,1), \\ u(0) &= 0, \quad u(1) = 0, \end{aligned} \tag{6.1}$$

where $a = a(x)$ is a given **positive** coefficient and $f = f(x)$ is a given source term. This is a model for stationary heat flow in a rod with conduction coefficient $a(x)$ and heat production $f(x)$, and $u = u(x)$ is the unknown temperature distribution. The variational formulation of (6.1), resulting from integration by parts as above, takes the form

$$\int_I au'v'\,dx = \int_I fv\,dx, \tag{6.2}$$

for all test functions v with $v(0) = v(1) = 0$.

The cG(1) method for (6.1) based on a partition T_h of $I = (0,1)$ with V_h^0 the corresponding finite element space defined above, reads as follows: find U in V_h^0 such that

$$\int_I aU'v'\,dx = \int_I fv\,dx, \qquad \text{for all } v \text{ in } V_h^0. \tag{6.3}$$

Subtracting (6.3) from (6.2), the Galerkin orthogonality is expressed in the relation

$$\int_I a(u-U)'v'\,dx = 0 \qquad \text{for all } v \text{ in } V_h^0. \tag{6.4}$$

Expressing the finite element solution in the form

$$U(x) = \sum_{j=1}^{M} \xi_j \varphi_j(x), \tag{6.5}$$

where $\{\varphi_j\}_{j=1}^M$ are the nodal basis functions associated with the interior nodes, we see that U is determined when the ξ_j are all determined. Setting the test function v to be each one of the basis functions φ_i in turn, the relation (6.3) is found to be equivalent to the following matrix equation for the vector $\xi = (\xi_j)$:

$$A\xi = b, \tag{6.6}$$

where $A = (a_{ij})$ is the a $M \times M$ stiffness matrix with coefficients

$$a_{ij} = \int_I a\varphi_j'\varphi_i'\,dx, \tag{6.7}$$

and $b = (b_j)$ the load vector with elements

$$b_j = \int_I f\varphi_j\,dx. \tag{6.8}$$

The system matrix A is positive definite and sparse (in the present case tridiagonal) and is easily solved by Gaussian elimination to give the approximate solution U.

The basic question now concerns the size of the error $u - U$. We first prove an a posteriori error estimate and then an a priori error estimate for

$u - U$.

6.1 An a posteriori error estimate

To measure the size of the error $e = u - U$, we use the *energy norm* $\| \cdot \|_E$ defined for functions v with $v(0) = v(1) = 0$ by

$$\|v\|_E \equiv \|v'\|_a \equiv \left(\int_I a(x)(v'(x))^2 \, dx \right)^{\frac{1}{2}}.$$

Note that for this definition to make sense, $a(x)$ must be strictly positive for all x, in which case $\| \cdot \|_a$ is called a *weighted* L_2 norm. If a is bounded above and below by strictly positive constants, then two functions are close in the weighted L_2 norm if and only if they are close in the standard L^2 norm. Moreover, if two functions v_1 and v_2 with $v_i(0) = v_i(1) = 0$, $i = 1, 2$, are close in the energy norm — i.e., if $\|(v_1 - v_2)'\|_a$ is small — then because $v_1(0) = v_2(0) = 0$, $\|v_1 - v_2\|$ is small as well.

Remark Recall that the error of the computation in Example 1 is 0.25 in the energy norm, while the error in the maximum norm is 0.01. In general, we expect the error in norms including derivatives, such as the energy norm, to be larger than the error in norms without derivatives.

We use several variations of Cauchy's inequality with the weight a present

$$\left| \int_I a v' w' \, dx \right| \leq \|v'\|_a \|w'\|_a, \tag{6.9}$$

$$\left| \int_I v w \, dx \right| \leq \|v\|_a \|w'\|_{\frac{1}{a}}. \tag{6.10}$$

To prove an a posteriori error estimate in the energy norm, we use the Galerkin orthogonality (6.3), choosing $v = \pi_h e$, where $\pi_h w$ in V_h^0 denotes the nodal interpolant of w, and obtain

$$\begin{aligned} \|e'\|_a^2 = \int_I a e' e' \, dx = \int_I a u' e' \, dx - \int_I a U' e' \, dx = \int_I f e \, dx - \int_I a U' e' \, dx \qquad (6.11) \\ = \int_I f \times (e - \pi_h e) \, dx - \int_I a U' (e - \pi_h e)' \, dx \\ = \int_I f \times (e - \pi_h e) \, dx - \sum_{i=1}^{M+1} \int_{I_j} a U' (e - \pi_h e)' \, dx. \end{aligned}$$

Now, we integrate by parts over each subinterval I_j in the last term and use the fact that all the boundary terms disappear because $(e - \pi_h e)(x_j) = 0$

to get

$$\|e'\|_a^2 = \int_I R(U)(e - \pi_h e) \leq \|hR(U)\|_{\frac{1}{a}} \|h^{-1}(e - \pi_h e)\|_a,$$

where

$$R(U) \equiv f + (aU')', \quad \text{on each subinterval } I_j$$

is the residual. In analogy to (5.10), the interpolation error measured in the weighted L_2 norm $\|\cdot\|_a$ satisfies

$$\|h^{-1}(e - \pi_h e)\|_a \leq C_i \|e'\|_a,$$

where C_i is an interpolation constant only depending on the maximum of the ratios $\max_{I_n} a / \min_{I_n} a$. We have now proved an a posteriori error estimate.

Theorem 6.1 *The finite element solution U in V_h^0 satisfies*

$$\|u - U\|_E \equiv \|u' - U'\|_a \leq C_i \|hR(U)\|_{\frac{1}{a}}. \tag{6.12}$$

6.2 An adaptive algorithm

We build an adaptive algorithm for automatic control of the energy norm error $\|u - U\|_E$ using the a posteriori error estimate (6.12) as follows

(1) Choose an initial mesh $T_{h^{(0)}}$ of mesh size $h^{(0)}$

(2) Compute the corresponding cG(1) finite element solution $U^{(0)}$ in $V_{h^{(0)}}$

(3) Given a computed solution $U^{(m)}$ in $V_{h^{(m)}}$ on a mesh with mesh size $h^{(m)}$, stop if

$$C_i \|h^{(m)} R(U^{(m)})\|_{\frac{1}{a}} \leq TOL. \tag{6.13}$$

(4) If not, determine a new mesh $T_{h^{(m+1)}}$ with mesh function $h^{(m+1)}$ of maximal size such that

$$C_i \|h^{(m+1)} R(U^{(m)})\|_{\frac{1}{a}} = TOL \tag{6.14}$$

and continue. We note that (6.13) is the stopping criterion and (6.14) defines the mesh modification. By Theorem 6.1, it follows that the error $\|u - U\|_E$ is bounded by the tolerance TOL, if the stopping criterion (6.13) is reached with $U = U^{(m)}$. The relation (6.14) defines the new mesh size $h^{(m+1)}$ by maximality, that is we seek a mesh function $h^{(m+1)}$ as large as possible (to maintain efficiency) such that (6.14) holds. In general, maximality is obtained by the "equidistribution" of error so that the error contributions from the individual intervals I_j are kept equal.

Equidistribution of the error results in the equation

$$a(x_j)^{-1}(h_j^{(m)} R(U^{(m-1)}))^2 h_j^{(m)} = \frac{TOL^2}{M^{(m)} + 1}, \qquad j = 1, .., M^{(m)} + 1, \tag{6.15}$$

where $M^{(m)} + 1$ is the number of intervals in $T_{h^{(m)}}$. The equation reflects the fact that the total error is given by the sum of the errors from each interval, and so the error on each interval must be a fraction of the total error. In practice, this nonlinear equation is simplified by replacing $M^{(m)}$ by $M^{(m-1)}$.

Remark The non-uniform mesh used in Example 1 is determined by the principle of equidistribution.

Note that in the Femlab implementation of this adaptive algorithm, the interpolation constant C_i is computed in an auxiliary computation and does not have to be supplied by the user.

6.3 An a priori error estimate

We now prove the basic a priori error estimate for (6.3) by comparing the Galerkin discretization error to the interpolation error. We use the notation

$$a(f,g) := \int_I a(x)f(x)g(x)\,dx.$$

Recalling the Galerkin orthogonality $a(u-U,v) = 0$ that holds for all v in V_h, we choose $v = \pi_h u$ and obtain from (6.3),

$$\int_I a(u-U)'(u-U)'\,dx = \int_I a(u-U)'(u-\pi_h u)'\,dx \leq \|u'-U'\|_a \|(u-\pi_h u)'\|_a,$$

so that

$$\|u' - U'\|_a \leq \|(u - \pi_h u)'\|_a \tag{6.16}$$

This inequality shows that the energy norm error of the Galerkin approximation is less than the energy norm error of the interpolant, and thus the Galerkin approximation is optimal in the present case. Together with the interpolation estimate

$$\|(u - \pi_h u)'\|_a \leq C_i \|hu''\|_a,$$

where C_i is an interpolation constant only depending on $\max_{I_j} a/\min_{I_j} a$, $j = 1, \cdots, M$, which is analogous to (5.11), this proves the following a priori error estimate:

Theorem 6.2 *The finite element solution U in V_h satisfies*

$$\|u' - U'\|_a \leq C_i \|hu''\|_a. \tag{6.17}$$

Remark

It is easy to prove that

$$\|hR(U)\|_{\frac{1}{a}} \leq CC_i \|hu''\|_a$$

with C a constant depending on a, indicating that the a posteriori error estimate is optimal in the same sense as the a priori estimate.

6.4 Dirichlet and Neumann boundary conditions

If the boundary condition $u(1) = 0$ in (3.1) (or (6.1)) is replaced by the boundary condition $u'(1) = 0$, then the functions in both the trial and test space need only satisfy the boundary condition at $x = 0$. In this case, the computation (3.3) is still valid because the term $-u'(1)v(1)$ now vanishes since $u'(1) = 0$. Further, by letting the test function v vary at $x = 1$, the boundary condition at $x = 1$ is found to be satisfied approximately in the discrete case. A boundary condition of the form $u = 0$ is called a *Dirichlet condition* and of the form $u' = 0$ a *Neumann condition*. We sum up as follows: the trial and test spaces should be forced to satisfy Dirichlet conditions, but not Neumann conditions.

6.5 Modeling and data errors

We discuss briefly how the a posteriori approach can also be used to analyze errors in data and modeling. Suppose that the correct coefficient $a(x)$ and data $f(x)$ in (6.3) are replaced by approximations $\hat{a}(x)$ and $\hat{f}(x)$. We use $\hat{U}$ in V_h^0 to denote the modified Fem approximation defined by the orthogonality relation,

$$\int_I \hat{a}\hat{U}'v'\,dx = \int_I \hat{f}v\,dx, \qquad \text{for all } v \text{ in } V_h^0. \tag{6.18}$$

It is natural to ask about the effects of the modeling and data errors on $u - \hat{U}$. To determine these, we start from a modified form of the error representation (6.11) obtained by using (6.18) instead of (6.3);

$$\|(u-\hat{U})'\|_a^2 = \int_I f \times (e - \pi_h e)\,dx - \sum_{i=1}^{M+1}\int_{I_j} a\hat{U}'(e-\pi_h e)'\,dx$$

$$+ \int_I (f-\hat{f})\pi_h e - \sum_{i=1}^{M+1}\int_{I_j}(a-\hat{a})U'(\pi_h e)'\,dx \equiv I + II - III$$

with the obvious definition of I, II, and III. The first term I is analysed above and just changes the residual to $R(\hat{U})$. For the new term III, we have using an estimate of the form $\|(\pi_h e)'\|_a \le C_i\|e'\|_a$,

$$III \le C_i\|(\hat{a}-a)\hat{U}'\|_{\frac{1}{a}}\|e'\|_a.$$

Similarly, integration by parts gives

$$II \le \|(\hat{F}-F)\|_{\frac{1}{a}}\|e'\|_a,$$

where $F' = f$, $\hat{F}' = \hat{f}$ and $F(0) = \hat{F}(0) = 0$. Altogether, the modified form of the a posteriori error estimate of Theorem 6.1 estimating the total error from contributions from Galerkin discretization, data and modeling errors is

Theorem 6.3 *The finite element solution $\hat{U}$ defined by (6.18) satisfies*

$$\|u' - \hat{U}'\|_a \leq C_i(\|hR(\hat{U})\|_{\frac{1}{a}} + \|(\hat{F} - F)\|_{\frac{1}{a}} + \|(\hat{a} - a)\hat{U}_x\|_{\frac{1}{a}}). \tag{6.19}$$

This estimate allows the mesh to be adapted to control the first term while simultaneously taking into account the effects of data and modeling errors.

7 A "parabolic" model problem

As a first example of a time-dependent problem, we consider the linear initial value problem: find $u = u(t)$ such that

$$\begin{aligned} u' + a(t)u &= f(t), \qquad \text{for } t > 0, \\ u(0) &= u_0, \end{aligned} \tag{7.1}$$

where $a(t)$ is a given coefficient, $f(t)$ a given source term, and $v' = \frac{dv}{dt}$ now denotes the time derivative of v. The exact solution $u(t)$ of this problem is given by the variation of constants formula

$$u(t) = e^{-A(t)}u_0 + \int_0^t e^{-(A(t)-A(s))} f(s)\, ds \tag{7.2}$$

where $A(t)$ is a primitive function to $a(t)$, that is $A' = a$ and $A(0) = 0$. The formula can be used to determine u if the integral can be evaluated, otherwise, we have to use a numerical method.

Before going into the numerics, we draw some conclusions concerning the nature of the exact solution u from (7.2). In general, the exponential factors are expected to grow with time. However, note that if $a(t) \geq 0$ for all t, then $A(t) \geq 0$ and $A(t) - A(s) \geq 0$ for all $t \geq s$. Hence, both u_0 and f are multiplied by quantities that are less than or equal to one, and if $a(t) > 0$, then these factors tend to zero as time passes. This decay of influence or *stability* is typical of parabolic problems. As we shall see, for such problems, Galerkin discretization errors accumulate in such a way that accurate long-time computation is possible.

We use the dG(0)-method for (7.1). Recall that T_k is a subdivision of $(0, \infty)$ into time intervals $I_n = (t_{n-1}, t_n)$ of length k_n and W_k denotes the space of discontinuous piecewise constant functions on T_k. The dG(0)-method for (7.1) reads as follows: find U in W_k such that for all polynomials v of degree 0 on I_n,

$$\int_{I_n} (U' + a(t)U)v\,dt + [U_{n-1}]v_{n-1}^+ = \int_{I_n} fv\,dt, \tag{7.3}$$

where $[v_n] = (v_n^+ - v_n^-)$, $v_n^\pm = \lim_{s\to\pm 0} v(t_n + s)$, and $U_0^- = u_0$. We need the last notation to distinguish the two values of a function v in T_k at each time node. We note that (7.3) in variational form expresses that the "sum" of the residual $U_t + a(t)U - f$ in I_n, and the "jump" $[U_{n-1}]$ is orthogonal to all discrete test functions. Since U is a constant on I_n, we have in fact that $U' \equiv 0$.

Letting U_n denote the constant value of U on the time interval I_n, the dG(0)-method (7.3) takes the form

$$U_n - U_{n-1} + U_n \int_{I_n} a\,dt = \int_{I_n} f\,dt, \qquad \text{for} \quad n = 1, 2, .., \tag{7.4}$$

where $U_0 = u_0$. This is a variant of the classical backward Euler method with exact evaluation of the integrals involving the given functions a and f. We assume that if $a(t)$ is negative, then the time step is so small that $|\int_{I_n} a\,dt| < 1$, in which case (7.4) defines U_n uniquely.

7.1 An a posteriori error estimate

To derive an a posteriori error estimate for the error $e_N \equiv u(t_N) - U_N$ for a given $N \geq 1$, we introduce the continuous dual "backward" problem,

$$\begin{aligned} -\varphi' + a(t)\varphi &= 0, \qquad \text{in} \quad (0, t_N), \\ \varphi(t_N) &= e_N, \end{aligned} \tag{7.5}$$

with solution given by $\varphi(t) = e^{A(t)-A(t_N)}e_N$. Integrating by parts over each subinterval I_n gives

$$\begin{aligned} e_N^2 &= e_N^2 + \sum_{n=1}^{N} \int_{I_n} e(-\varphi' + a\varphi)\,dt \\ &= \sum_{n=1}^{N} \int_{I_n} (e' + ae)\varphi\,dt + \sum_{n=1}^{N-1} [e_n]\varphi_n^+ + (u_0 - U_0^+)\varphi_0^+ \\ &= \sum_{n=1}^{N} \Big(\int_{I_n} (f - aU)\varphi\,dt - [U_{n-1}]\varphi_{n-1}^+ \Big), \end{aligned}$$

where, in the last step, we use the facts that $U_t = 0$ on each I_n and $U_0^- = u_0$. Now, we use the Galerkin orthogonality (7.3) by taking $v = \pi_k\varphi$, where we recall that

$$\int_{I_n} (\varphi - \pi_k\varphi)\,dt = 0, \quad n = 1, .., N,$$

to obtain the error representation formula

$$e_N^2 = \sum_{n=1}^{N} \Big(\int_{I_n} (f - aU))(\varphi - \pi_k\varphi)\,dt - [U_{n-1}](\varphi - \pi_k\varphi)_{n-1}^+ \Big).$$

Recalling the interpolation error estimate

$$\|\varphi - \pi_k\varphi\|_{I_n} \le \int_{I_n} |\varphi'|\, dt, \tag{7.6}$$

we obtain

$$\begin{aligned} e_N^2 &\le \left(\int_0^{t_N} |\varphi'| dt\right) \max_{n=1,..,N} (|[U_{n-1}]| + \|k(f - aU)\|_{I_n}) \\ &\le S(t_N) \max_{n=1,..,N} (|[U_{n-1}]| + \|k(f - aU)\|_{I_n}) |e_N|, \end{aligned} \tag{7.7}$$

where $S(t_N)$ is a stability factor defined by

$$S(t_N) = \frac{\int_0^{t_N} |\varphi_t| dt}{|e_N|}. \tag{7.8}$$

To complete the proof of the a posteriori error estimate, we need to estimate the stability factor $S(t_N)$. The following lemma presents such a stability estimate in both the general case and the dissipative case when $a(t) \ge 0$ for all t. We also state an estimate for the dual solution φ itself.

Lemma 7.1 *If $|a(t)| \le \mathcal{A}$ for t in $(0, t_N)$, then φ satisfies for all t in $(0, t_N)$:*

$$|\varphi(t)| \le \exp(\mathcal{A} t_N)|e_N|, \tag{7.9}$$

and

$$S(t_N) \le t_N \mathcal{A} \exp(\mathcal{A} t_N). \tag{7.10}$$

If $a(t) \ge 0$ for all t, then φ satisfies for all t in $(0, t_N)$:

$$|\varphi(t)| \le |e_N|, \tag{7.11}$$

and

$$S(t_N) \le 1. \tag{7.12}$$

Proof The first and second estimate directly follow from the boundedness assumption on a, and we leave the details as an exercise. The third estimate follows directly from the fact that $A(t_N) - A(t)$ is non-negative for $t \le t_N$. Further, since a is non-negative,

$$\begin{aligned} \int_I |\varphi'| dt &= |e_N| \int_I a(t) \exp(A(t_N) - A(t)) dt \\ &= |e_N|(1 - \exp(A(0) - A(t_N))) \le |e_N|, \end{aligned}$$

which completes the proof. □

We now insert the strong stability estimates (7.10) or (7.12) into (7.7), and obtain the following a posteriori error estimate for the parabolic model problem (7.1).

Theorem 7.2 *The finite element solution U satisfies for $N = 1, 2, ..,$*

$$|u(t_N) - U_N| \leq S(t_N) \max_{n=1,..,N} (|U_n - U_{n-1}| + k_n \|f - aU\|_{I_n}).$$

Remark It is important to compare the general result when a is only known to be bounded to the result for dissipative problems with $a \geq 0$. In the first case, the errors can accumulate at an "exponential" rate, and, depending on $\mathcal{A}$, the stability factor $S(t_N)$ can quickly become so large that controlling the error is no longer possible. In the case $a \geq 0$, there is no accumulation of error so that accurate computation is possible over arbitrarily long times.

Example 2 Consider the dissipative problem $u' + u = 0$, $y(0) = 1$ with solution $u(t) = e^{-t}$. We compute with dG(0) and plot the solution and the approximation in Fig. 9a. The approximation is computed with an error tolerance of .001. In Fig. 9b, we plot the stability factor versus time. Note that $S(t)$ tends to 1 as t increases, indicating that the numerical error does not grow significantly with time, and accurate computations can be made over arbitrarily long time intervals.

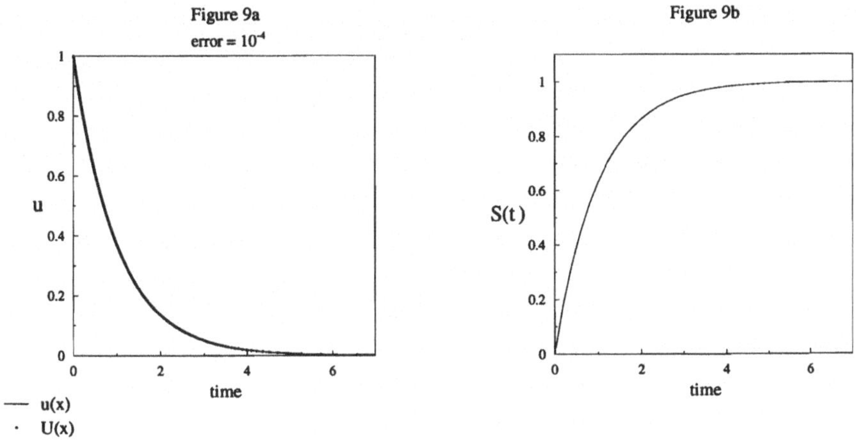

FIG. 9. Solution and approximation for Example 2.

Example 3 We now consider the problem $u' - u = 0$, $y(0) = 1$ with solution $u(t) = e^t$. We compute with dG(0) keeping the error below .025. Since the problem is not dissipative, we expect to see the error grow. The difference $U(t) - u(t)$ is plotted in Fig. 10a and the exponential growth rate is clearly visible. Given a certain amount of computational power — for example, a fixed precision or a fixed amount of computing time — there is some value of t at which accurate computation is no longer possible. The stability factor is plotted in Fig. 10b, and we note that it reflects the instability rate precisely.

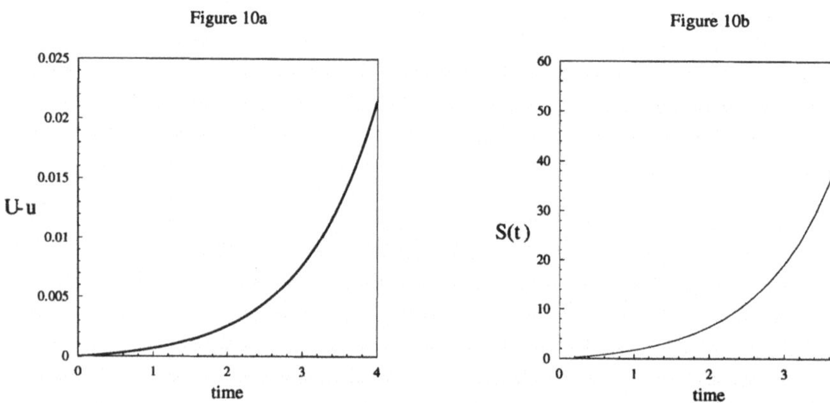

FIG. 10. Solution and approximation for Example 3.

7.2 Adaptive error control

An adaptive algorithm based on the above a posteriori error estimate takes the form: determine the time steps k_n so that

$$\hat{S}(t_N)(|U_n - U_{n-1}| + k_n\|f - aU\|_{I_n}) = TOL, \qquad n = 1, 2, ..., N,$$

where $\hat{S}(t_N) = \max_{1\leq n\leq N} S(t_n)$, which guarantees the following error control:

$$\|u(t_n) - U(t_n)\| \leq TOL, \qquad n = 1, \cdots N.$$

The stability constant $\hat{S}(t_N)$ is computed in an auxiliary computation solving the backward problem with chosen initial data, see [8] for more details.

Example 4 We consider a more complicated problem,

$$u' + (.25 + 2\pi\sin(2\pi t))u = 0, \qquad t > 0,$$
$$u(0) = 1,$$

with solution,

$$u(t) = \exp(-.25t + \cos(2\pi t) - 1).$$

The solution oscillates as time passes, but the oscillations dampen. In Fig.

11a, we plot the solution together with the dG(0) approximation computed with error below .12. In Fig. 11b. we plot the time steps used for the computation. We see that the steps are adjusted for each in oscillation and in addition that there is an overall trend to increasing the steps as the size of the solution decreases.

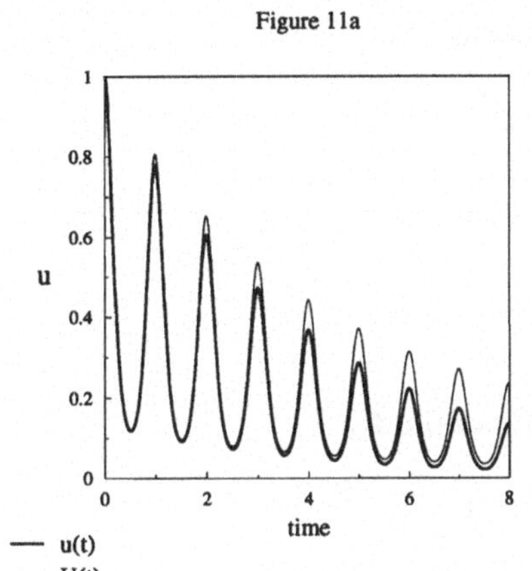

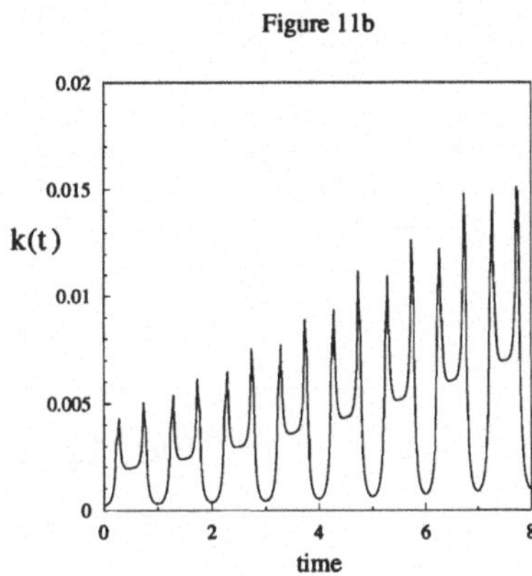

FIG. 11. Plots for Example 4.

In addition, the solution has changing stability characteristics. In Fig. 12a, we plot the stability factor versus time, and it is evident that the numerical error decreases and increases in alternating periods of time. If a crude "exponential" bound on the stability factor had been used instead of a computational estimate, then the error would have been greatly overestimated with the consequence that the computation could only be done over a much smaller interval. To demonstrate the effectiveness of the a posteriori estimate for error control, we plot the ratio of the true error to the computed bound versus time. The ratio quickly settles down to a constant, which means that the bound is predicting the behavior of the error in spite of the fact that the error oscillates a good deal.

7.3 An a priori error estimate

The a priori error estimate for (7.3) reads

Theorem 7.3 *If $|a(t)| \leq \mathcal{A}$ for all t, then there is a constant $C > 0$ such that U satisfies for $N = 1, 2, ..,$*

$$|u(t_N) - U_N| \leq C\mathcal{A}t_N \exp(C\mathcal{A}t_N)\|ku'\|_I,$$

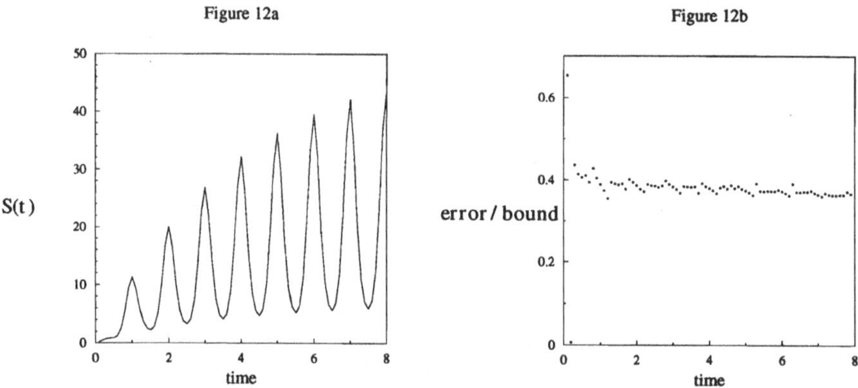

FIG. 12. Plots for Example 4.

and if $a(t) \geq 0$ *for all* t, *then for* $N = 1, 2, ..,$

$$|u(t_N) - U_N| \leq \|ku'\|_I. \tag{7.13}$$

We note the optimal nature compared to interpolation in the case $a(t) \geq 0$.

Proof We introduce the discrete dual backward problem: find Φ in W_k such that for $n = N, N-1, .., 1$,

$$\int_{I_n} (-\Phi' + a(t)\Phi)v dt - [\Phi_n]v_n^- = 0, \quad \text{for all } v \text{ in } W_k, \tag{7.14}$$

where $\Phi_N^+ = (\pi_k u - U)_N^-$. It suffices to estimate the "discrete" error $\bar{e} \equiv \pi_k u - U$ in W_k since $u - \pi_k u$ is already known. Choosing $v = \bar{e}$, the Galerkin orthogonality allows U to be replaced by u and we obtain the following representation

$$|\bar{e}_N^-|^2 = \sum_{n=1}^{N} \int_{I_n} (-\Phi' + a(t)\Phi)\bar{e} dt - \sum_{n=1}^{N-1} [\Phi_n]\bar{e}_n^- + \Phi_N^- \bar{e}_N^-$$

$$= \sum_{n=1}^{N} \int_{I_n} (-\Phi' + a(t)\Phi)(\pi_k u - u) dt - \sum_{n=1}^{N-1} [\Phi_n](\pi_k u - u)_n^- + \Phi_N^-(\pi_k u - u)_N^-$$

$$= - \int_I a\Phi(u - \pi_k u) dt + \sum_{n=1}^{N-1} [\Phi_n](u - \pi_k u)_n^- - \Phi_N^-(u - \pi_k u)_N^-,$$

where we used that $\Phi_t = 0$ on each time interval. Recalling the interpolation estimate (7.6), the desired result now follows from the following lemma expressing the weak and strong stability of the discrete dual problem (7.14). □

Lemma 7.4 *When* $|a(t)| \leq \mathcal{A}$ *for all* t*, then there is a constant* $C > 0$ *such that the solution of the discrete dual problem (7.14) satisfies*

$$|\Phi_n^-| \leq \exp(C\mathcal{A}t_N)|\bar{e}_N^-|, \tag{7.15}$$

$$\sum_{n=1}^{N-1} |[\Phi_n]| \leq C\mathcal{A}t_N \exp(C\mathcal{A}t_N)|\bar{e}_N^-|, \tag{7.16}$$

$$\sum_{n=1}^{N} |\int_{I_n} a|\Phi_n|dt| \leq C\mathcal{A}t_N \exp(C\mathcal{A}t_N)|\bar{e}_N^-|. \tag{7.17}$$

If $a(t) \geq 0$ *for all* t*, then*

$$|\Phi_n^-| \leq |\bar{e}_N^-|, \tag{7.18}$$

$$\sum_{n=1}^{N-1} |[\Phi_n]| \leq |\bar{e}_N^-|, \tag{7.19}$$

$$\sum_{n=1}^{N} |\int_{I_n} a|\Phi_n|dt| \leq |\bar{e}_N^-|. \tag{7.20}$$

Proof The discrete dual problem (7.14) takes the form

$$-\Phi_{n+1} + \Phi_n + \Phi_n \int_{I_n} a(t)dt = 0, \quad n = N, N-1, ...1,$$
$$\Phi_{N+1} = \bar{e}_N^-,$$

where Φ_n denotes the value of Φ on I_n, so that

$$\Phi_n = \prod_{j=n}^{N} (1 + \int_{I_j} a\, dt)^{-1} \Phi_{N+1}.$$

In the case that a is bounded, the results follow from standard estimates. When a is nonnegative, this immediately proves (7.18). To prove (7.19), we assume without loss of generality that Φ_{N+1} is positive, then the sequence Φ_n decreases when n decreases, so that

$$\sum_{n=1}^{N} |[\Phi_n]| = \sum_{n=1}^{N} [\Phi_n] = \Phi_{N+1} - \Phi_1 \leq |\Phi_{N+1}|.$$

Finally (7.20) follows from the discrete equation. □

We note that the a priori error estimate (7.13) is optimal compared to interpolation.

7.4 The effect of quadrature

The dG(0)-method (7.3) contains integrals involving the given functions a and f. If these integrals are computed approximately using quadrature, then the *quadrature error* also contributes to the total error. Quadrature is a subject familiar from Calculus since integration is defined as the limit of quadrature formulas such as the rectangle or Simpson's rule. The important fact to recall is that there are quadrature formulas of different accuracy on a given interval, for example Simpson's rule is more accurate than the rectangle rule. It is natural to control the quadrature error on the same tolerance level as the Galerkin discretization error, which is controlled by (7.2). We focus on the quadrature error arising from the integration of f. For the quadrature, we could use the rectangle rule or the more accurate Simpson's rule, though it turns out there is a reason to use the more accurate formula. We could also use the special "Gauss" quadrature given by the midpoint rule:

$$\int_{I_n} f\,dt \approx k_n f(t_{n-\frac{1}{2}}), \qquad t_{n-\frac{1}{2}} = \frac{1}{2}(t_{n-1} + t_n), \tag{7.21}$$

since this has the same accuracy as Simpson's rule while only requiring one function evaluation per element as opposed to two for Simpson's rule. The effect on the total error from using quadrature can be estimated by the sum of the quadrature error over each time interval by invoking a weak stability estimate. One can show that the Gauss rule has local error

$$|\int_{I_n} f\,dt - k_n f(t_{n-\frac{1}{2}})| \leq C_q \int_{I_n} |k^2 f''|\,dt, \tag{7.22}$$

where C_q is a quadrature constant, so summing the errors from each element shows that the total error resulting from quadrature is bounded by

$$C_q \int_I |k^2 f''|\,dt \leq S_q \|k^2 f''\|_I, \tag{7.23}$$

where $S_q = C_q T$ is the stability factor related to quadrature. We note that this estimate includes a linear growth in the length of the time interval T, even in the dissipative case when there would be no accumulation of error if the integrals were computed exactly. This is the reason that we choose a more accurate quadrature than the rectangle rule, since this growth may be compensated for somewhat by the two powers of the time step k. If we instead used the rectangle rule,

$$\int_{I_n} f\,dt \approx k_n f(t_n), \tag{7.24}$$

which gives the classic backward Euler scheme, then we get a linear growth in time that is not compensated for by two powers of k since the quadrature error in this case has the form

$$C_q \int_I |kf'|\,dt \le C_q T\|kf'\|_I. \tag{7.25}$$

Example 5 We consider the approximation of $u' - 0.1u = t^3$, $u(0) = 1$. We compute using the dG(0), backward Euler (rectangle rule quadrature), and the one-point Gauss rule, with accuracies plotted below. The function u is plotted in Fig. 13a; the problem is not dissipative, so we expect error accumulation. We plot the errors of the three computations in Fig. 13b. The dG(0) and one-point Gauss rule are very close in accuracy, while the backward Euler computation errors accumulate at a much faster rate.

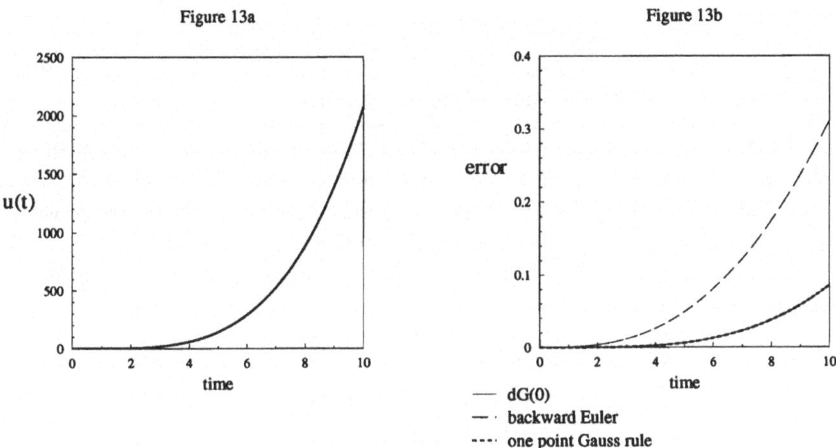

FIG. 13. Plots for Example 5.

Since the computational cost of the quadrature is usually small compared to the Galerkin computational work (which requires the solution of a system of equations), the precision of the quadrature may be increased if needed without significantly increasing the overall work. This illustrates the importance of separating Galerkin discretization and quadrature errors since they accumulate differently and they involve different quantities. These errors should not be confused as happens in the classical analysis of the backward Euler method, leading to non-optimal performance.

8 A "hyperbolic" model problem

We finally consider the ode model for a "hyperbolic" problem: find $u = (u_1, u_2)$ such that

$$\begin{aligned} u_1' + au_2 = f_1, &\quad \text{for} \quad t > 0, \\ u_2' - au_1 = f_2, &\quad \text{for} \quad t > 0, \\ u_1(0) = u_{10}, &\quad u_2(0) = u_{20}, \end{aligned} \tag{8.1}$$

where the $a = a(t)$ is a given bounded coefficient with $|a| \leq \mathcal{A}$, and the f_i and u_{i0} are given data.

We study the application of the cG(1)-method to (8.1), with V_k the set of continuous piecewise linear functions $v = (v_1, v_2)$ on a partition T_k. This method takes the form: find $U = (U_1, U_2)$ in V_k such that for $n = 1, 2, ..,$

$$\begin{aligned} \int_{I_n} (U_1' + aU_2)\, dt &= \int_{I_n} f_1\, dt, \\ \int_{I_n} (U_2' - aU_1)\, dt &= \int_{I_n} f_2\, dt, \\ U_1(0) = u_{10}, \quad U_2(0) &= u_{20}, \end{aligned} \tag{8.2}$$

corresponding to using piecewise constant test functions on each interval I_n. We use piecewise constant test functions because there are only first order derivatives in (8.1). In the elliptic problem discussed above, there were second order derivatives, hence we used piecewise linear test functions. In the case that a is constant, the method (8.2) reduces to the following using the notation $U_{i,n} = U_i(t_n)$:

$$\begin{aligned} U_{1,n} - U_{1,n-1} + k_n a(U_{2,n} + U_{2,n-1})/2 &= \int_{I_n} f_1\, dt, \\ U_{2,n} - U_{2,n-1} - k_n a(U_{1,n} + U_{1,n-1})/2 &= \int_{I_n} f_2\, dt, \\ U_1(0) = u_{10}, \quad U_2(0) &= u_{20}, \end{aligned} \tag{8.3}$$

which is a variant of the classical Crank–Nicolson method. The method cG(1) has less dissipation and better accuracy than dG(0), and it is advantageous to use it the case when the solution is smooth.

8.1 An a posteriori error estimate

To derive an a posteriori error estimate for the error $e_N = u(t_N) - U_N$, $U_N \equiv U(t_N)$, we introduce the dual problem: find $\varphi = (\varphi_1, \varphi_2)$ such that

$$\begin{aligned} -\varphi_1' + a\varphi_2 &= 0, \quad \text{for} \quad 0 < t < t_N, \\ -\varphi_2' - a\varphi_1 &= 0, \quad \text{for} \quad 0 < t < t_N, \\ \varphi(t_N) &= e_N. \end{aligned} \tag{8.4}$$

Again by using Galerkin orthogonality, we obtain an error representation formula

$$|e_N|^2 = -\int_0^{t_N} R(\varphi - \pi_k \varphi)\, dt,$$

where

$$R_1 = U_1' + a_1U_2 - f_1, \quad R_2 = U_2' - a_2U_1 - f_2,$$

and π_k is the nodal interpolation operator into V_k. Multiplying the equations by φ_1 and φ_2 respectively and using the cancellation of the terms $\pm a_1\varphi_1\varphi_2$, we obtain the following stability estimate:

$$\max_{0\le t\le t_N} |\varphi(t)| \le |e_N|,$$

$$\max_{0\le t\le t_N} |\varphi'(t)| \le \mathcal{A}|e_N|.$$

Combining the error representation with the strong stability estimate and using the interpolation estimate (5.10) we arrive at the following a posteriori error estimate:

Theorem 8.1 *The solution U of the cG(1)-method (8.2) satisfies for $N = 1, 2, ..,$*

$$|u(t_N) - U_N| \le \mathcal{A}\int_0^{t_N} k|R|\,dt.$$

8.2 An a priori error estimate

The corresponding a priori error estimate takes the form

Theorem 8.2

$$|u(t_N) - U_N| \le \mathcal{A}\int_0^{t_N} |k^2u''|\,dt \le \mathcal{A}t_N \max_{0\le t\le t_N} |k^2u''|.$$

We note the linear growth in time that is typical of a hyperbolic problem.

Example 6 We compute on the problem

$$\begin{aligned} &u_1' + 2u_2 = \cos(\pi t/3), \\ &u_2' - 2u_1 = 0, \\ &u_1(0) = 0, \quad u_2(0) = 1, \end{aligned}$$

using the cG(1)-method with error below .07. The two components of the approximation are plotted in Fig. 14a. This demonstrates that different components of a system of equations may behave differently at different times. The error control must choose the time steps to maintain accuracy in all components simultaneously. In Fig. 14b, we plot the stability factor and the linear growth of error is clear.

Remark Generally, when the solution is smooth, then the computation is more efficient as the *order* of the approximation is increased. This is due to the fact that the cost of computing the approximation is given by a polynomial function of the order q, while the accuracy depends exponentially on q, (i.e. k^q). For the problem in Example 6, we plot the step sizes used for the dG(0) and the cG(1) approximations computed with the same accuracy

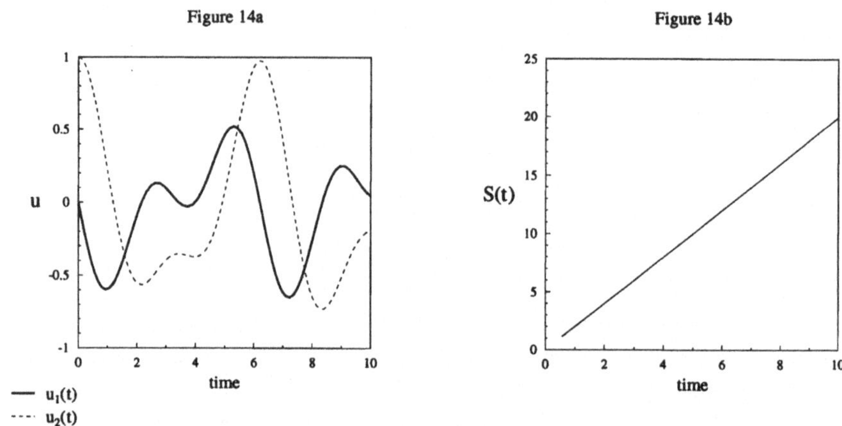

FIG. 14. Plots for Example 6.

of .07. Note that the cG(1) step sizes are about 10 times larger. Viewed in

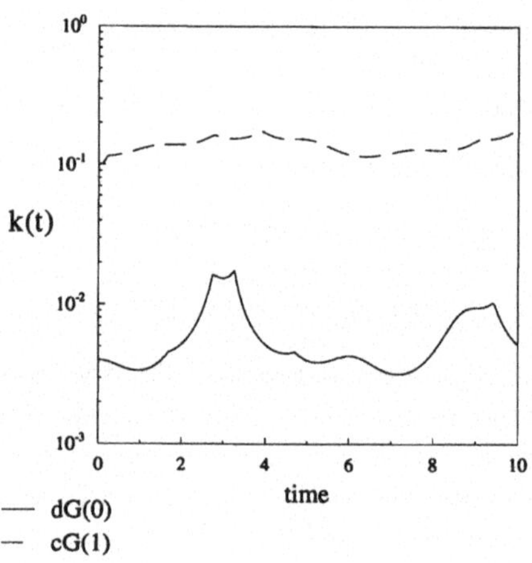

FIG. 15. Step sizes for dG(0) and cG(1).

a different way, this means that maximum interval over which the cG(1) approximation is accurate to a given tolerance is "$1/k$" times longer than the corresponding maximum interval for the dG(0)-method. However, the solution must have the required smoothness, i.e. q derivatives to obtain q^{th} order convergence. If the solution is not smooth, then all order methods use the same step sizes, and the lower order methods are then cheaper.

9 The general initial value problem

Next, we show how the a posteriori error control presented above extends directly to a general nonlinear initial value problem for ODEs in $\mathbb{R}^d, d \geq 1$ without difficulty. This is surprising because nonlinear differential equation may be very complex. However, the a posteriori character of the error analysis means that the difficulties are handled through computation involving the approximation rather than through mathematical analysis.

We consider the following initial value problem: find $u = u(t)$ such that

$$\begin{aligned} u' + f(t,u) &= 0, \qquad \text{for } t > 0, \\ u(0) &= u_0, \end{aligned} \tag{9.1}$$

where $f(t,\cdot) : \mathbb{R}^d \to \mathbb{R}^d$ is a given function for $t > 0$. The dG(0)-method for (9.1) reads as follows: find U in W_k such that for all constant d-vectors v,

$$\int_{I_n} (U' + f(t,U))v dt + [U_{n-1}]v_{n-1}^+ = 0, \tag{9.2}$$

where $[v_n] = (v_n^+ - v_n^-)$, $v_n^\pm = \lim_{s\to\pm 0} v(t_n + s)$, and $U_0^- = u_0$. Note that the generalization to $d > 1$ dimensions is completely natural using vector notation. Recalling the notation $U_n \equiv U|_{I_n}$, the dG(0)-method (9.2) takes the form

$$U_n - U_{n-1} + \int_{I_n} f(t, U_n)\, dt = 0, \qquad \text{for} \quad n = 1, 2, ..., \tag{9.3}$$

where $U_0 = u_0$. Again, this is an improved variant of the classical backward Euler method with exact evaluation of the integral with integrand $f(t,U)$.

9.1 An a posteriori error estimate

To derive an a posteriori error estimate for the error $e_N \equiv u(t_N) - U_N$ for a given $N \geq 1$, we introduce the continuous dual "backward" problem

$$\begin{aligned} &-\varphi' + A(t)^*\varphi = 0 \quad \text{in} \quad (0, t_N), \\ &\varphi(t_N) = e_N, \end{aligned} \tag{9.4}$$

where A^* is the transpose of A and

$$A(t) \equiv \int_0^1 f_u(t, su + (1-s)U)\, ds, \tag{9.5}$$

where $f_u(t,\cdot)$ is the Jacobian of $f(t,\cdot)$, so that

$$A(t)e = \int_0^1 f_u(t, su + (1-s)U)e\, ds$$

and

$$\int_0^1 \frac{d}{ds} f(t, su + (1-s)U)\, ds = f(t,u) - f(t,U).$$

Integrating by parts, we get

$$\begin{aligned}
|e_N|^2 &= |e_N|^2 + \sum_{n=1}^{N} \int_{I_n} e(-\varphi' + A^*\varphi)\, dt \\
&= \sum_{n=1}^{N} \int_{I_n} (e' + A(t)e)\varphi\, dt + \sum_{n=1}^{N-1} [e_n]\varphi_n^+ + (u_0 - U_0^+)\varphi_0^+ \\
&= -\sum_{n=1}^{N} (\int_{I_n} (U' + f(t,U))\varphi\, dt + [U_{n-1}]\varphi_{n-1}^+).
\end{aligned}$$

Now the Galerkin orthogonality (9.2) can be used by choosing $v = \pi_k\varphi$ in W_k defined above, and we obtain, since $U' = 0$ on I_n, the following error representation formula

$$\begin{aligned}
|e_N|^2 &= -\sum_{n=1}^{N} \left(\int_{I_n} (U' + f(t,U)(\varphi - \pi_k\varphi) \right) dt + [U_{n-1}](\varphi - \pi_k\varphi)_{n-1}^+) \\
&= -\int_0^{t_N} f(t,U))(\varphi - \pi_k\varphi)\, dt - \sum_{n=0}^{N-1} [U_n](\varphi - \pi_k\varphi)_n^+.
\end{aligned}$$

Recalling the interpolation estimate (5.5), we see that

$$|e_N|^2 \leq \int_0^{t_N} |\varphi'| dt \max_{n=0,1,..,N} (|[U_{n-1}]| + k_n \|f(\cdot, U_n)\|_{I_n}).$$

Finally, we define the strong stability factor $S(t_N)$ by φ of (9.4)

$$S(t_N) = \max \varphi(t_N) \frac{\int_0^{t_N} |\varphi'| dt}{|\varphi(t_N)|}. \tag{9.6}$$

We thus arrive at the following "abstract" a posteriori error estimate for the initial value problem (9.1):

Theorem 9.1 *The solution U of the dG(0)-method (9.2) satisfies for $N = 1, 2, ..,$*

$$|u(t_N) - U_N| \leq S(t_N)(\max_{n=1,..,N} (|U_n - U_{n-1}| + k_n \|f(\cdot, U_n)\|_{I_n}).$$

Remark There is a corresponding a priori result.

We remark that the a posteriori error estimate is proved in the same way for the general ode (9.1) as in the case of the model problem (7.1). This is remarkable since (9.1) is a much more complex class of problems. The difficulty is hidden in the definition of the strong stability factor $S(t_N)$. As in the model case, we can prove an "exponential" bound on $S(t_N)$. In practice, this is found to overestimate the error too much to be useful. Instead, in the adaptive algorithm, the stability factor $S(t_N)$ is estimated computationally by solving the dual problem numerically with suitable data at $t = t_N$, and as a result quantitative error control is also possible for complex problems.

Example 7 In the early 1960s, the meteorologist E. Lorenz derived a simple ODE model in order to explain why weather forecasts are unreliable over more than a couple of days. The model is derived by taking a three element Fem space discretization of the Navier–Stokes equations for fluid flow (the "fluid" being the atmosphere in this case) and simply ignoring the discretization error. This gives a three-dimensional system of ODEs in time:

$$\begin{aligned} x' &= -\sigma x + \sigma y, \\ y' &= -rx - y - xz, \\ z' &= -bz + xy, \\ x(0) &= x_0, y(0) = y_0, z(0) = z_0, \end{aligned} \tag{9.7}$$

where σ, r, and b are positive constants. These were determined originally as part of the physical problem, but the interest among mathematicians quickly shifted to studying (9.7) for different values of the parameters because this makes the problem *chaotic*.

A precise definition of chaotic behavior seems difficult to give, but we point out two distinguishing features: firstly, while confined to a fixed region in space, the solutions do not "settle down" into a steady state or periodic state; secondly, the solutions are *data sensitive*, which means that perturbations of the initial data of a given solution eventually causes large changes in the solution. This corresponds to the so-called "butterfly effect" in meteorology suggesting that small causes may sometimes have large effects on the evolution of the weather. In such a situation, numerical approximations always become inaccurate after some time. An important issue is to determine this time, which in this context is related to the maximal length of time of accurate weather forecasts.

We choose standard values $\sigma = 10$, $b = 8/3$, and $r = 28$, and we compute with the dG(1)-method. In Fig. 16, we plot two views of the solution corresponding to initial data $(1, 0, 0)$ computed with an error of .5 up to time 30. The solutions always behave similarly: after some short initial time, they begin to "orbit" around one of two points, with an occasional "flip" back and forth between the points. The chaotic nature of the solutions

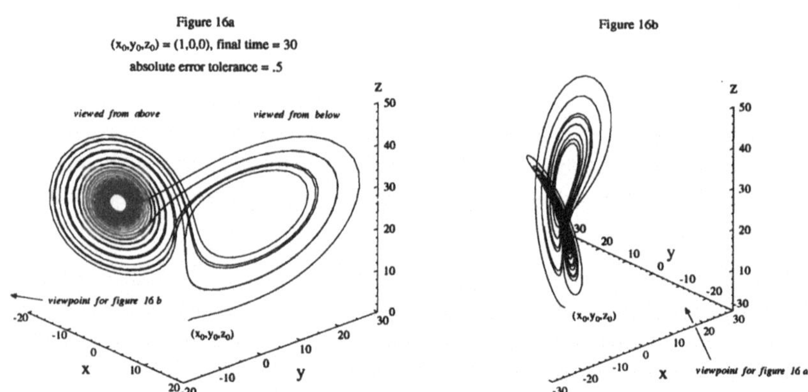

FIG. 16. Two views of a solution of the Lorenz system.

is this flipping that occurs at apparently random times. In fact, accurate computation can reveal much detail about the behavior of the solutions, see [21].

Here, we settle for demonstrating the quality of the error control explained in these notes. In Fig. 17a, we plot the stability factor $S(t)$ on a logarithmic scale. The data sensitivity of this problem is reflected in the overall exponential growth of $S(t)$, and it is clear that any computation becomes inaccurate at some point. The error control allows us to determine this time. Note, however, that $S(t)$ does not grow uniformly rapidly and there are periods of time with different data sensitivity. It is important for the error control to detect these to avoid grossly overestimating the error. To test this, we do an experiment. We compute using two error tolerances, one 10^{-5} smaller than the other, then we subtract the least accurate computation from the more accurate computation. This should be a good approximation to the true error (which is unknown of course). In Fig. 17b, we plot this approximate error together with the error bound predicted by the error control based on a posteriori estimates as we have described. There is remarkable agreement.

10 An elliptic PDE: Poisson's equation

10.1 Introduction

In this chapter, we consider the finite element method based on piecewise linear approximation applied to an elliptic PDE, the Poisson equation on a polygonal domain Ω with boundary Γ. We describe optimal a priori and a

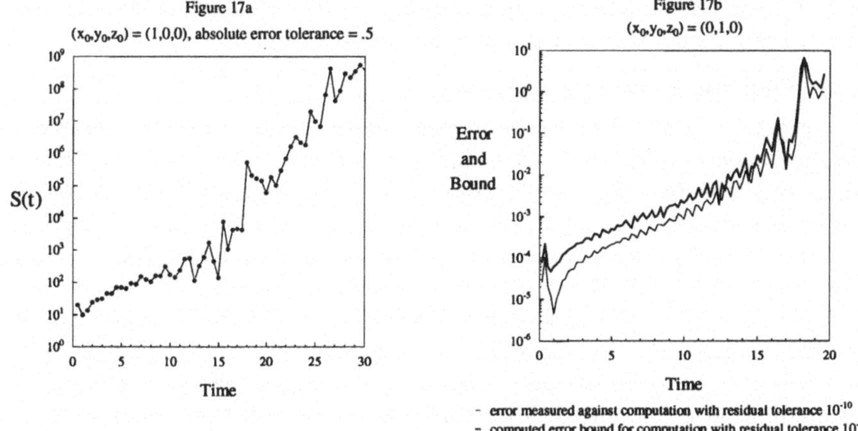

FIG. 17. $S(t)$, actual error and error bound for the Lorenz system.

posteriori error estimates in the energy norm and present some numerical results using Femlab. We use the notation $\|\cdot\|_2 = \|\cdot\|_{L_2(\Omega)}$.

10.2 Poisson's equation

Consider the Poisson equation with homogeneous Dirichlet boundary conditions: find $u = u(x)$ such that

$$\begin{aligned} -\Delta u &= f, \quad \text{in } \Omega, \\ u &= 0 \quad \text{on } \Gamma, \end{aligned} \tag{10.1}$$

where Ω is a bounded polygonal domain in $\mathbb{R}^2$ with boundary Γ, $x = (x_1, x_2)$, $\Delta = (\partial/\partial x_1)^2 + (\partial/\partial x_2)^2$ is the Laplace operator, $f(x)$ is a given source term and $g(x)$ given boundary data. This boundary value problem is the basic model problem of elliptic type with Dirichlet boundary conditions, and plays a fundamental role in most areas of physics and mechanics. One may replace the Dirichlet boundary conditions on a part $\Gamma_1 \subset \Gamma$ by Neumann or, more generally, Robin boundary conditions of the form

$$\frac{\partial u}{\partial n} + \lambda u = g \qquad \text{on } \Gamma_1, \tag{10.2}$$

where $\lambda \geq 0$. For the sake of simplicity, we consider the case of homogeneous Dirichlet boundary conditions with $g = 0$ in (10.1).

The variational form of (10.1) reads: find u such that

$$(\nabla u, \nabla v) = (f, v), \qquad \text{for all } v \text{ in } V, \tag{10.3}$$

where V is the set of test functions v such that $v = 0$ on Γ, $(w, v) =$

$\int_\Omega wv\,dx$, and $(\nabla w, \nabla v) = \int_\Omega \nabla w \cdot \nabla v\,dx$. This results from multiplying (10.1) by the test function v, integrating over Ω, and then integrating by parts using Green's theorem to move derivatives from u to v.

10.3 Fem for Poisson's equation

We consider the simplest finite element method for (10.1) by applying Galerkin's method to the variational formulation (10.3) using a finite-dimensional subspace V_h based on piecewise linear approximation on triangles. To this end, let $T_h = \{K\}$ be a triangulation of Ω into closed triangles K of diameter h_K, with the associated set of internal nodes $N_h = \{N\}$, where each node N is the corner of at least one triangle, and the set of internal edges $S_h = \{S\}$ of length h_S. We require that the intersection of any two triangles K_1 and K_2 in T_h be either empty or a triangle side or a node. We note that the size and shape of the triangles may vary as long as a triangulation is generated. We define the "mesh function" $h(x)$ so that $h(x) = h_K$ for x in K.

The "isotropy" of the finite element triangulation T_h may be measured by the positive constant c_1 defined by

$$c_1 h_K^2 \leq 2 \int_K dx, \qquad \text{for all } K \text{ in } T_h. \tag{10.4}$$

This inequality expresses a "minimal angle" condition stating that the angles of all triangles in T_h are bounded below. If the constant $c_1 \sim 1$, then all triangles are more or less isosceles and T_h is "isotropic", while if c_1 is small, then some triangles are thin with one or two small angles. Clearly, for each T_h a corresponding constant c_1 always exists, small or not.

We use the finite element space V_h of continuous piecewise linear functions that vanish on Γ, that is

$$V_h = \{v \in V : v \text{ is linear in } x = (x_1, x_2) \text{ on each } K \text{ in } T_h\}. \tag{10.5}$$

A basis $\{\varphi_j\}$ for V_h is given by the "tent functions" φ_j, where the basis function φ_j in V_h is defined by $\varphi_j(N_i) = 1$ if $i = j$ and $\varphi_j(N_i) = 0$ if $i \neq j$.

The finite element method for (10.1) reads: find U in V_h such that

$$(\nabla U, \nabla v) = (f, v), \qquad \text{for all } v \text{ in } V_h. \tag{10.6}$$

Problem (10.6) is equivalent to the linear system of equations

$$A\xi = b, \tag{10.7}$$

where $U = \sum_i \xi_i \varphi_i$, $\xi = (\xi_i)$ with $\xi_i = U(N_i)$, $A = (a_{ji})$ is the $M \times M$ stiffness matrix with elements $a_{ji} = (\nabla\varphi_i, \nabla\varphi_j)$ and $b = (b_j)$ with $b_j = (f, \varphi_j)$ is the load vector. The stiffness matrix A is symmetric and positive definite, and thus in particular (10.7) has a unique solution U.

The Galerkin orthogonality for (10.6), resulting from (10.3) and (10.6) takes the form:

$$(\nabla(u-U),\nabla v)=0, \qquad \text{for all } v \text{ in } V_h. \tag{10.8}$$

10.4 Estimates of the interpolation error

We use the following error estimate for interpolation by piecewise linear functions, where the piecewise linear interpolant $\pi w \equiv \pi_h w$ in V_h of a given continuous function w is defined by $\pi w(N_i) = w(N_i)$, for all i.

Theorem 10.1 *For $r=0,1,\ s=0,1,$ there is a constant C_i depending only on c_1, such that the piecewise linear interpolant πw in V_h of w satisfies*

$$\|h^r D^s(w-\pi w)\|_2 \quad \leq C_i\|h^{2+r-s}D^2 w\|_2 \tag{10.9}$$

10.5 A priori error estimates in the energy norm

The basic energy norm a priori error estimate reads

Theorem 10.2 *There exists a constant C_i only depending on c_1, such that if u and U are the solutions of (10.3) and (10.6), respectively, then*

$$\|\nabla(u-U)\|_2 \leq \|\nabla(u-\pi u)\|_2 \leq C_i\|hD^2u\|_2. \tag{10.10}$$

Proof Using (10.8) with $v=U-\pi u$, we get using Cauchy's inequality,

$$\begin{aligned}\|\nabla e\|_2^2 &= (\nabla e,\nabla(u-U)) = (\nabla e,\nabla(u-U)) + (\nabla e,\nabla(U-\pi u))\\ &= (\nabla e,\nabla(u-\pi u)) \leq \|\nabla e\|_2\,\|\nabla(u-\pi u)\|_2.\end{aligned}$$

This proves the desired result recalling the interpolation estimate of Theorem 10.1. □

10.6 A posteriori error estimates in the energy norm

We now prove an a posteriori error estimate for (10.6) following the same argument used in the case of the two-point boundary value problem. Using first (10.3) and then (10.6), we have

$$\begin{aligned}\|\nabla e\|_2^2 &= (\nabla(u-U),\nabla e) = (\nabla u,\nabla e) - (\nabla U,\nabla e)\\ &= (f,e) - (\nabla U,\nabla e) = (f,e-\pi_h e) - (\nabla U,\nabla(e-\pi_h e).\end{aligned}$$

Now, integrating by parts over each element in the second term, we get

$$\begin{aligned}\|\nabla e\|_2^2 &= \sum_K \int_K (f+\Delta U)(e-\pi_h e)dx + \sum_K \int_{\partial K} \frac{\partial U}{\partial n_K}(e-\pi_h e)ds\\ &= \sum_K \int_K (f+\Delta U)(e-\pi_h e)dx + \sum_{S \text{ in } S_h} \int_S h_S^{-1}[\frac{\partial U}{\partial n_S}](e-\pi_h e)h_S ds\end{aligned}$$

where $[\frac{\partial v}{\partial n_S}]$ is the jump in the normal derivative of the function v in V_h. Using an interpolation estimate analogous to (5.10), we obtain the following a posteriori error estimate

Theorem 10.3 *The solution U of (10.6) satisfies*

$$\|\nabla u - \nabla U\|_2 \leq C_i \|hR(U)\|_2,$$

where

$$R(U) = R_1(U) + R_2(U),$$

$$R_1(U)|_K = |f + \Delta U| \quad \text{on } K \text{ in } T_h,$$

$$R_2(U)|_K = \max_{S \subset \partial K} h_S^{-1} \left| \left[\frac{\partial U}{\partial n_S} \right] \right|.$$

Remark If $c_1 \sim 1$, then $C_i \sim 1$.

We note that the total residual $R(U)$ has one contribution $R_1(U)$ from the interior of the elements K, and one contribution from the element sides S involving the jump in the normal derivative of U divided by the local mesh size. In the one dimensional problem considered above, the second contribution vanishes, because the interpolation error may be chosen to be zero at the node points. In two dimensions, the interpolation error does not vanish on the element sides, and so the jump term has to be taken into account.

11 A parabolic PDE: the heat equation

We briefly consider the heat equation, a standard model problem of parabolic type: find $u = u(x,t)$ such that

$$\begin{aligned} u_t - \Delta u &= f, \quad && \text{in } \Omega \times I, \\ u &= 0, && \text{on } \Gamma \times I, \\ u &= u_0, && \text{in } \Omega, \end{aligned} \tag{11.1}$$

where Ω is a bounded polygonal domain in R^2 with boundary $\Gamma = \partial\Omega$, $x = (x_1, x_2)$, $\Delta = (\partial/\partial x_1)^2 + (\partial/\partial x_2)^2$, $I = (0,T)$ is a time interval, $u_t = \partial u/\partial t$, and the functions f and u_0 are given data.

For discretization of (11.1) in time and space we use the dG(r)-method based on a partition $0 = t_0 < t_1 < \cdots < t_n < \cdots < t_N = T$ of I and associate with each time interval $I_n = (t_{n-1}, t_n]$ of length $k_n = t_n - t_{n-1}$, a triangulation T_n of Ω and the corresponding space $V_n \in H_0^1(\Omega)$ of piecewise linear functions as in Section 5. Note that we allow the space discretizations to change with time. We define

$$V_{rn} = \{v : v = \sum_{j=0}^{r} t^j \varphi_j, \quad \varphi_j \text{ in } V_n\},$$

and discretize (11.1) as follows: find U such that for $n = 1, 2, \ldots$, $U|_{\Omega \times I_n} \in$

V_{rn} and

$$\int_{I_n} \{(U_t, v) + (\nabla U, \nabla v)\} dt + ([U]_{n-1}, v_{n-1}^+) = \int_{I_n} (f, v) dt, \quad \text{for all } v \text{ in } V_{rn}, \tag{11.2}$$

where $[w]_n = w_n^+ - w_n^-$, $w_n^{+(-)} = \lim_{s \to 0+(-)} w(t_n + s)$, and $U_0^- = u_0$.

As above, if $r = 0$, then (11.2) reduces to a variant of the Euler backward method and for $r = 1$ to a variant of the subdiagonal Padé scheme of order (2,1) which is third-order accurate in U_n^- at the nodal points t_n.

The a posteriori error estimate in the case $r = 0$ has essentially the form

$$\|u(t_N) - U_N\|_2 \le L_N \max_{n=1,\ldots N} (\|R(U_n)\|_2 + \|[U_{n-1}]\|_2), \tag{11.3}$$

where we use the notation of Sections 8 and 11, and $L_N = \max_{n=1,\ldots N}(1 + \log(\frac{t_n}{k_n}))^{\frac{1}{2}}$ is a logarithmic factor entering in the present case. Clearly, one may design an adaptive algorithm based on (11.3). For more details see [10].

12 Concluding remarks. References

First of all the Ptolemaic system is exploded. For who will believe that there are as many theories for the Sun as there are planets. It seems that a single theory of the Sun is enough. (Kepler)

The approach presented here differs radically from a classical one. For example, we do not discuss such concepts as truncation error, consistency, finite difference method, difference quotient, multistep methods, Runge–Kutta methods, B-stability, one-sided Lipschitz bounds, logarithmic norms, local error or extrapolation, that form the foundation of the classical tradition for numerical methods for differential equations. This is because we do not use these concepts in our work. In some cases of course, the concepts used in these notes have a connection to classical concepts. For instance the interpolation error is connected to the "truncation error" and thus to "consistency" but the concepts are not identical. Several of the new basic concepts, however, have no natural classical counterparts, such as Galerkin discretization error, quadrature error, strong stability or a posteriori error estimate.

We summarize some features of the proposed approach

- integration: mathematics-computation-application
- simplicity: conceptual as well as algorithmic
- generality: the same basic methodology for all problems
- reliability: automatic adaptive error control
- efficiency: adaptive algorithms applicable to complex problems
- interplay between theory and software.

We give now a brief account of the current status in the development of the framework as reflected in our own work now available in the literature. We also give some references into the large literature on adaptive methods, which are of particular relevance for our work. A general introduction to our work is given in [9] and a complete account will be given in [8]. Introductory educational material, partly coinciding with the present note, is given in [10].

Adaptive methods for linear elliptic problems with energy norm control where first developed by Babuska et al. (see [4] and references therein), and Bank et al. (see [2]). In both cases, a posteriori error estimates are obtained by solving local problems with the residual acting as data. Residual-based a posteriori energy norm error estimates were also derived for Stokes's equations in [37].

The basic approach we use for adaptive methods for linear elliptic problems, including a priori and a posteriori error estimates in H^1, L_2 and L_∞ norms, is presented in [11] and [7]. Extensions to adaptive control of the discrete solution error using multigrid methods are developed in [3]. Nonlinear elliptic problems including obstacle and plasticity problems are considered in [26], [30] and [29]. Recently, applications to eigenvalue problems were given in [36].

Early a posteriori error analysis for ordinary differential equations was used in [6] and [38]. These approaches are quite different from ours. We develop adaptive global error control for systems of ordinary differential equations in [24], [19], [20], [21], and [22]. An influence to our early work came from [35].

The series [11]-[16] develops adaptive Fem for a class of parabolic problems in considerable generality including space-time discretization that is variable in space-time, and applications to nonlinear problems.

Adaptive Fem for linear convection-diffusion problems are considered in [25], [17] and [18]. Extensions to the compressible Euler equations are given in [23] and [34]. Extensions to the Navier–Stokes equations for incompressible flow are given in [31], [33] and [32]. Second-order wave equations are considered in [28].

The framework presented also applies to Galerkin methods for integral equations. An application to integral equations is given in [1]. The potential of the framework is explored in [5].

13 How to obtain Femlab and Introduction to Computational Methods for Differential Equations

Femlab contains software for solving (i) two-point boundary value problems (Femlab-1d), (ii) initial value problems for general systems of ODEs (Femlab-ode) and (iii) 2d boundary value problems (Femlab-2d). Femlab together with the educational material [9] which is closely related to the present notes, may be obtained over Internet as follows.

To obtain Femlab-ode, use anonymous ftp to
ftp.math.gatech.edu.
Change to directory */pub/users/estep* and get *femlabode.tar*. This tar file contains the codes and a brief users manual. In that same directory is *intro.ps.Z*, a compressed postscript version of [9].

To obtain the Femlab-1d and Femlab-2d, open to the WWW (World Wide Web) address
http://www.math.chalmers.se/ kenneth
using (for instance) the Mosaic program. There is a README file located there that gives further instructions.

Femlab-1d consists of a number of Matlab script and M-files. You can import these files using the "save as .." command under the "file" menu. To run the code, you then just start your local Matlab program and give the command adfem, calling the script file adfem.m. Femlab-2d is a more complex Fortran program. For more details, see the README file.

Bibliography

1. Asadzadeh, M. and Eriksson, K. An adaptive finite element method for a potential problem. To appear in *M3AS*.
2. Bank, R. (1986). Analysis of local a posteriori error estimate for elliptic equations. In *Accuracy Estimates and Adaptive Refinements in Finite Element Computations*, edited by I. Babuška, O. C. Zienkiewicz, J. Gago, and E. R. de A. Oliveira. Wiley, New York.
3. Becker, R., Johnson, C. and Rannacher, R. (1994). An error control for multigrid finite element methods. Preprint #1994-36, Department of Mathematics, Chalmers University of Technology, Göteborg.
4. Babuška, I. (1986). Feedback, adaptivity and *a posteriori* estimates in finite elements: aims, theory, and experience. In *Accuracy Estimates and Adaptive Refinements in Finite Element Computations*, edited by I. Babuška, O. C. Zienkiewicz, J. Gago, and E. R. de A. Olivieira, Wiley, New York.
5. Carstensen, C. and Stephan, E. (1993). *Adaptive boundary element methods.* Institute für Angewandt Mathematik, Universität Hannover.
6. Cooper, G. (1971). Error bounds for numerical solutions of ordinary differential equations. *Num. Math.*, **18**, 162–170.
7. Eriksson, K. An adaptive finite element method with efficient maximum norm error control for elliptic problems. To appear in *M3AS*.
8. Eriksson, K., Estep, D., Hansbo, P. and Johnson, C. *Adaptive Finite Element Methods.* North Holland, in preparation.
9. Eriksson, K., Estep, D., Hansbo, P. and Johnson, C. (1994). *Introduction to Computational Methods for Differential equations.* Department of Mathematics, Chalmers University of Technology.

10. Eriksson, K., Estep, D., Hansbo, P. and Johnson, C. Adaptive Methods for Differential equations. To appear in *Acta Numerica*.
11. Eriksson, K. and Johnson, C. (1991). Adaptive finite element methods for parabolic problems I: A linear model problem. *SIAM J. Numer. Anal.*, **28**, 43–77.
12. Eriksson, K. and Johnson, C. Adaptive finite element methods for parabolic problems II: Optimal error estimates in $L_\infty L_2$ and $L_\infty L_\infty$. To appear in *SIAM J. Numer. Anal.*
13. Eriksson, K. and Johnson, C. Adaptive finite element methods for parabolic problems III: Time steps variable in space. In preparation.
14. Eriksson, K. and Johnson, C. Adaptive finite element methods for parabolic problems IV: Non-linear problems. To appear in *SIAM J. Numer. Anal.*
15. Eriksson, K. and Johnson, C. Adaptive finite element methods for parabolic problems V: Long-time integration. To appear in *SIAM J. Numer. Anal.*
16. Eriksson, K., Johnson, C. and Larsson, S. (1994). *Adaptive finite element methods for parabolic problems VI: Analytic semigroups.* Preprint, Department of Mathematics, Chalmers University of Technology.
17. Eriksson, K. and Johnson, C. (1993). Adaptive streamline diffusion finite element methods for stationary convection-diffusion problems. *Math. Comp.*, **60**, 167–188.
18. Eriksson, K. and Johnson, C. Adaptive streamline diffusion finite element methods for time-dependent convection-diffusion problems. To appear in *Math. Comp.*
19. Estep, D. A posteriori error bounds and global error control for approximations of ordinary differential equations. To appear in *SIAM J. Numer. Anal.*
20. Estep, D. and French, D. Global error control for the continuous Galerkin finite element method for ordinary differential equations. To appear in *RAIRO M.M.A.N.*
21. Estep, D. and Johnson, C. (1994). *The computability of the Lorenz system.* Preprint #1994-33, Department of Mathematics, Chalmers University of Technology.
22. Estep, D. and Williams, R. The structure of an adaptive differential equation solver. In preparation.
23. Hansbo, P. and Johnson, C. (1991). Adaptive streamline diffusion finite element methods for compressible flow using conservation variables. *Comput. Methods Appl. Mech. Engrg.*, **87**, 267–280.
24. Johnson, C. (1988). Error estimates and adaptive time step control

for a class of one step methods for stiff ordinary differential equations. *SIAM J. Numer. Anal.*, **25**, 908–926.

25. Johnson, C. (1990). Adaptive finite element methods for diffusion and convection problems. *Comput. Methods Appl. Mech. Engrg.*, **82**, 301–322.
26. Johnson, C. (1992). Adaptive finite element methods for the obstacle problem. *Math. Models Methods Appl. Sci.*, **2**, 483–487.
27. Johnson, C. (1992). A new approach to algorithms for convection problems based on exact transport + projection. *Comput. Methods Appl. Mech. Engrg.*, **100**, 45–62.
28. Johnson, C. (1993). Discontinuous Galerkin finite element methods for second order hyperbolic problems. *Comput. Methods Appl. Mech. Engrg.*, **107**, 117-129.
29. Johnson, C. and Hansbo, P. (1992). Adaptive finite element methods for small strain elasto-plasticity. In *Finite Inelastic Deformations - Theory and Applications*, edited by D. Besdo and E. Stein, Springer, Berlin.
30. Johnson, C. and Hansbo, P. (1992) Adaptive finite element methods in computational mechanics. *Comput. Methods Appl. Mech. Engrg.*, **101**, 143–181.
31. Johnson, C. Rannacher, R. and Boman, M. Numerics and hydrodynamic stability: Towards error control in CFD. To appear in *SIAM J. Numer. Anal.*
32. Johnson, C. Rannacher, R. and Boman, M. (1994). *On transition to turbulence and error control in CFD.* Preprint #1994-26, Department of Mathematics, Chalmers University of Technology.
33. Johnson, C. and Rannacher, R. On error control in CFD. To appear in Proc. of. Conf. on Navier–Stokes Equations, Oct. 93, Vieweg.
34. Johnson, C. and Szepessy, A. Adaptive finite element methods for conservation laws based on a posteriori error estimates. To appear in *Comm. Pure Appl. Math.*
35. Lippold, G. (1988). *Error estimates and time step control for the approximate solution of a first order evolution equation.* Preprint, Akademie der Wissenschaften der Karl-Weierstrass-Institut für Mathematik, Berlin.
36. Nystedt, C. *Adaptive finite element methods for eigenvalue problems.* Licenciate Thesis, Department of Mathematics, Chalmers University of Technology, in preparation.
37. Verfürth, R. (1989). A posteriori error estimators for the Stokes equations. *Numer. Math.*, **55**, 309–325.
38. Zadunaisky, P. (1976). On the estimation of errors propagated in the

numerical integration of ordinary differential equations. *Numer. Math.* **27**, 21–39.

Numerical Solution of Differential-Algebraic Equations

Linda R. Petzold

Department of Computer Science,
University of Minnesota,
Minneapolis 55455,
USA

Abstract

Many physical systems are naturally described as systems of differential-algebraic equations (DAEs). These types of systems occur in the modeling of electrical networks, flow of incompressible fluids, control, mechanical systems subject to constraints, and in many other applications. This class of problems includes systems which are in many ways quite different from ODEs. In these notes we outline some of the theory and available software for the numerical solution of DAE systems, and explore some challenging applications.

1 Introduction

We will focus on the numerical solution of systems of differential-algebraic equations (DAEs). These are systems of the form

$$0 = F(t, y, y'), \tag{1.1}$$

where $\frac{\partial F}{\partial y'}$ may be singular. Initial value problems in DAEs arise in a wide variety of applications, including circuit and control theory, chemical process simulation, modeling of constrained mechanical systems, fluid dynamics and robotics. DAEs arise as boundary value problems in parameter estimation and optimal control of the above-mentioned systems. We will examine the formulation and numerical solution of DAE systems, describe the software which is available, and explore some challenging applications.

The basic idea of using a numerical method for solving DAE systems was introduced by Gear [20], and consists of replacing y and/or y' in (1.1) by a difference approximation, and then solving the resulting equation for an approximation to y. The simplest example of a numerical ODE method for (1.1) is the implicit Euler method. Using this approach, the derivative $y'(t_n)$ at time t_n is approximated by a backward difference of $y(t)$. The resulting system of nonlinear equations is solved for y_n,

$$0 = F\left(t_n, y_n, \frac{y_n - y_{n-1}}{h}\right), \tag{1.2}$$

where $h = t_n - t_{n-1}$. In this way the solution is advanced from time t_{n-1} to t_n. Although it is sometimes possible to rewrite DAE systems as standard ODEs $y' = f(t, y)$, it is not always desirable to do this because it may lead to a loss of sparsity of the system, may be difficult to implement, and the resulting ODE may not share the properties of the original system with respect, for example, to preserving the constraints in the presence of numerical errors. A difficulty with the general form (1.1) is that there are problems which can be written in this form which are not solvable by standard numerical ODE methods. There are also problems for which some numerical ODE methods are appropriate but not others. To distinguish these classes of problems requires the definition of some basic concepts and fundamental structures in DAEs.

2 Structure of DAE systems

The class of DAEs which we consider has the property that solutions exist and are in some sense unique. We note that solutions to a DAE do not in general exist for all initial values. More precisely, we define the class of *solvable* DAEs as follows.

Definition 2.1 *The DAE (1.1) is solvable for t in a finite interval I, if for each t, there is a manifold M_t, through every point of which there is a smooth solution which exists over all of I, solutions never bifurcate, and two solutions beginning at distinct initial values never intersect in I.*

The concept which to date has been most important in understanding the structure and numerical solution of DAEs is that of the *index*. Roughly speaking, the index is a measure of the amount of singularity in the system. In general, the higher the index the more difficulties we are likely to encounter in trying to solve the system by a numerical ODE method. We define the index as follows.

Definition 2.2 *The index of a solvable DAE (1.1) is the smallest nonnegative integer m such that F has m continuous derivatives and the nonlinear system*

$$\begin{aligned} F(t, y, y') &= 0 \\ \frac{dF}{dt}(t, y, y', y'') &= \frac{\partial F}{\partial y} y' + \frac{\partial F}{\partial y'} y'' + \frac{\partial F}{\partial t} = 0 \\ &\vdots \\ \frac{d^m F}{dt^m}(t, y, y', y'', \ldots, y^{(m+1)}) &= 0 \end{aligned} \tag{2.1}$$

can be solved pointwise for y' uniquely in terms of y and t: $y' = \varphi(y, t)$.

We note that there are a number of definitions of index, including (local) index, (global) index (which is what we have defined here), and (perturbation) index. The relationship between these definitions is discussed in [26]; they all coincide for linear constant-coefficient DAEs but may differ for more complex problems. Basic existence and uniqueness for differential-algebraic systems is studied by Rheinboldt et al. [44, 45, 42].

To put these definitions into perspective, consider a simple index two problem

$$\begin{aligned} y_1 &= g(t) \\ y_2 &= y_1' \end{aligned}$$

and note that for systems of index m, $(m \geq 1)$, in contrast to an ODE,

(1) The solution involves derivatives of order $m-1$ of the forcing function. For the above example, the solution, $y_1 = g(t)$, $y_2 = g'(t)$ depends on the first derivative of $g(t)$, as the system is index two.

(2) Not all initial conditions admit a smooth solution if $m \geq 1$. Those that do admit smooth solutions are called *consistent initial conditions.* For the above example, the consistent initial condition is given by $y_1(t_0) = g(t_0)$, $y_2(t_0) = g'(t_0)$.

(3) Higher index DAEs can have hidden algebraic constraints. For the above example, the constraint $y_2 = g'(t)$ is in some sense hidden.

There are two important classes of index one systems. The first is the general *fully-implicit* index one system (1.1). If we assume that the rank of $\partial F/\partial y'$ is constant, and that the index is identically equal to one in a neighborhood of the solution, then we will refer to these systems as *uniform index one.* The second class of index one systems is that of *semi-explicit* index one systems. These are systems which are written in the special form

$$\begin{aligned} 0 &= F_1(x, x', y, t) \\ 0 &= F_2(x, y, t), \end{aligned} \tag{2.2}$$

where $\partial F_1/\partial x'$ is nonsingular. The system index one iff $\partial F_2/\partial y$ is nonsingular. These semi-explicit index one systems arise frequently in applications. It is important to distinguish them from the general index one system for (2.2) but poorly for (1.1).

For high index systems in the general form (1.1), numerical ODE methods such as the implicit Euler method can be unstable for small step sizes [23]. It is often possible to reduce the index of a semi-explicit system by analytically differentiating the constraints [23]. Campbell [15] has devised an algorithm for solving higher-index systems which repeatedly differentiates the entire system and solves the resulting overdetermined system for y' in terms of y and t in the spirit of Definition 2.1. Algorithms of this type require the extensive use of symbolic or automatic differentiation [25], however they are the only known means for dealing with the most general high-index DAE numerically.

Despite the discouraging results on the instability of numerical ODE methods when they are applied directly to the most general high-index linear systems, we can identify important classes of high-index nonlinear problems for which the numerical methods are stable and accurate. These systems arise for example in the modeling of electrical networks and constrained mechanical systems and in the solution of the equations of fluid flow, and in fact in practically all applications known to this author. For index two, these are semi-explicit DAE systems of the form (2.2). The system is index two if $(\partial F_2/\partial x)(\partial F_1/\partial y)$ is nonsingular. For index three systems, the ones we can solve are semi-explicit systems which can be written in a *Hessenberg form* which includes the class of index three constrained mechanical systems described in the next section. In general, a system is *Hessenberg of index m* if it can be written in the form

$$\begin{aligned} x_1' &= F_1(x_1, x_2, \ldots, x_m, t) \\ x_2' &= F_2(x_1, x_2, \ldots, x_{m-1}, t) \\ &\vdots \\ x_i' &= F_i(x_{i-1}, x_i, \ldots, x_{m-1}, t) \\ 0 &= F_m(x_{m-1}, t), \end{aligned}$$

where the matrix $(\partial F_m/\partial x_{m-1})(\partial F_{m-1}/\partial x_{m-2}) \cdots (\partial F_2/\partial x_1)(\partial F_1/\partial x_m)$ is nonsingular.

Gear [21] has noted a simple relationship between the fully-implicit index m system and the semi-explicit index $m+1$ system with which it is often possible to transfer methods and convergence results from one class of problems to the other. It is also possible to convert a higher-index system to a lower-index system by differentiating the constraints. If the original constraints are then discarded, the resulting system may be subject to 'drift' away from satisfying the original constraints, because of numerical errors introduced at each step. There are various means of avoiding or minimizing the effects of this drift. Gear *et al.* [22] have shown how to write the system which includes the original constraint as well as the differentiated constraint by introducing some extra variables which are Lagrange multipliers to the system. Various regularizations have also been proposed to stabilize the differentiated system [6, 28, 29]. Ascher and Petzold [5] have analysed these formulations with respect to stability, and found that almost all of the formulations which are widely used in practice retain the stability of the original high-index system. However, some formulations lead to numerical methods with better stability than others for practical step sizes.

3 Multistep methods

Historically, the first numerical methods to be implemented for DAEs were the backward differentiation formulae. These methods are well suited for a wide variety of DAE problems, and are the basis for the most widely used production codes for DAEs. Here we will describe the results on the order and stability of BDF methods. Later in this section we will state some results for general multistep methods.

To solve (1.1) by a BDF method, we replace $y'(t_n)$ by a k-step backward differentiation formula,

$$\rho y_n = \sum_{i=0}^{k} \alpha_i y_{n-i} \tag{3.1}$$

to obtain the system of nonlinear equations

$$F\left(t_n, y_n, \frac{\rho y_n}{h}\right) = 0. \tag{3.2}$$

Then we have the following result for BDF applied to index one systems [23].

Theorem 3.1 *If F is uniform index one and is differentiable with respect to y and y', the solution of (1.1) by the k-step BDF method with fixed step size h for $k < 7$ converges to order $O(h^k)$ if all initial values are correct to order $O(h^k)$.*

An extension of this theorem to variable step sizes is given in [22].

While the BDF methods converge as expected for index one problems, there are still practical difficulties in implementing these methods for this class of problems [39] [40]. It should be noted that systems whose index is less than or equal to one are the problems that general-purpose codes [39] are designed to handle.

For higher index systems, we have noted earlier that it is not possible to obtain convergence even of backward Euler, in general. Therefore, attention has focused on semi-explicit index two systems and higher index systems of Hessenberg form. Hessenberg index three systems arise in the solution of constrained mechanical systems of the form

$$\begin{aligned} M(q)q'' &= f(q, q', t) + G(q)\lambda \\ \varphi(q) = 0, &\quad \frac{\partial \varphi}{\partial q} = G^T, \end{aligned} \tag{3.3}$$

where q are the positions and λ the Lagrange multipliers. The index of the mechanical systems (3.3) is three.

For semi-explicit index two systems and for index three constrained mechanical systems of the form (3.3), the k-step constant-step-size BDF method converges to order of accuracy $O(h^k)$ if the initial values are sufficiently accurate [10, 32]. Gear *et al.* [22] generalized this result to show that variable-step-size BDF methods converge for semi-explicit index two systems. In contrast to the situation for semi-explicit index two systems, the $O(h^k)$ convergence results for the k-step BDF method applied to index three systems cannot be extended to hold for variable-step-size meshes because each time the step size is changed, a new boundary layer of reduced convergence rates is initiated. In particular, the first order BDF (backward Euler method) fails to converge at the end of the first step following a change in the step size. Gear and Keiper [30] have shown that variable step size BDF methods converge for Hessenberg index three systems if the order of the method is greater than one, and for Hessenberg index four systems if the order of the method is greater than three.

It is possible to use a general purpose code based on backward differentiation formulas to solve these special high-index nonlinear systems [40]. However, there are some practical difficulties which must be dealt with. It can be shown that for an index m system, the iteration matrix which the code uses in the Newton iteration for solving the nonlinear equation (3.2) has a condition number which is $O(1/h^m)$. This difficulty can be remedied by scaling the equations and the variables. The convergence test and error test must also be modified to allow a variable-step-size BDF code to solve these types of problems. In our experience, it has not been possible to attain a robust multistep method for DAEs of index higher than two. Therefore we recommend reducing the index of the system first, by one of the *projected invariants* methods described in [5].

For general multistep methods, it is easy to see that the methods are stable and convergent for semi-explicit index one systems. In general, there is an additional set of order conditions which must be satisfied for the methods to attain a given order for fully-implicit index one systems. These order conditions are already satisfied by the BDF formulae. Convergence results for multistep methods applied to a class of fully-implicit index one systems are given in [24] for several different formulations of the multistep methods. Multistep methods for DAEs are studied in much detail in [1].

4 Runge–Kutta methods

Runge–Kutta methods are potentially advantageous over multistep methods for some systems (for example, for systems with frequent discontinuities). However, some care must be taken in choosing a Runge–Kutta method which is appropriate for DAEs, as these methods do not in general attain the same order of accuracy for DAEs as they do for ODEs.

In the case of implicit Runge–Kutta methods (IRK), an M-stage IRK method applied to the system of DAEs (1.1) is written as

$$F\left(t_{n-1} + c_i h, y_{n-1} + h\sum_{j=1}^{M} a_{ij}Y'_j, Y'_i\right) = 0$$

$$y_n = y_{n-1} + h\sum_{i=1}^{M} b_i Y'_i. \tag{4.1}$$

The method can be written in the shorthand notation which displays the matrix of coefficients,

$$\begin{array}{c|cccc}
c_1 & a_{11} & a_{12} & \cdots & a_{1M} \\
c_2 & a_{21} & a_{22} & \cdots & a_{2M} \\
\vdots & \vdots & \vdots & \ddots & \vdots \\
c_M & a_{M1} & a_{M2} & \cdots & a_{MM} \\
\hline
 & b_1 & b_2 & \cdots & b_M.
\end{array}$$

We will assume that the matrix $\mathcal{A} = (a_{ij})$ is nonsingular.

We will first give some results on the properties of implicit Runge–Kutta methods applied to different classes of index one systems, and then we will discuss the properties of these methods applied to nonlinear semi-explicit index two systems. First we need to define some terminology.

The Runge–Kutta method is called *strictly stable* if the difference between a perturbed Runge–Kutta step

$$F\left(t_{n-1} + c_i h, z_{n-1} + h\sum_{j=1}^{M} a_{ij} Z'_j + \delta_n^{(i)}, Z'_i\right) = 0 \quad i = 1, 2, \ldots, M$$

$$z_n = z_{n-1} + h\sum_{i=1}^{M} b_i Z'_i + \delta_n^{(M+1)} \tag{4.2}$$

where $z_0 = y_0 + \delta_0^{(M+1)}$, $\|\delta_n^{(i)}\| \leq \Delta$, and an unperturbed Runge–Kutta step (4.1) satisfies $\|y_n - z_n\| \leq K_0\Delta$, where $0 < h \leq h_0$ and K_0, h_0 are constants depending only on the method and the DAE.

Defining the *stability constant*

$$r = 1 - b^T \mathcal{A}^{-1} \epsilon_M,$$

where $\epsilon_M = (1, 1, \ldots, 1)^T$, it is easy to show for index one DAEs that the IRK method (4.1) is stable if and only if the method coefficients satisfy the *strict stability condition* $|r| < 1$. This stability condition for DAEs is related to the stability criterion $|R(z)| \leq 1$, $R(z) = 1 + zb^T(I - z\mathcal{A})^{-1}\epsilon_M$, where $z = h\lambda$, for the stiff model problem $y' = \lambda y$ because $\lim_{|z|\to\infty} R(z) = 1 - b^T\mathcal{A}^{-1}\epsilon_M$.

One class of IRK methods which appears to be particularly promising for the solution of DAEs is that of *stiffly accurate* methods. These are methods whose coefficient matrices satisfy $c_M = 1$, $a_{Mj} = b_j$, $j = 1, \ldots, M$ and $\mathcal{A}$ is nonsingular. For semi-explicit index one systems, these methods have the property, analogous to BDF, that the constraint equations are satisfied exactly at the end of each time step. There is no order reduction for stiffly accurate methods applied to semi-explicit index one systems. Other IRK methods can suffer an order reduction even for semi-explicit index one systems, unless (in this case) the constraints are enforced on each internal stage.

The IRK methods are, in general, even less accurate for fully-implicit nonlinear index one systems than for semi-explicit systems. The additional loss of accuracy comes about because of mixing which can occur between the errors in the differential and singular parts of the system. Defining

$$C(q): \sum_{j=1}^{M} a_{ij} c_j^{k-1} = c_i^k/k, \qquad i = 1,2,\ldots,M \quad k = 1,2,\ldots,q$$

$$B(q): \sum_{j=1}^{M} b_j c_j^{k-1} = 1/k, \qquad k = 1,2,\ldots,q \tag{4.3}$$

We have the following result[13].

Theorem 4.1 *Suppose that (1.1) is uniform index one and linear in y', the Runge–Kutta method satisfies the stability condition $|r| \le 1$, the errors in the initial conditions are $O(h^G)$ and the errors in terminating the Newton iterations are $O(h^{G+\delta})$, where $\delta = 1$ if $|r| = 1$ and $\delta = 0$ otherwise, and $G \ge 2$. Then the global errors satisfy $\|e_n\| = O(h^G)$ where*

$$G = \begin{cases} q, & \text{if } C(q) \text{ and } B(q) \\ q+1, & \text{if } C(q),\ B(q+1) \text{ and } -1 \le r < 1 \\ q+1, & \text{if } C(q),\ B(q+1),\ A_1(q+1) \text{ and } r = 1. \end{cases}$$

$A_1(q)$ are the order conditions for IRK methods applied to index one systems with constant coefficients, and are given by

$$A_1(q): \quad b^T \mathcal{A}^{-1} c^j = 1 \qquad j = 1,2,\ldots,q. \tag{4.4}$$

For nonlinear semi-explicit index two systems, because of the close relationship between fully implicit index one systems and semi-explicit index two systems, we have that the global error in the "differential" variable x is given by Theorem 4.1. For methods satisfying the strict stability condition, a lower bound on the global errors in the y variable is $O(h^{G_y})$, where G_y [11] is given by

$$G_y = \begin{cases} q & \text{if } C(q),\ B(q),\ A_1(q) \\ q+1 & \text{if } C(q),\ B(q+1),\ A_1(q+1),\ A_2(q+1), \end{cases}$$

where A_2 are the order conditions for Runge–Kutta methods applied to constant-coefficient index two systems and are given by

$$A_2(q): \qquad \begin{aligned} & b^T \mathcal{A}^{-1} \epsilon_M = b^T \mathcal{A}^{-2} c \\ & b^T \mathcal{A}^{-2} c^i = i, \quad i = 2,3,\ldots,q. \end{aligned}$$

These results give only a lower bound for the order. A complete set of order results, which is based on a nontrivial extension of Butcher's theory of rooted trees, is given in [26], for index one and Hessenberg index two and index three systems.

We note that these order conditions assume: (1) $\mathcal{A}$ nonsingular and (2) the methods satisfy the strict stability condition. Order results for the case where $\mathcal{A}$ is singular, and to Rosenbrock and extrapolation methods are given in [26]. The importance of the strict stability condition deserves special mention. In particular, this condition excludes symmetric methods such as midpoint. These methods are important especially in the solution of boundary value problems, where a well-conditioned problem may have both stable increasing and decreasing components. If we define the symmetric methods as in (4.1), they suffer from a severe order reduction [26], but also from some even worse problems, namely a persistent oscillation and instability[2]. A cure for this problem of instability, which also cures the order reduction and oscillation, for Hessenberg index two systems, is given by *Projected Implicit Runge–Kutta Methods* [3], and generalized to higher-index systems in [4]. The stability is recovered by means of an extra projection onto the constraints which is done on the final stage. The stability problem will also be present for certain fully implicit index one systems. These systems can be more difficult to stabilize; one way which is natural for many physically based systems is to write them in a conservative form [17].

For nonstiff semi-explicit DAEs, it is natural to solve the ODE with an explicit method (the constraints are still handled implicitly). These types of methods are called half-explicit and are defined and analysed in [26].

5 Software for DAEs

Available codes

There are several codes which are available for the solution of various forms of index one DAEs. DASSL [9] is designed for solving initial value problems of the fully implicit form $F(t, y, y') = 0$ which are index zero or one. LSODI [27] is similar to DASSL in that it is based on BDF methods. This code is written for *linearly implicit* DAEs of the form $A(t, y)y' = f(t, y)$. LSODI differs from DASSL in that it uses a fixed coefficient implementation of the BDF formulae, as compared to the fixed leading coefficient implementation in DASSL. The strategies for selecting the order and step size are also quite different. The SPRINT code [7] also uses BDF methods for the solution of linearly implicit DAEs. The LIMEX code of Deuflhard and Nowak [18] is based on extrapolation of the semi-implicit Euler method. The theory for this method is limited to linearly implicit DAEs and is given mostly in [34]. The Radau 5 code [26] solves non stiff linearly implicit DAEs which may be of higher index. The MEXX code of Lubich [33, 35] solves nonstiff mechanical systems in an index one or index two formulation. The methods are based on extrapolation of half-explicit midpoint or half-explicit Euler methods. There is also a version of this code for stiff mechanical systems.

A great deal of useful software for solving ODEs and DAEs and a wide variety of other numerical and non-numerical problems is available freely on the Internet via Netlib [19]. This includes DASSL and DASRT. One can obtain an index of Netlib ODE software by

```
mail netlib@ornl.gov
Subject: send index from ode
```

The netlib system will then mail back an index of ODE solvers and descriptions. To obtain one of these solvers (for example, to obtain DDASSL — double precision DASSL), send the following message

```
mail netlib@ornl.gov
send ddassl from ode
```

On many X-window systems, an interactive version of netlib called Xnetlib is available.

Extensions to DASSL

Although DASSL can handle a wide variety of problems, some problems require capabilities which are not implemented in the standard version of DASSL or which can be performed more efficiently with a code based on DASSL but oriented towards a more specific task. These problems include root-finding, sensitivity analysis, and the solution of very large systems of DAEs whose iteration matrices do not fit naturally into a banded structure.

The standard version of DASSL solves a DAE system from time T to time TOUT, where TOUT is specified by the user. In some problems it is more natural to stop the code at the root of some function $g(t, y)$. The code DASRT [9] has this capability. A vector of functions $g_i(t, y)$, $i = 1, 2, \ldots, NG$, may be supplied to DASRT such that the root of any of the NG functions g_i is desired. If there are several roots in a given output interval, DASRT returns them one at a time, in the order in which they occur along the solution. An integer array tells the user which g_i, if any, were found to have a root on any given return.

One of the obvious deficiencies of the standard DASSL is the lack of options for dealing with Jacobian matrices which do not fit naturally into a dense or banded category. Marquardt [37] has implemented a version of DASSL which makes use of the HARWELL sparse direct linear system solver.

For very large DAE systems, such as those arising from the method of lines solution of partial differential equations in two and three dimensions, preconditioned iterative methods are combined with the time-stepping methods of DASSL. Here we will briefly indicate how this works.

At each time step, a nonlinear system of equations $G(y)$ must be solved. This is done by Newton iteration. At each iteration, a linear system $Ax = b$ must be solved. Instead of relegating this problem to a standard linear-system solution algorithm as is done in DASSL, for large problems we use iterative methods. Many iterative methods for linear systems are known, but some are much more appealing than others in the setting of large DAE systems. Such methods are known as Krylov subspace iteration methods. Their crucial property is that at each iteration they require only the value of the matrix-vector product Av for a given vector v. That is, if m iterations have been done, so that one has $x_0, x_1, \ldots, x_m$ (or some equivalent set of vectors), a vector v is generated as a linear combination of these vectors, and the next iterate x_{m+1} is a linear combination of Av and the older vectors. Many methods of this type (such as conjugate gradient iteration, for example) are known to work well when A has certain special properties (such as symmetry), but only a few are good candidates when no such assumptions about A are made. These are the most useful choices, because no special properties can be assumed about the function G from which A is obtained.

Given a suitable Krylov method, it can be exploited to best advantage by finding an efficient way to calculate products Av that does not entail calculating the matrix A itself. To do this, we note that A is just the matrix of partial derivatives of $G(y)$. This implies that for a suitably small constant ϵ, $[G(y+\epsilon v)-G(y)]/\epsilon$ is a good approximation to Av. The Krylov iteration proceeds until convergence of the iterates is achieved to within a suitable tolerance. The resulting method involves no explicit construction or storage of the matrix A.

Krylov methods are not powerful enough, by themselves, to handle with acceptable efficiency the wide variety of matrices A that can occur. However, they can be assisted greatly by a technique known as *preconditioning*. Suppose we can find a matrix P (the preconditioner matrix) that resembles A to some extent but is much easier to construct and operate with. In particular, suppose that we can solve linear systems $Px = b$ reasonably efficiently. To solve $Ax = b$, we write an equivalent system, $(P^{-1}A)x = P^{-1}b$, with a different matrix $A' = AP^{-1}$ and a different solution vector $x' = Px$, and apply the Krylov method to the problem $A'x' = b$. Each iteration requires the evaluation of a product $A'v = P^{-1}Av$, but that is achieved by evaluating Av and then solving $Pw = Av$. Convergence is more likely to occur now, because A' is closer to the identity matrix, depending on how close P is to A.

Brown, Hindmarsh and Petzold have written a variant of DASSL called DASPK [12], which includes the GMRES Krylov method with user-supplied preconditioning as an option. DASPK actually includes the direct methods of DASSL as well. Because the choice of preconditioner can best be made by exploiting the structure of the problem, the user of DASPK must supply the preconditioner. Preconditioners for DAEs arising from reaction-diffusion type PDE systems have been studied in [12].

For use on massively parallel machines, two modified versions of DASPK have been written — one using Fortran 90 (with data-parallelism), and one using message-passing [36].

In current versions of DASSL and DASPK, the user must supply a consistent set of initial conditions. That is, the initial conditions must satisfy the DAE. Improvements in DASSL and DASPK which will include options for computing consistent initial conditions for several important classes of DAEs is currently under development by Brown, Hindmarsh and Petzold.

An important problem which is related to solving DAEs is sensitivity analysis. Given a DAE

$$\begin{aligned} F(t,y,y',p) &= 0, \\ y(t_0,p) &= y_0(p) \end{aligned} \tag{5.1}$$

where y, $y' \in \mathbb{R}^N$, whose solution depends on a vector of M time independent parameters p, the sensitivity analysis problem is to compute the (N, M) matrix $W(t)$ of sensitivity functions

$$W(t) = \frac{\partial y(t)}{\partial p} \tag{5.2}$$

which describes how the solution components change as a result of changes in the parameters. DASAC [16] has been developed to handle sensitivity analysis for DAEs.

A new sensitivity code based on DASSL and DASPK is under development by Petzold and Maly. This code uses a directional derivative approximation to the sensitivity equations. The equations to be solved for a first order difference approximation are

$$\begin{aligned} F(t,y,y',p) &= 0, \\ (F(t,y+\delta_i s_i, y'+\delta_i s_i', p+\delta_i e_i) - F(t,y,y',p))/\delta_i &= 0, \; i = 1,\cdots,M \end{aligned} \tag{5.3}$$

Optionally, the user can input the sensitivities directly. The sensitivity equations can be solved very efficiently simultaneously with the original DAE because the Jacobian matrix for each of the sensitivity equations is the same as for the original system. Thus in a direct method or preconditioner, the matrix can be decomposed once and used over and over again. For the iterative method, there will be many clusters of identical eigenvalues. This approach differs from DASAC in several ways. The directional difference is used rather than directly approximating the Jacobian matrix in the sensitivity equations because it allows for a 'matrix free' formulation of the equations for DASPK and does not require recomputation of the matrix at each time step in DASSL. We do not iterate the Newton method to convergence before solving the sensitivity equations on each step; instead we solve the system of nonlinear equations arising from (5.3) simultaneously; this is advantageous for parallel computation. We have proved a theorem which shows that the terms involving partials of F with respect to the original variables can be neglected in the sensitivity equations without adversely effecting the convergence, and our tests have confirmed this result. The resulting codes are very easy to use. They have the same parameter lists as DASSL and DASPK; the only additional information required from the user is a list of parameters.

6 Applications

Although DAEs arise in a wide variety of problems from engineering and science, there are a number of applications in which there has recently been or is likely to be much interest because of the importance of the problems and the challenges for their numerical solution. For initial value problems, there has recently been much research on numerical methods for constrained mechanical systems. These systems arise in real-time vehicle simulation and design, in computer-aided design of mechanical systems, and in robotics. For boundary-value problems, there are challenging numerical difficulties in parameter estimation and optimal control of chemical, electrical and mechanical systems. Here we briefly describe some of the computational difficulties in these areas. We note that similar difficulties are also encountered in the related problems of electrical network simulation, power systems, and chemical process simulation.

Multibody system simulation

The Euler–Lagrange equations of motion for mechanical systems are usually posed initially in the form of a system of differential equations (Newton's laws of motion)

$$\begin{aligned} M(q)q'' &= f(q, q', t) + G(q)\lambda \\ \varphi(q) = 0, &\quad \frac{\partial \varphi}{\partial q} = G^T, \end{aligned} \tag{6.1}$$

coupled with nonlinear constraints which are enforced via a Lagrange multiplier. Direct discretization of this index three system yields numerical methods which are often not very robust because of well known[9] difficulties with error estimation and step size control, as well as severe ill-conditioning of linear systems at each time step and other problems. A wide variety of reformulations of the problem and associated numerical methods have been suggested in an attempt to find a system of equations describing the system which can be effectively solved numerically. However, each has some apparent disadvantage in terms of speed and/or robustness. Because the constraints are sometimes highly nonlinear and have a strong physical relevance, it is generally considered important that the constraints, and sometimes the time derivative of the constraints, be satisfied very accurately. In addition, there are other potential difficulties: the constraints can become rank-deficient or nearly rank-deficient, the solution may have components which are oscillating at a high frequency, and there is the possibility of frequent discontinuities which are especially troublesome because the solution of a high-index DAE can be less continuous than its input. A number of formulations, called projected invariants, which are advantageous from the point of view of numerical stability for realistic step sizes, have been proposed in [5].

Frequent discontinuities are possible in a multibody system. Some of these discontinuities can be located very efficiently by a root-finder such as in DASRT. However, others may arise from user-defined functions or other unanticipated situations, and need to be located automatically and handled efficiently. In the case of a collision, conservation properties of the solution should be preserved across the interface. The situation for DAEs presents difficulties in addition to the ODE case because the solution of a high-index system can be less continuous than the input. Thus, impulsive solutions are possible.

It can sometimes happen that the constraint matrix G^T becomes rank-deficient or nearly rank-deficient. In such a case, the problem becomes poorly conditioned, and numerical methods can experience serious difficulties. There are a number of possibilities for dealing with this, including regularizing the system [41], or eliminating the redundant constraints. However, there are difficulties with either alternative.

Often in multibody systems the solution may have components which are oscillating at a high frequency. This may arise, for example, from components which are rotating, or from the natural frequencies of the system. In a numerical method such as multistep or Runge–Kutta, which are based on approximating the solution locally, the step size must be chosen very small to resolve the oscillation in the solution. In some cases this could ruin the possibility for obtaining a solution in real-time. Usually, the details of the oscillating solution are not so important as the long-term solution behavior. The oscillations might be damped by the numerical method if the amplitude is small enough, however this leads to problems with Newton convergence for the Lagrangian formulation. Recently a coordinate-splitting formulation has been developed [46] which effectively handles this problem. Methods for oscillations which cannot be damped or neglected need to be developed.

Real-time simulation imposes severe requirements on the solution method. The solution must be computed extremely rapidly, necessitating the use of massively parallel computers. The most efficient serial method for multibody systems, which allows the numerical solution for each step to be computed in $O(n)$ time where n is the number of independent variables, solves the linear system at each time step by expressing the elimination as a recursion. Unfortunately, this method is quite difficult to parallelize, especially in the case of massively parallel computers. A number of other alternatives have been suggested for parallelization; however it remains to be seen after much experience and testing which if any of these methods are viable. In addition to parallelization, efficient solution methods will treat the nonstiff parts of the system explicitly, and automatically switch to fully or partially implicit methods when necessary. All methods will need to exploit the structure of the systems extensively.

Parameter estimation and optimal control for singular systems
Differential-algebraic equation (DAE) models are common in all aspects of chemical and process engineering, as well as many other important areas of engineering design[31, 38, 43]. Typically, these models contain operating or design parameters which can be adjusted so as to improve the efficiency of the process or the initial design (lower cost, higher yields, etc.). Thus an inherent part of the design of such complex systems is optimization of the design parameters to give improved system performance. A closely related problem is the optimal (or near-optimal) control of such a process after a chemical plant has been built. Typically this requires real-time control of all, or part, of the process based on a DAE model which includes control variables.

Both of these situations lead to large-scale constrained nonlinear optimization problems with both continuous and discrete variables and many nonlinear constraints. Because of the combination of discrete variables and many nonlinear constraints, these are very difficult problems to solve and typically require a large amount of computation for each specific case. In addition, the optimal control model must often be solved in real time to be effective in the actual process control.

The DAE optimization model can be represented in the following general form

$$\min_{x,u(t),z(t)} \varphi(x, u(t), z(t)) \tag{6.2}$$

subject to

$$\begin{aligned} c(x, u(t), z(t)) &= 0 \\ g(x, u(t), z(t)) &\geq 0 \\ \dot{z} &= f(x, u(t), z(t)), \quad t \in [0, T] \end{aligned}$$

where in addition, the parameter vector x, the control profile vector $u(t)$ and the state profile vector $z(t)$ are constrained by specific lower and upper bounds for all $t \notin [0, T]$. To solve this continuous system numerically, the time interval $[0, T]$ is discretized and the continuous vectors $u(t)$ and $z(t)$ are approximated by the finite-dimensional vectors w and y. Also, $\dot{z}$ is approximated by Ay, where A is a suitable finite difference operator. Typically the matrix A will be sparse and nonsingular.

The system (6.2) poses challenges for both the discretization and the optimization. The optimization leads to a boundary value problem DAE which may be high-index. There is a problem of stability for DAE boundary value problems, and especially for stiff problems which occur for example in control of chemical processes. In order to maintain stability for the differential part of the system, symmetric discretizations such as Gauss collocation are often necessitated. These discretizations have been shown to have severe limitations, including instability, oscillation and loss of accuracy when applied to differential-algebraic systems. This is true even for index one systems; the problems are even more severe for higher-index systems. Recently a new class of methods called *projected implicit Runge–Kutta methods* [3, 4], has been developed which overcome the stability and accuracy limitations above, and allow the use of symmetric discretizations to maintain stability in the differential part. This is done in part by exploiting the structure of the DAE system. A practically important subset of the projected implicit Runge–Kutta methods is given by projected collocation methods. These collocation methods achieve the nonstiff superconvergence order at the mesh points, for stable nonlinear DAE systems. One of the most interesting applications of these methods is in the area of singular optimal control problems. These problems arise in chemical process control. A difficulty in many of these problems is that there are regions of the solution called singular arcs, where the problem may for a time change its index and become a higher-index system. To effectively deal with this complication, methods will need to be developed which can treat the different parts of the subsystem of different index in a unified way. There are other practical difficulties. For example, since many of the control problems are embedded in an optimization problem, discontinuous changes in the solution from one iteration to the next due to the adaptive mesh procedure could cause a problem.

Following discretization, the system becomes a large but finite nonlinear programming problem in the vectors x, w, and y as

$$\min_{x,w,y} \Phi(x,w,y) \tag{6.3}$$

subject to

$$\begin{aligned} c(x,w,y) &= 0 \\ g(x,w,y) &\geq 0 \\ Ay &= f(x,w,y) \end{aligned}$$

where the vectors x, w, and y are bounded as before.

Bibliography

1. Arevalo, C. (1993). *Matching the Structure of DAEs and Multistep Methods.* Ph.D. thesis, Department of Computer Science, Lund University, Sweden.
2. Ascher, U. (1989). On Symmetric Schemes and Differential-Algebraic

Equations. *SIAM J. Sci. Stat. Comp.*, **10**, 937–949.

3. Ascher, U. and Petzold, L.R. (1991). Projected Implicit Runge–Kutta methods for Differential-Algebraic Equations. *SIAM J. Numer. Anal.*, **28**, 1097-1120.
4. Ascher, U. and Petzold, L.R. (1991). Projected Collocation for Higher-Order Higher-Index Differential-Algebraic Equations. To appear in *Comp. Appl. Math.*
5. Ascher, U. and Petzold, L.R. (1993). Stability of Computational Methods for Constrained Dynamics Systems. *SIAM J. Sci. Comput.*, **14**, 95–120.
6. Baumgarte, J. (1972). Stabilization of Constraints and Integrals of Motion in Dynamical Systems. *Comp. Math. Appl. Mech. Engrg.*, **1**, 1–16.
7. Berzins, M. and Furzeland, R.M. (1985). *A User's Manual for SPRINT: Part 1.* Dept. of Computer Studies Report 199, Leeds University.
8. Bock, H.G., Eich, E. and Schlöder, J.P. (1987). Numerical Solution of Constrained Least Squares Boundary Value Problems in Differential-Algebraic Equations. Universität Heidelberg, 1987.
9. K. Brenan, E., Campbell, S.L. and Petzold, L.R. (1989). *The Numerical Solution of Initial Value Problems in Differential-Algebraic Equations.* Elsevier Science Publishing Company.
10. Brenan, K.E. and Engquist, B. (1988). Backward Difference Approximations of Nonlinear Differential-Algebraic Equations. *Math. Comp.*, **51**, 659–676.
11. Brenan, K.E. and Petzold, L.R. (1989). The Numerical Solution of Higher Index Differential/Algebraic Equations by Implicit Runge–Kutta Methods. *SIAM J. Numer. Anal.*, **26**, 976–996.
12. Brown, P.N., Hindmarsh, A.C. and Petzold, L.R. (1994). Using Krylov Methods in the Solution of Large-Scale Differential-Algebraic Systems. To appear in *SIAM J. Sci. Comp.*, **15**.
13. Burrage, K. and Petzold, L.R. (1991). On Order Reduction for Runge–Kutta Methods Applied to Differential/Algebraic Systems and to Stiff Systems of ODEs. *SIAM J. Numer. Anal.*, **28** 205–226.
14. Burrage, K., Hundsdorfer, W.H. and Verwer, J.G. (1986). A Study of B-Convergence of Runge–Kutta Methods Applied to Stiff Systems. *Computing*, **36**, 17–34.
15. Campbell, S.L. (1987). A General Form for Solvable Linear Time Varying Singular Systems of Differential Equations. *SIAM J. Math. Anal.*, **18**, 1101–1115.
16. Caracotsios, M. and Stewart, W.E. (1985). Sensitivity Analysis of Ini-

tial Value Problems with Mixed ODEs and Algebraic Equations. *Computers and Chemical Engineering*, **9**, 359–365.

17. Clark, K.D. and Petzold, L.R. (1990). Numerical Methods for Differential-Algebraic Equations in Conservative Form. Lawrence Livermore National Laboratory UCRL-JC-103423.
18. Deuflhard, P. and Nowak, U. (1987). Extrapolation Integrators for Quasilinear Implicit ODEs. In *Large Scale Scientific Computing*, edited by P. Deuflhard and B. Engquist. Progress in Scientific Computing, **7**, Birkhäuser.
19. Dongarra, J. and Grosse, E. (1987). Distribution of Mathematical Software via Electronic Mail. *Comm. ACM*, **30**, 403–407.
20. Gear, C.W. (1971). Simultaneous Numerical Solution of Differential/Algebraic Equations. *IEEE Trans. Circuit Theory*, **CT-18**, 89–95.
21. Gear, C.W. (1988). Differential-Algebraic Equation Index Transformations. *SIAM J. Sci. Stat. Comp.*, **9**, 39–47.
22. Gear, C.W., Leimkuhler, B. and Gupta, G.K. (1985). Automatic Integration of Euler-Lagrange Equations with Constraints. *J. Comp. Appl. Math.*, **12** and **13**, 77–90.
23. Gear, C.W. and Petzold, L.R. (1984). ODE Methods for the Solution of Differential/Algebraic Systems. *SIAM J. Numer. Anal.*, **21**, 367–384.
24. Griepentrog, E. and März, R. (1986). *Differential- Algebraic Equations and their Numerical Treatment.* Teubner-Texte, Leipzig.
25. Griewank, A., Juedes, D. and Srinivasan, J. (1991). *ADOL-C, a Package for the Automatic Differentiation of Algorithms Written in C/C++.* Preprint, Argonne National Laboratory, Argonne, Illinois.
26. Hairer, E. and Wanner, G. (1991). *Solving Ordinary Differential Equations II: Stiff and Differential-Algebraic Problems.* Springer-Verlag.
27. Hindmarsh, A.C. (1981). ODE Solvers for use with the Method of Lines. In *Advances in Computer Methods for Partial Differential Equations IV*, edited by R. Vichnevetsky and R. S. Stepleman. IMACS, New Brunswick, NJ, 1981.
28. Kalachev, L. and O'Malley, R.E. (1991). *The Regularization of Linear Differential-Algebraic Equations.* University of Washington Dept. of Applied Math. report 91-10.
29. Kalachev, L. and O'Malley, R.E. (1991). *Regularization of Nonlinear Differential-Algebraic Equations.* University of Washington Dept. of Applied Math., report 92-4.
30. Keiper, J.B. (1989). *Generalized BDF Methods Applied to Hessenberg Form DAEs.* Ph.D. Thesis, Department of Computer Science, Univer-

sity of Illinois, UIUCDCS-R-89-1534.

31. Logsdon, J.S. and Biegler, L.T. (1989). Accurate Solution of Differential-Algebraic Optimization Problems. *I&EC Research*, **28**, 1628–1639.
32. Lötstedt, P. and Petzold, L.R. (1986). Numerical Solution of Nonlinear Differential Equations with Algebraic Constraints I: Convergence Results for Backward Differentiation Formulas. *Math. Comp.*, **46**, 491–516.
33. Lubich, Ch. (1989). h^2 - Extrapolation Methods for Differential-Algebraic systems of Index 2. *Impact Comput. Sci. Eng.*, **1**, 260–268.
34. Lubich, Ch. (1989). Linearly Implicit Extrapolation Methods for Differential-Algebraic Systems. *Numer. Math.*, **55**, 197–211.
35. Lubich, Ch. (1991). Extrapolation Integrators for Constrained Multibody Systems. *Impact Comput. Sci Eng.*, **3**, 213–234.
36. Maier, R.S., Petzold, L.R. and Rath, W. (1994). Solving Large-Scale Differential-Algebraic Equations via DASPK on the CM5. Submitted to *Concurrency: Practice and Experience.*
37. Marquardt, W. (1988). Universität Stuttgart, personal communication.
38. Morison, K.R. and Sargent, R.W.H. (1986). *Optimization of Multistage Processes Described by Differential-Algebraic Equations.* Lecture Notes in Mathematics No. 1230, pp. 86–102. Springer-Verlag, Berlin.
39. Petzold, L.R. (1982). A Description of DASSL: A Differential/Algebraic System Solver. *IMACS Trans. on Scientific Computation, Vol. 1*, edited by R. S. Stepleman.
40. Petzold, L.R, and Lötstedt, P. (1986). Numerical Solution of Nonlinear Differential Equations with Algebraic Constraints II: Practical Implications. *SIAM J. Sci. Stat. Comput.*, **7**, 720–733.
41. Petzold, L.R., Ren, Y. and Maly, T. (1993). *Regularization of Higher-Index Differential-Algebraic Equations with Rank-Deficient Constraints.* Department of Computer Science, University of Minnesota.
42. Rabier, P.J. and Rheinboldt, W.C. (1991). A General Existence and Uniqueness Theorem for Implicit Differential-Algebraic Equations. *J. Diff. and Integral Equations*, **4**, 563–582.
43. Renfro, J.G., Morshedi, A.M. and Asbjornsen, A.O. (1987). Simultaneous Optimization and Solution of Systems Described by Differential-Algebraic Equations. *Comput. Chem. Engrg.*, **11**, 503–517.
44. Rheinboldt, W.C. (1984). Differential-Algebraic Systems as Differential Equations on Manifolds. *Math. Comp.*, **43**, 473–482.
45. Rheinboldt, W.C. (1991). On the Existence and Uniqueness of Solu-

tions of Nonlinear Semi-Implicit Differential Algebraic Equations. *J. Nonlinear Analysis: Theory, Methods and Applications*, **16**, 647–661.

46. Yen, J. and Petzold, L.R. (1994). *On the Numerical Solution of Constrained Multibody Dynamic Systems.* Army High Performance Computing Research Center, University of Minnesota.

Boundary Element Methods

Ian H. Sloan

Department of Applied Mathematics,
University of New South Wales,
P.O. Box 1,
Kensington,
New South Wales 2033,
Australia.

1 A first brief tour of boundary element methods

1.1 Introduction

What are boundary element methods? They are methods for the approximate solution of two- or three-dimensional boundary value problems, which rely on casting the problem into the form of an integral equation over the boundary; then chopping the boundary into pieces ('elements'), on each of which a low-degree polynomial approximation is assumed; and then approximating the resulting discrete problem numerically. Approximate solutions to the original problem can then be obtained by carrying out a suitable approximate integration over the boundary.

Boundary element methods (BEM) have a rich relationship with finite-element methods (FEM). Since boundary element methods were developed later, they have borrowed many of the ideas and many of the techniques. But there are marked differences. Because we work only on the boundary, the dimension is lower for BEM than for FEM: thus a two-dimensional physical problem is reduced to a one-dimensional problem over the bounding curve while a three-dimensional problem is reduced to a two-dimensional problem over a surface. On the other hand, the BEM approach always, as we shall see, gives dense matrices, whereas FEM matrices are always sparse. Boundary methods have a special advantage when the physical problem is an **exterior** problem, for example one posed in the region outside a sphere, became the main computations are carried out on the two-dimensional spherical surface, instead of for an **infinite** three-dimensional region.

On the down side, BEM rely for their formation, as we shall see, on **linearity** of the problem (or at least of the main part of the problem), whereas FEM, being local, can cope easily with the nonlinear aspects of, for example, fluid flow.

These days there is little debate about whether BEM or FEM is the better approach. Rather, it is recognized that each has its role. Sometimes the best way is to use both, perhaps using BEM in an unbounded exterior part of the physical domain in which the physical processes are simple, and FEM in the relatively small region in which dissipative or nonlinear processes take place.

1.2 A model problem

Think of the problem of externally heating an object (either two-dimensional or three-dimensional) made of a uniform classical conducting material. In appropriate units, the heat flow $\mathbf{q}$ is equal to minus the gradient of the temperature φ,

$$\mathbf{q} = -\nabla\varphi,$$

while if a steady state has been established the principle of conservation of energy gives

$$\nabla \cdot \mathbf{q} = 0.$$

Together they give

$$-\Delta\varphi = 0,$$

where Δ is the two- or three-dimensional Laplacian. Perhaps the temperature φ is prescribed on part of the boundary (Dirichlet boundary condition), while on another part the normal derivative $\frac{\partial\varphi}{\partial n}$, which is related to the heat flux density, may be prescribed (Neumann boundary condition). The special case $\frac{\partial\varphi}{\partial n} = 0$ arises if part of the boundary is insulated, or if symmetry is used to reduce by half the domain of a symmetric problem. Newton's law of cooling introduces a boundary condition that relates both φ and $\frac{\partial\varphi}{\partial n}$.

The same equation also arises in electrostatics problems, with φ now being the electrical potential. Related but more complex equations arise from acoustics and other wave scattering problems (in this case the Laplace equation is replaced by the Helmholtz equation), transmission problems, linear elasticity, and many others. In these lectures we shall save time by concentrating on the Laplace equation.

1.3 A first boundary integral equation

Let us consider then the Laplace equation

$$-\Delta\varphi = 0 \quad \text{in } \Omega, \tag{1.1}$$

for a two-dimensional bounded, open domain Ω with piecewise-smooth boundary Γ, and Dirichlet boundary condition

$$\varphi = f \quad \text{on } \Gamma. \tag{1.2}$$

The simplest boundary integral formulation of this problem is via the 'single-layer' representation of the potential φ; that is, we seek a representation of φ in the form

$$\varphi(t) = -\frac{1}{\pi}\int_\Gamma \log|t-s|z(s)dl_s, \quad t\in\Omega, \tag{1.3}$$

where $|t-s|$ is the Euclidean distance between t and s, dl_s is the element of arc length with respect to s, and z is an unknown function, the 'single-layer density', or 'charge density'. The motivation is easily stated: because $\log|t-s|$ is the fundamental solution of the Laplace equation, (1.3) yields a solution of the Laplace equation, no matter how z is chosen; thus all that remains is to satisfy the boundary condition (1.2). Let t approach the boundary. We shall see later that the right side of (1.3) is continuous onto the boundary, so the limit of equation (1.3) is

$$f(t) = -\frac{1}{\pi}\int_\Gamma \log|t-s|z(s)dl_s, \quad t\in\Gamma.$$

This is our first boundary integral equation, or BIE. This particular BIE is an integral equation of the 'first kind' (which merely means that the unknown z occurs **only** under the integral sign). If we introduce the single-layer 'integral operator' V defined by

$$Vv(t) := -\frac{1}{\pi}\int_\Gamma \log|t-s|v(s)dl_s, \quad t\in\Gamma, \tag{1.4}$$

then we may write this integral equation as

$$Vz = f. \tag{1.5}$$

We shall refer to V as the **single-layer operator**. Once we have solutions of this integral equation, we can recover values of the original unknown φ (which we shall call the 'potential') at any point $t\in\Omega$ by appeal to (1.3).

That the integral on the right of (1.3) is continuous as t approaches the boundary has been shown by Gaier [14], under the assumptions that $z\in L_p(\Gamma)$ for some $p>1$, and that the curve is piecewise smooth and has no cusps.

Before we think of numerical solution of our equation (1.5), there are other issues we should consider. Does a solution exist? Is it unique? Is the problem well posed? We shall touch on these issues later, both for this and other boundary integral equations.

1.4 Two important numerical methods — the Galerkin and collocation methods

Suppose our boundary integral equation is of the form

$$Lu = g, \tag{1.6}$$

where L is a linear operator which maps functions on Γ (real or complex valued, as appropriate) to other real or complex functions on Γ.

We assume that Γ is the piecewise-smooth boundary of a simply connected open domain in two- or three-dimensional Euclidean space and that the measure on Γ is $d\Gamma$, denoting the element of arc length or surface area, as appropriate. (If the surface or curve Γ is parametrised in a particular way then a different measure might be more appropriate, but the necessary changes are usually obvious.)

How should we choose the domain and co-domain of L? That depends on L (more about this later), on the nature of Γ (for example, does it have corners?) and on the approximation scheme. For the moment, let us assume just that L maps $L_2(\Gamma)$ into $L_2(\Gamma)$. In this setting it is natural to introduce the inner product

$$(v, w) := \int_\Gamma v\overline{w}d\Gamma, \tag{1.7}$$

with the complex conjugate $\overline{w}$ replaced by w if the functions are real. The corresponding norm is the L_2 norm,

$$\|v\|_{L_2} := (v, v)^{1/2}. \tag{1.8}$$

Now let S_h be a finite-dimensional space within which the approximate solution is to be sought. Typically, S_h is defined by partitioning Γ into a finite number of pieces with simple geometry (e.g. plane or curved triangles) and maximum diameter h, on each of which the restriction of S_h is a piecewise polynomial space with respect to an appropriate local parametrization (e.g. one in which the element boundary is a triangle). If Γ is a curve in the plane, then each piece is just a smooth arc of length $\leq h$. Continuity conditions across elements may or may not be imposed, depending on the circumstances. (For further details see, for example, Brebbia, Telles and Wrobel, [7].) Then the **Galerkin method** for this problem and this choice of elements is: find $u_h \in S_h$ such that

$$(Lu_h, \chi_h) = (g, \chi_h) \quad \forall \chi_h \in S_h. \tag{1.9}$$

We shall take up the theory of the Galerkin method later.

While the Galerkin method is the theorist's favourite, it is in truth not easy to implement. Let $\{\varphi_1, \ldots, \varphi_N\}$ be a basis for S_h. Then we may write

$$u_h = \sum_{j=1}^{N} a_j \varphi_j, \tag{1.10}$$

so that the equations to be solved in practice are

$$\sum_{j=1}^{N} (L\varphi_j, \varphi_k) a_j = (g, \varphi_k), \quad k = 1, \ldots, N, \tag{1.11}$$

in which each matrix element $(L\varphi_j, \varphi_k)$, even in the two-dimensional case, is a two-dimensional integral — one integral for the integral operator, and one for the inner product. In the three-dimensional case four levels of integration are needed for each matrix element. And the difficulty is compounded by the fact that if L is an integral operator the matrix $\{(L\varphi_j, \varphi_k)\}$ is invariably dense. The point is that an integral operator (see, for example, (1.4)), quite unlike a differential operator, is nonlocal.

Often practical people prefer the **collocation method**, because it is much easier to implement than the Galerkin method. Like the Galerkin method, the collocation method starts with a choice of a space S_h within the solution is to be approximated. (Note that choosing S_h involves choosing each of the following: the mesh into which Γ is to be partitioned, the degree of the piecewise polynomials, and the continuity properties to be imposed at the inter-element boundaries.) In the collocation method one chooses also a set of "collocation points" $t_1, \ldots, t_N$, where $N = N_h$ is the dimension of S_h. Then the collocation method is: find $u_h \in S_h$ such that

$$Lu_h(t_k) = g(t_k), \quad k = 1, \ldots, N. \tag{1.12}$$

Needless to say, the choice of the collocation points is a very important question, to which we shall return later for particular boundary integral equations.

With u_h again written in the form (1.10), the equations to be solved in practice are now

$$\sum_{j=1}^{N} L\varphi_j(t_k)a_j = g(t_k), \quad k = 1, \ldots, N. \tag{1.13}$$

Clearly, the labor involved in setting up the matrix is much less than in the Galerkin method: in the two-dimensional case each matrix element requires just one integration, and even in the three-dimensional case a matrix element needs just two integrations. This is to be compared to four levels of integration for the Galerkin method.

However, saving time in the construction of the matrix is only one of many issues we have to worry about in selecting a numerical method. For example, given that the exact problem (1.6) is well posed and has a unique solution, can we be sure that the approximate system has a solution? (At best we usually know this only for h sufficiently small.) Is the solution unique? Is the numerical method stable? And, most importantly, if all these hold, what is the rate of convergence? Even that question on its own has many answers, because the rate of convergence depends on the norm with which we choose to measure the end. And numerical analysts often ask questions such as: can we obtain "superconvergence" for some integral of the solution, or for its value at special points? (Superconvergence, roughly speaking, is higher-than-usual convergence, which can be seen only in special circumstances.)

We shall see later that the Galerkin method has a more robust theory than the collocation method and that a satisfying theory of the collocation method exists only in special cases. We shall also meet other methods (such as the **qualocation method** and its friends) for which the theoretical development is just beginning.

This brings to an end our first brief glimpse of boundary element methods. Now we are ready to dig deeper.

2 Basic boundary integral equations

2.1 Classical BIE formulations

Let us return again to the two-dimensional Laplace equation (1.1) with Dirichlet boundary condition (1.2) for a bounded planar region with smooth or piecewise-smooth boundary Γ. The classical way of setting up a boundary integral equation (BIE) for this problem is to look for a solution not in the form of a single-layer potential, as in (1.3), but instead as a "double-layer" potential of the form

$$\begin{aligned} \varphi(t) &= \frac{1}{\pi}\int_\Gamma \frac{\partial}{\partial n_s}(\log|t-s|)z(s)dl_s \\ &= \frac{1}{\pi}\int_\Gamma \frac{n(s)\cdot(s-t)}{|t-s|^2}z(s)dl_s, \quad t\in\Omega. \end{aligned} \tag{2.1}$$

Here the derivative is the normal derivative (with respect to s), in the direction of the outward unit normal $\mathbf{n}$ (i.e., the normal directed away from Ω). The double-layer potential satisfies the Laplace equation for every choice of z, just as the single-layer potential (1.3) does, became for $t \notin \Gamma$ the differentiation of $\log|t-s|$ with respect to s still leaves us with a function of t that satisfies the Laplace equation. The double-layer potential may be thought of as arising from a distribution of **dipoles** over Γ, with density z. This may be contrasted with the **charge** distribution associated with the single-layer potential.

Why the classical preference for the double-layer potential? The point is that this approach leads to a quite different kind of integral equation, because the double-layer potential is **not** continuous onto Γ. The following theorem is proved by Mikhlin [28] for the case of a Lyapunov curve, and by Hackbusch [16] and Wendland [39] for a curve with corners. (A Lyapunov curve or surface has a normal at each point, and the angles of the normals at points s_1 and s_2 satisfy a Hölder continuity condition $|\theta_1-\theta_2| \le C|s_1-s_2|^\alpha$ for some $\alpha > 0$.)

Theorem 2.1 *Let Γ be piecewise Lyapunov without cusps, and let $z \in$*

$C(\Gamma)$. *Then the integral*

$$\frac{1}{\pi}\int_\Gamma \frac{\partial}{\partial n_s}(\log|t-s|)z(s)dl_s, \quad t \in \Omega, \tag{2.2}$$

has a limit as t approaches Γ *from* Ω. *If* $t' \in \Gamma$ *is a point at which* Γ *has a tangent then the limit as* $t \to t'$ *is*

$$\frac{1}{\pi}\int_\Gamma \frac{\partial}{\partial n_s}(\log|t'-s|)z(s)dl_s + z(t'). \tag{2.3}$$

The proof proceeds by representing $z(s)$ in (2.2) as $z(t') + (z(s) - z(t'))$. It can be shown, because $z(s) - z(t) \to 0$ as $t \to \Gamma$, that the integral corresponding to the second term is continuous as $t \to \Gamma$. On the other hand in the first term $z(t')$ can be taken outside the integral, so is sufficient to prove the result for $z \equiv 1$. Here is the argument for the case $z \equiv 1$. For t a fixed point in Ω or on Γ (but not a corner point of Γ), and s a point running along Γ, let ρ, θ be polar coordinates of $s - t$, and let ψ be the angle between the outward normal $n(s)$ and the vector $s - t$. Then

$$\frac{\partial}{\partial n_s}(\log|t-s|) = \frac{n(s)\cdot(s-t)}{\rho^2} = \frac{\cos\psi}{\rho}, \tag{2.4}$$

and

$$dl_s = \frac{\rho d\theta}{\cos\psi},$$

from which we see that

$$\frac{1}{\pi}\int_\Gamma \frac{\partial}{\partial n_s}(\log|t-s|)dl_s = \frac{1}{\pi}\int d\theta = \frac{1}{\pi}\begin{cases} 2\pi & \text{if } t \in \Omega, \\ \pi & \text{if } t \in \Gamma. \end{cases} \tag{2.5}$$

That is, if $z \equiv 1$ the integral (2.2) has the value 2, while (2.3) has the value $1 + 1 = 2$, which is the same.

Motivated by this theorem, we define

$$\begin{aligned} Kz(t) &:= \frac{1}{\pi}\int_\Gamma \frac{\partial}{\partial n_s}(\log|t-s|)z(s)dl_s \\ &= \frac{1}{\pi}\int_\Gamma \frac{n(s)\cdot(s-t)}{|t-s|^2}z(s)dl_s, \quad t \in \Gamma, \end{aligned} \tag{2.6}$$

the **double-layer operator** on Γ. The limit (2.3) may now be replaced by

$$Kz(t') + z(t'). \tag{2.7}$$

Now we return to the double-layer representation (2.1) of the interior Dirichlet problem (1.1), (1.2). By taking the limit as t approaches a point on the boundary and using the theorem (and swapping the two sides of the equation), we see that z satisfies

$$Kz(t) + z(t) = f(t), \ t \in \Gamma, \ t \text{ not a corner point.} \tag{2.8}$$

This is an integral equation of the *second* kind, in the nomenclature introduced by Fredholm. The crucial difference between this equation and the first kind integral equation (1.4) is that the unknown z now appears outside the integral sign, as well as under the integral sign.

Now in the classical Fredholm theory of second kind equations (see, for example Kress [23]) the central assumption is that K is a 'compact' operator in an appropriate Banach space. (This assumption can be weakened: it is enough that some power of K be compact.) Under the compactness assumption the equation $z + Kz = f$ has the kind of nice properties we associate with square systems of linear equations. For a start, if the corresponding homogeneous equation $z + Kz = 0$ has only the trivial solution, then a solution of the inhomogeneous equation exists for **every** choice of f from the appropriate Banach space. And if the homogeneous equation has nontrivial solutions then the dimension of the solution space of $z + Kz = 0$ is equal to the number of (integral) side conditions to be satisfied by f before a solution of $z + Kz = f$ can exist.

The compactness of K also plays a key role in numerical approximation schemes for second-kind equations (see Atkinson [5]).

Is the integral operator K defined by (2.6) compact in some appropriate space? If Γ is a Lyapunov curve then the kernel of K, given by (2.4), turns out to be weakly singular (i.e. to have at worst an integrable singularity). If Γ is a C^2 curve then the kernel $\cos\psi/\rho$ is even continuous, because in this case ψ approaches $\pi/2$ as $s \to t$, and, as is easily seen,

$$\cos\psi = C\rho + o(\rho) \quad \text{as } |s - t| = \rho \to 0.$$

In both cases K is a compact operator in $C(\Gamma)$, the space of continuous functions on Γ with uniform norm, so the classical Fredholm theory applies.

On the other hand, for a region with a corner the operator K is **not** compact. We shall come back to this question shortly.

2.2 The classical BIE quartet

For the moment let us assume that the boundary Γ of the planar domain Ω is a Lyapunov curve. So far we have considered just the Dirichlet problem for the interior domain Ω. Now we introduce also the **exterior** Dirichlet problem, i.e. we consider the Laplace equation (1.1) for the region $\Omega_c :=$ complement of $\overline{\Omega}$, with a Dirichlet boundary condition on Γ, and with φ assumed bounded at infinity. We consider also the interior and exterior **Neumann** problems. The four resulting second-kind boundary integral equations make a nicely matched foursome, which we may call the classical quartet.

For the interior Dirichlet problem, recall that we used a double-layer representation of the potential, and used its discontinuity property at the boundary to obtain the BIE

$$z + Kz = f, \tag{2.9}$$

where K is given by (2.6). For the exterior problem the double-layer representation (2.1) can again be used, but this time the last term in (2.3), expressing the discontinuity onto the boundary, has the opposite sign. (This follows from the fact that for $t \in \Omega_c$ the integral (2.5) has the value 0.) As a result the BIE for the exterior Dirichlet problem is

$$-z + Kz = f. \tag{2.10}$$

Now we turn to the interior and exterior Neumann problems

$$-\Delta\varphi = 0, \;\; t \in \Omega \text{ or } \Omega_c, \tag{2.11}$$

with

$$\frac{\partial \varphi}{\partial n} = h \text{ on } \Gamma, \tag{2.12}$$

and in the case of the exterior problem, with φ bounded at infinity. (Here, and below, the normal derivative is always taken to be outward from Ω, and into Ω_c.) It is easily verified that a necessary condition for a solution to exist is

$$\int_\Gamma h d\ell = 0, \tag{2.13}$$

and that if φ is a solution so is φ + constant.

The classical approach to the Neumann problem is via the single-layer representation (1.3) of φ. It can be shown (Mikhlin [28]) that if z is integrable on Γ and $t \notin \Gamma$ that the potential φ can be differentiated under the integral sign, giving for the two-dimensional case

$$\begin{aligned} \nabla\varphi(t) &= -\frac{1}{\pi}\int_\Gamma (\nabla_t \log|t-s|) z(s) dl_s \\ &= -\frac{1}{\pi}\int_\Gamma \frac{t-s}{|t-s|^2} z(s) dl_s, \quad t \notin \Gamma. \end{aligned}$$

Now let t' denote a point on Γ, and let $\frac{\partial \varphi}{\partial n}$ be the directional derivative of φ in the direction of the (outward) normal at t'. Then the last equation gives

$$\frac{\partial \varphi}{\partial n}(t) = -\frac{1}{\pi}\int_\Gamma \frac{n(t')\cdot(t-s)}{|t-s|^2} z(s) dl_s, \quad t \notin \Gamma. \tag{2.14}$$

This normal derivative has jump discontinuities analogous to those in Theorem 2.1 as $t \to t'$. By the same kind of arguments as before, the limits as

$t \to t' \in \Gamma$ from Ω and Ω_c are (Mikhlin [28])

$$-K^*z(t') \pm z(t'), \tag{2.15}$$

where the upper and lower signs hold for $t \in \Omega$ and Ω_c respectively, and

$$\begin{aligned} K^*z(t) &:= \frac{1}{\pi}\int_\Gamma \left(\frac{\partial}{\partial n_t}\log|t-s|\right) z(s)dl_s, \\ &= \frac{1}{\pi}\int_\Gamma \frac{n(t)\cdot(t-s)}{|t-s|^2} z(s)dl_s, \quad t \in \Gamma, \end{aligned} \tag{2.16}$$

the **normal derivative of the single-layer operator**. Note that the normal derivative in this case is with respect to t, whereas in the double-layer operator (2.2) it is with respect to s. In fact K and K^* are **adjoints** of each other, because in the respective kernels s and t are interchanged. It is well known that compactness of an integral operator implies compactness of its adjoint, so in the case of a Lyapunov curve K^* is compact.

It now follows from the jump discontinuity relations (2.15) that the two-dimensional interior and exterior Neumann problems are characterized by the equation

$$-K^*z(t) \pm z(t) = h(t), \quad t \in \Gamma, \tag{2.17}$$

with the upper and lower signs holding for the interior and exterior problems respectively.

Thus we obtain the classical quartet. Written in an order that emphasizes a natural mathematical pairing, the four members of the quartet are:

$$z + Kz = f \quad \text{(interior Dirichlet)}, \tag{2.18}$$
$$z + K^*z = -h \quad \text{(exterior Neumann)}, \tag{2.19}$$
$$z - Kz = -f \quad \text{(exterior Dirichlet)}, \tag{2.20}$$
$$z - K^*z = h \quad \text{(interior Neumann)}. \tag{2.21}$$

The first and second equations are mutually adjoint, as are the third and fourth.

It turns out that the first two of these have nicer properties than the last two. For the first two, the interior Dirichlet and exterior Neumann BIEs, it is known (Mikhlin [28], Section 13) that a unique $z \in L_2$ exists for each $f \in L_2$ or $h \in L_2$. (The Fredholm theory, see e.g. Kress [23], tells us that this holds for both equations if it holds for either of them.) In contrast, the homogeneous versions of the last two equations each have a one-dimensional solution space, and solutions of $z - Kz = -f$ exist in L_2 if and only if $\int_\Gamma fw = 0$ for every solution w of the adjoint homogeneous equation, and vice versa. (In fact it should not be a surprise that the homogeneous version of (2.20) has a nontrivial solution, because (2.5) tells us that $K\mathbf{1} = \mathbf{1}$, where **1** denotes the constant function with value 1. The corresponding condition for (2.21) to have a solution is just (2.13).)

The classical arguments considered so far in this section assume considerable regularity of the boundary, and in particular do not allow corners. However, it is now known that the results hold also in the much more general setting of Lipschitz curves. (Roughly speaking, a curve is Lipschitz if there exist local coordinate systems in which short sections of the curve form Lipschitz functions. Thus ordinary corners are allowed, whereas cusps are not.)

To understand why corners cause difficulties, one need only look again at the definition of the double-layer operator

$$Kz(t) = \frac{1}{\pi}\int_\Gamma \frac{n(s)\cdot(s-t)}{|t-s|^2} z(s)d\ell_s, \quad t \in \Gamma,$$

when t is near a corner. You can see that the kernel is no longer continuous by allowing s and t to approach the corner from the two different arcs that join at a corner: the kernel is not even bounded. It turns out that K is still a bounded operator in L_2, but it is no longer compact. For a thorough examination of the double-layer operator near a corner, see Atkinson and De Hoog [6].

2.3 Direct method

The classical BIE formulations described above are **indirect**, in that the solution φ of the original problem has been expressed as a double-layer or single-layer potential, in which the unknown quantity z is not of direct interest. Actually, most practical applications of boundary integral equations nowadays employ **direct** formulations, based on Green's theorem or its relatives. In the direct methods the unknown quantity **is** of direct interest: in the simple example of the Laplace equation the unknown function is the value on the boundary of φ itself, or of its normal derivative.

Green's theorem for a bounded two-dimensional region Ω with smooth boundary Γ, and for a function $\varphi \in C^2(\overline{\Omega})$ that satisfies the Laplace equation (1.1), is (Mikhlin [28], p.224)

$$\varphi(t) = \frac{1}{2\pi}\int_\Gamma \left[\left(\frac{\partial}{\partial n_s}\log|t-s|\right)\varphi(s) - \log|t-s|\frac{\partial\varphi(s)}{\partial n_s}\right] dl_s, \quad t \in \Omega. \tag{2.22}$$

An equation on the boundary may now be obtained by letting $t \to \Gamma$ and using the continuity properties of the single- and double-layer potentials discussed above. In this way we obtain

$$\varphi(t) = \frac{1}{2}(K\varphi(t) + \varphi(t)) + \frac{1}{2}V\frac{\partial\varphi}{\partial n}(t), \quad t \in \Gamma,$$

where K and V are defined by (2.6) and (1.4). Written concisely, the equation on Γ is

$$\varphi = K\varphi + V\frac{\partial \varphi}{\partial n}. \tag{2.23}$$

This equation is an identity, not an integral equation. It holds whenever φ satisfies the Laplace equation in Ω.

Thus far we have assumed stringent conditions on Γ and φ, but these can be relaxed significantly (see, for example, Costabel [11]), to allow curves that are merely Lipschitz, and hence may have corners.

Now let us introduce boundary conditions. Consider first the case of the Dirichlet boundary condition (1.2). Then (2.23) gives an integral equation of the first kind for $z := \frac{\partial \varphi}{\partial n}$, namely

$$Vz = f - Kf. \tag{2.24}$$

In fact we see that this is the same integral equation as the one we obtained from the indirect method with the single-layer potential, namely (1.5), except that now the right hand side is different. Of course the solution this time has quite a different meaning: now it is the normal derivative of φ on the boundary.

The biggest advantage of direct methods is that they allow **mixed boundary conditions** to be handled in a natural way. Suppose that the boundary conditions prescribe the value of φ on one part of the boundary, denoted by Γ_1, and the normal derivative $\frac{\partial \varphi}{\partial n}$ on the remaining part, denoted by Γ_2. Now we need to employ not only (2.23), but also the equation obtained by taking its normal derivative, namely

$$\frac{\partial \varphi}{\partial n} = W\varphi - K^*\frac{\partial \varphi}{\partial n}, \tag{2.25}$$

where K^* is given by (2.16), and W, the so-called **hypersingular operator**, is defined by

$$W\varphi(t) = \frac{1}{\pi}\frac{\partial}{\partial n_t}\int_\Gamma \frac{\partial}{\partial n_s}(\log|t-s|)\varphi(s)dl_s, \quad t \in \Gamma. \tag{2.26}$$

To obtain a closed set of equations, all we need to do now is to split each integral operator in (2.23) and (2.26) into two parts, one corresponding to the integral over Γ_1 and one to the integral over Γ_2; and then look at the values on Γ_1 and Γ_2 separately; and finally substitute the given boundary data, namely the Dirichlet data on Γ_1, and the Neumann data on Γ_2. In this way we obtain a pair of **coupled** equations, for the unknowns $\frac{\partial \varphi}{\partial n}$ on Γ_1 and φ on Γ_2. We will not pursue the details any further.

2.4 Three-dimensional problems

In this chapter we choose, for simplicity, to concentrate on two-dimensional problems, but in fact almost everything we have said applies with only small changes to three-dimensional problems, if we define the layer operators appropriately:

$$
\begin{aligned}
Vz(t) &= \frac{1}{2\pi}\int_\Gamma \frac{1}{|t-s|}z(s)dS_s, \quad t\in\Gamma, && (2.27)\\
Kz(t) &= -\frac{1}{2\pi}\int_\Gamma \left(\frac{\partial}{\partial n_s}\frac{1}{|t-s|}\right)z(s)dS_s, \quad t\in\Gamma, && (2.28)\\
K^*z(t) &= -\frac{1}{2\pi}\int_\Gamma \left(\frac{\partial}{\partial n_t}\frac{1}{|t-s|}\right)z(s)dS_s, \quad t\in\Gamma, && (2.29)\\
Wz(t) &= -\frac{1}{2\pi}\frac{\partial}{\partial n_t}\int_\Gamma \left(\frac{\partial}{\partial n_s}\frac{1}{|t-s|}\right)z(s)dS_s, \quad t\in\Gamma, && (2.30)
\end{aligned}
$$

where dS is the element of surface area. For more details see, for example, Sloan [33].

3 Smoothing properties, Sobolev spaces, and so forth

3.1 Introduction

Many numerical methods for boundary integral equations rely on the fact that we know in a rather precise way the mapping properties of the common boundary integral operators. Suppose we write our boundary integral equation in the form

$$Lu = g. \tag{3.1}$$

Given u, what can we say about Lu? More importantly, given $g = Lu$, what can we say about u? And following on from that, what is the "right" function-space setting?

For the main boundary integral operators such questions have nice answers, provided the boundary Γ is reasonably smooth. So for the present we shall assume that Γ is the smooth boundary of a planar domain Ω.

3.2 The single-layer operator

We start with the single-layer operator (1.4). This is often said to be a 'pseudodifferential operator (or ψdo) of order –1'. Let's see if we can explain what that means.

We shall assume for the present that Γ is parameterized by $t = \nu(x)$, where

$$\nu : [0,1] \to \Gamma,\ \nu \text{ is 1-periodic},\ \nu \in C^\infty,\ |\nu'(x)| \neq 0.$$

Then from the definition (1.4) we have

$$Vz(\nu(x)) = -\frac{1}{\pi}\int_0^1 \log|\nu(x)-\nu(y)|z(\nu(y))|\nu'(y)|dy \tag{3.2}$$

$$= -2\int_0^1 \log|\nu(x)-\nu(y)|u(y)dy,$$

where we have introduced a new unknown function

$$u(y) := \frac{1}{2\pi} z(\nu(y))|\nu'(y)| \tag{3.3}$$

which incorporates the Jacobian $|\nu'(y)|$ and a convenient normalization factor. Thus we now have $V(z(\nu(x)) = Lu(x)$, where

$$Lu(x) := -2\int_0^1 \log|\nu(x)-\nu(y)|u(y)dy, \tag{3.4}$$

and our equation (1.5) becomes $Lu = g$ if we define $g(x) := f(\nu(x))$.

The simplest result, and at the same time the key to understanding the properties of the operator L for general smooth curves, is for the case in which Γ is a circle. Let A denote the operator L for the specific case of a circle of radius a. With the circle parameterized by $t = (t_1, t_2) = a(\cos 2\pi x, \sin 2\pi x)$, the definition becomes

$$Au(x) := -2\int_0^1 \log|2a\sin\pi(x-y)|u(y)dy. \tag{3.5}$$

Since we are working in the context of 1-periodic functions, it is natural to ask: what does A do to the Fourier mode $e^{2\pi ikx}$, where $k \in \mathbb{Z}$? The answer is surprisingly simple.

Proposition 3.1 *If $\varphi_k(x) = e^{2\pi ikx}$, with $k \in \mathbb{Z}$, then*

$$A\varphi_k(x) = \begin{cases} -2\log a, & \text{if } k = 0, \\ \frac{1}{|k|}\varphi_k(x), & \text{if } k \neq 0. \end{cases}$$

This follows from the well known Fourier cosine series, valid for $x \neq 0$,

$$-\log(2|\sin\pi x|) = \sum_{k=1}^{\infty} \frac{1}{k}\cos 2\pi kx = \frac{1}{2}\sum_{k\neq 0} \frac{1}{|k|} e^{2\pi ikx},$$

from which follows, using only the property of the logarithm of a product,

$$-2\log(2a|\sin\pi x|) = -2\log a + \sum_{k\neq 0} \frac{1}{|k|} e^{2\pi ikx}.$$

The proposition tells us that the effect of the operator A on the k-th Fourier component of v, if $k \neq 0$, is just to multiply that component by $1/|k|$. Notice that the effect of the operator A on the Fourier mode is like that of indefinite integration, if we disregard a constant factor, except that here we have $|k|^{-1}$ as the multiplier, instead of k^{-1}. It is the exponent -1 here that makes this a pseudodifferential operator of order -1; for definitions and discussion of ψdo's see Agranovich [1], Wendland [38] and McLean [25].

The operator A is invertible, we see, unless $a = 1$. In the exceptional case $a = 1$, on the other hand, A is not invertible, since we then have $A1 = 0$. In fact it is known that a similar situation occurs for the single-layer equation for curves Γ of any shape: very briefly, to every curve Γ (and, more generally, every point set in the plane) there is associated a real number called the 'transfinite diameter', or 'logarithmic capacity', a number which grows linearly as the curve is magnified in scale. For the definitions of these terms, and the connections to potential theory, see Hille [18]. The single-layer operator fails to be invertible when the transfinite diameter is equal to 1, and in no other case. (Jaswon and Symm [21] call the exceptional curve a 'Γ-contour'.) For a circle the transfinite diameter equals the radius, so that, consistently with the proposition, invertibility fails when the radius is 1. For a fuller discussion of transfinite diameter in the context of the single-layer equation, see for example Sloan [33] or Sloan and Spence [35].

The result in Proposition 3.1 is simplest for the case $a = e^{-1/2}$, since with that choice we have $A1 = 1$. From now on we make that choice of a, unless stated otherwise.

Now we come to the question of a function-space setting. Given the simple effect of the operator A on the Fourier modes, it is not surprising that we would want to use Fourier ideas in defining our function spaces. Remembering the 1-periodic setting, any reasonable function (or distribution) can be represented as a Fourier series,

$$v \sim \sum_{k \in \mathbb{Z}} \hat{v}(k) e^{2\pi i k x},$$

where

$$\hat{v}(k) = \int_0^1 e^{-2\pi i k x} v(x)\,dx, \quad k \in \mathbb{Z}.$$

For any real number s we define the Sobolev norm $\|v\|_s$ of v by

$$\|v\|_s^2 = |\hat{v}(0)|^2 + \sum_{k \neq 0} |k|^{2s} |\hat{v}(k)|^2. \tag{3.6}$$

When $s = 0$ the norm $\|v\|_0$ is just the L_2 norm. The norm $\|v\|_s$ also has a simple enough interpretation when s is a positive integer: if we recall that the sth derivative of v has the Fourier series

$$v^{(s)} \sim \sum_{k \in \mathbb{Z}} (2\pi i k)^s \hat{v}(k) e^{2\pi i k x},$$

we see that, apart from an unimportant constant factor, $\|v\|_s$ is essentially the L_2 norm of the sth derivative. (The term $|\hat{v}(0)|^2$ is included on the right of (3.6) to make this a norm, and not just a semi-norm.) Similarly, for negative integer values of s the norm is essentially the L_2 norm of the sth anti-derivative of v.

Corresponding to the norm $\|\cdot\|_s$ is the Sobolev space H^s: this may be defined as the closure with respect to the norm $\|\cdot\|_s$ of the space of 1-periodic C^∞ functions. The elements of H^s are 1-periodic functions (or more generally distributions) whose $\|\cdot\|_s$ norm is finite. The space H^s is a Hilbert space with respect to the inner product

$$(v, w)_s := \hat{v}(0)\overline{\hat{w}(0)} + \sum_{k \neq 0} |k|^{2s} \hat{v}(k)\overline{\hat{w}(k)}. \tag{3.7}$$

Note that for $s = 0$ we have $H^0 = L_2$, and the inner product reduces to

$$(v, w)_0 = (v, w) = \int_0^1 v(x)\overline{w(x)}dx,$$

the ordinary L_2 inner product.

An important inequality, holding for all real s and α, is

$$|(v, w)_s| \leq \|v\|_{s-\alpha}\|w\|_{s+\alpha}, \quad v \in H^{s-\alpha}, \quad w \in H^{s+\alpha}. \tag{3.8}$$

The proof is an easy application of the Cauchy–Schwarz inequality, starting from

$$(v, w)_s = \hat{v}(0)\overline{\hat{w}(0)} + \sum_{k \neq 0} |k|^{s-\alpha}\hat{v}(k)|k|^{s+\alpha}\overline{\hat{w}(k)}.$$

We also have the stronger result

$$\|v\|_{s-\alpha} = \sup_{w \in H^{s+\alpha}} \frac{|(v, w)_s|}{\|w\|_{s+\alpha}}, \quad v \in H^{s-\alpha}, \tag{3.9}$$

with the supremum being achieved by $\hat{w}(0) = \overline{\hat{v}(0)}$ and $\hat{w}(k) = |k|^{-2\alpha}\overline{\hat{v}(k)}$ for $k \neq 0$. In the jargon of the field, $H^{s-\alpha}$ and $H^{s+\alpha}$ provide a 'duality pairing' with respect to the inner product $(\cdot, \cdot)_s$.

By far the most important special case of a duality pairing is obtained by setting $s = 0$ in (3.9): we see that H^α **and** $H^{-\alpha}$ **are dual spaces with respect to the ordinary** L_2 **inner product.**

Now we shall see how beautifully adapted these Sobolev spaces are to the single-layer operator for the circle. From Proposition 3.1 (with $a = e^{-1/2}$) and the definition given in eqn (3.6) we see immediately that, for any 1-periodic function or distribution v,

$$Av = \hat{v}(0) + \sum_{k \neq 0} \frac{1}{|k|}\hat{v}(k)e^{2\pi ikx},$$

and hence

$$\|Av\|_s = \|v\|_{s-1}, \quad s \in \mathbb{R}.$$

That makes it obvious that A is a bounded operator from H^{s-1} to H^s, and that A^{-1} is a bounded operator from H^s to H^{s-1}; indeed, A is even an **isometric** operator from H^{s-1} to H^s.

Now we return to the single-layer operator L for a general smooth curve Γ, given by (3.4). Using nothing more than the properties of the logarithm function, we can write

$$L = A + B, \tag{3.10}$$

where A is the operator for a circle, as above, and

$$Bu(x) = -2\int_0^1 \log\left|\frac{\nu(x)-\nu(y)}{2e^{-1/2}\sin\pi(x-y)}\right| u(y)dy. \tag{3.11}$$

The kernel in the latter operator might look complicated, but in fact it is very well behaved, since the zeros in the numerator and denominator coincide. Using L'Hôpital's rule, the kernel is a C^∞ 1-periodic function in x and y.

Here is the standard way of studying the single-layer equation for the case of a smooth curve Γ. We can write (3.1) as

$$(A+B)u = g, \tag{3.12}$$

or, since A is invertible, after multiplying by A^{-1} on the left,

$$(I+M)u = A^{-1}g, \tag{3.13}$$

where

$$M = A^{-1}B. \tag{3.14}$$

Because B has a smooth kernel it is easy to see that so does M, thus M is a compact operator in H^{s-1}, no matter how we choose s. According to the Fredholm theory, that leaves just two possibilities: either $(I+M)$ is not invertible (which is the case if and only if L is not invertible, that is, if and only if the transfinite diameter of Γ equals 1); or $I+M$ **is** invertible, in which case a solution of (3.13) exists in H^{s-1} for each choice of $g \in H^s$. In the second case $I+M$ has a **bounded** inverse in H^{s-1}. (Of course we are talking here only of the properties of the exact equation rather than of any numerical approximation, but we shall see that the analysis above also plays a key role in the analysis of numerical methods.)

Thus we see that, provided we exclude the one bad choice for the scaling of Γ, the single-layer operator L is a bounded operator from H^{s-1} to H^s, and L^{-1} is a bounded operator from H^s to H^{s-1}.

There is one choice of s that is particularly natural for the single-layer operator L, namely $s = \frac{1}{2}$: in this setting L maps $H^{-1/2}$ to $H^{1/2}$, and

L^{-1} maps back in the other direction. The two spaces $H^{-1/2}$ and $H^{1/2}$ are dual to each other with respect to the ordinary L_2 inner product, a feature that makes this a natural setting when we come to talk about the Galerkin method in Section 4.

3.3 Other integral operators

In the case of the double-layer operator for the interior Dirichlet problem, the equation we have to consider is

$$(I + K)u = f,$$

where K is given of (2.6). So long as we stick to smooth curves Γ, the operator K, as we saw in Section 2, has a smooth kernel, so that K is a compact operator in any of our Sobolev spaces H^s, and the classical Fredholm theory applies; in the language of pseudodifferential operators, the operator $I + K$ is a ψdo of order 0. The natural setting is $H^0 = L_2$. Another operator of order 0 is the Hilbert transform

$$Hu(x) = 2\pi \int_0^1 \cot \pi(x-y)u(y)dy, \tag{3.15}$$

which is related to our ψdo A by $Hu = DAu$; that is, H is the first derivative of A.

The hypersingular operator W defined by (2.26), on the other hand, is rather more like a differential operator: it is a bounded map from H^{s+1} to H^s, and is a pseudodifferential operator of order 1.

3.4 Regions with corners

It turns out that at least some of the properties we have looked at for smooth curves still hold for Lipschitz curves. For example, the single-layer operator V and the hypersingular operator W are still defined, and have mapping properties which include (see, for example, Costabel [11])

$$V : H^{-1/2}(\Gamma) \rightarrow H^{1/2}(\Gamma), \quad \text{bounded}, \tag{3.16}$$

$$W : H^{1/2}(\Gamma) \rightarrow H^{-1/2}(\Gamma), \quad \text{bounded}. \tag{3.17}$$

How are the Sobolev spaces defined for a general Lipschitz curve Γ? Fortunately, there is one simple description: $H^{1/2}(\Gamma)$ is the 'trace' of $H^1(\Omega)$; and $H^{-1/2}(\Gamma)$ is then defined by duality with respect to the L_2 inner product. (An authority on the trace theorems is Lions and Magenes [24]).

Note, however, that we do **not** say as we did for the smooth curve, that V maps **onto** $H^{1/2}$.

Although most of our discussion in this chapter has been for two-dimensional regions, the properties (3.16) and (3.17) above hold just as well for three-dimensional or even n-dimensional regions; see Costabel [11] for details.

4 Strong ellipticity and the Galerkin method

4.1 Introduction

The Galerkin method, which we met briefly in Section 1, has a very flexible theoretical basis. The method is applicable to any **strongly elliptic** operator.

Before we move to define strong ellipticity, let's motivate this by noting that, except perhaps for smooth curves, we need more than we have now even to establish that a solution of $Lu = g$ exists. For smooth curves we know (at least for the single-layer operator) that the image of one Sobolev space under the boundary operator L is another Sobolev space; so existence is assured if u and f are chosen from the appropriate Sobolev spaces. But for a region with corners we know so far only that the single-layer operator maps $H^{-1/2}(\Gamma)$ **into** $H^{1/2}(\Gamma)$.

The first step is to adopt a more liberal notion of what we mean by 'solution': we will say that we have a **weak solution** of $Lu = g$ if

$$(Lu, \chi) = (g, \chi) \tag{4.1}$$

for all χ in an appropriate set of smooth functions. The inner product here is the L_2 inner product on Γ, or (for a smooth curve) on an appropriate parameterization of Γ.

4.2 Strong ellipticity

The strong ellipticity property is expressed in terms of the **bilinear form** (Lv, w), with v and w belonging to some Hilbert space H^α.

How should we choose α? We have to choose α so that the L_2 inner product of Lv and w exists. But we recall that H^α and $H^{-\alpha}$ are dual to each other with respect to the L_2 inner product. Thus it is natural to choose α so that

$$L : H^\alpha \to H^{-\alpha}, \ L \text{ bounded.} \tag{4.2}$$

Example 1 The single-layer operator, we recall, maps $H^{-1/2}$ into $H^{1/2}$, thus we may choose $\alpha = -1/2$ in this case. △

Example 2 The hypersingular operator maps $H^{1/2}$ into $H^{-1/2}$, thus in this case $\alpha = +1/2$ is appropriate. △

It follows from the assumption (4.2) and the duality of H^α and $H^{-\alpha}$ that

$$|(Lv, w)| \leq \|Lv\|_{-\alpha}\|w\|_\alpha \leq C\|v\|_\alpha\|w\|_\alpha, \tag{4.3}$$

thus the bilinear form (Lv, w) is **bounded** in the space H^α. But we need more than that. A very nice property, if it holds, is **coercivity**.

Definition 4.1 *The bilinear form* (Lv, w) *is* **coercive** *with respect to* H^α *if there exists* $\gamma > 0$ *such that*

$$\mathrm{Re}(Lv, v) \geq \gamma\|v\|_\alpha^2 \ \forall v \in H^\alpha. \tag{4.4}$$

Example 3 The single-layer operator A for the circle of radius a, defined by (3.5), is coercive with respect to $H^{-1/2}$ if and only if $a < 1$. This is because Proposition 3.1 gives

$$(Av, v) = -2\log a\ |\hat{v}(0)|^2 + \sum_{k\neq 0} \frac{1}{|k|}|\hat{v}(k)|^2. \tag{4.5}$$

For $a < 1$ we have $-\log a > 0$, and

$$\begin{aligned}(Av, v) &\geq \min(-2\log a, 1)[|\hat{v}(0)|^2 + \sum_{k\neq 0} \frac{1}{|k|}|\hat{v}(k)|^2] \\ &= \min(-2\log a, 1)\|v\|^2_{-1/2}.\end{aligned}$$

On the other hand for $a \geq 1$ the choice $v \equiv 1$ gives $(Av, v) = -2\log a \leq 0$, so coercivity certainly fails. △

If L is coercive then it must be one-to-one: for if there were to exist a nontrivial $v \in H^{-\alpha}$ such that $Lv = 0$ then (4.4) would be contradicted. But the really nice thing about the coervicity property comes from the Lax–Milgram theorem (Gilbarg and Trudinger [15] Theorem 5.8, Ciarlet [10]): adapted to our Sobolev space setting, it gives us

Theorem 4.2 *If $L : H^\alpha \to H^{-\alpha}$ is bounded, and if (Lv, w) is coercive with respect to H^α, then the equation*

$$(Lu, \chi) = (g, \chi),\ \ \chi \in H^\alpha, \tag{4.6}$$

has a unique solution $u \in H^\alpha$ for every choice of $g \in H^{-\alpha}$.

Coercivity, in other words, guarantees the existence and uniqueness of a weak solution to $Lu = g$.

Coercivity is an attractive property, but is very restrictive, as we see from the example of the single-layer operator for a circle: we know that the operator A is well behaved for $a > 1$, even if not coercive. The strong ellipticity property that now follows is more forgiving, and hence more useful.

Definition 4.3 *The bounded operator $L : H^\alpha \to H^{-\alpha}$ is* **strongly elliptic** *with respect to H^α if there exists a compact operator $T : H^\alpha \to H^{-\alpha}$ and a constant $\gamma > 0$ such that*

$$\text{Re}\ (Lv + Tv, v) \geq \gamma\|v\|^2_\alpha\ \forall v \in H^\alpha. \tag{4.7}$$

Example 4 The single-layer operator for a smooth curve, given by (3.4), is strongly elliptic with respect to $H^{-1/2}$, since from (3.10) and (4.5) we have

$$\begin{aligned}(Lv, v) &= (Av, v) + (Bv, v) \\ &= \|v\|^2_{-1/2} - (2\log a + 1)|\hat{v}(0)|^2 + (Bv, v)\end{aligned}$$

$$= \quad \|v\|^2_{-1/2} - (Tv, v),$$

where T, a compact operator in $H^{-1/2}$, is defined by

$$T = -B + (2\log a + 1)J,$$

with

$$Jv = \int_0^1 v(x)dx = \hat{v}(0). \qquad \triangle$$

A strongly elliptic operator need not be invertible (witness the operator A for a circle of radius 1). But if A **is** invertible then the main conclusion of Theorem 4.2 still holds, that (4.6) has a solution $u \in H^\alpha$ for every choice of $g \in H^{-\alpha}$. (This is proved, for example, by Hildebrandt and Wienholtz [17] Remark 3.)

How can we know, in more difficult situations, whether the operator is strongly elliptic? For smooth curves, and similarly for smooth three-dimensional surfaces the theory of pseudodifferential operators is a very powerful tool. (See Stephan and Wendland [36], Wendland [37],[38],[39] Hsiao and Wendland [20]). However, this approach breaks down as soon as there are corners. An alternative approach that works even for Lipschitz curves and surfaces (Costabel and Wendland [12], Costabel [11]) deduces the strong ellipticity from the ellipticity of the associated differential operator on Ω. In our Laplace equation context the original differential operator $-\Delta$ is certainly elliptic, so the strong ellipticity of the boundary integral operators is assured for all two- and three-dimensional Lipschitz domains.

4.3 The Galerkin method

Let us suppose that $L : H^\alpha \to H^{-\alpha}$ is strongly elliptic. The first step in the Galerkin method, we recall, is to choose a finite-dimensional space $S_h \in H^\alpha$ within which the approximate solution is to be sought. Then the Galerkin method for the equation $Lu = g$ is: find $u_h \in S_h$ such that

$$(Lu_h, \chi_h) = (g, \chi_h) \;\forall\; \chi_h \in S_h. \tag{4.8}$$

The theory of the Galerkin method is particularly simple if the bilinear form (Lv, w) is coercive.

Theorem 4.4 (**Céa's lemma**) *Suppose that $L : H^\alpha \to H^{-\alpha}$ is bounded, and that the bilinear form (Lv, w) is coercive with respect to H^α. Then the Galerkin equation (4.8) has a unique solution $u_h \in S_h$, which satisfies*

$$\|u_h - u\|_\alpha \leq C \inf_{\chi_h \in S_h} \|\chi_h - u\|_\alpha. \tag{4.9}$$

Proof. The existence and uniqueness of u_h follows from the Lax–Milgram theorem (see Theorem 4.2), now applied to S_h instead of H^α. Then from

(4.4), (4.6) and (4.8) we have, for any $\chi_h \in S_h$,

$$\begin{aligned}\gamma\|u_h - u\|_\alpha^2 &\leq |(L(u_h - u),\ u_h - u)| = |(L(u_h - u),\ \chi_h - u)| \\ &\leq C\|u_h - u\|_\alpha \|\chi_h - u\|_\alpha,\end{aligned}$$

and the result follows on cancelling $\|u_h - u\|_\alpha$. □

Céa's lemma tells us that the error in the Galerkin method in the H^α norm is **optimal** with respect to order. If L is merely strongly elliptic the argument is more delicate, because the compact operator T in the definition of strong ellipticity has to be dealt with. However, Hildebrandt and Wienholtz [17] show that the essential conclusion still holds.

Theorem 4.5 *Suppose that $L : H^\alpha \to H^{-\alpha}$ is strongly elliptic and one-to-one. Then there exists $h_0 > 0$ such that the Galerkin equation (4.8) has a unique solution $u_h \in S_h$ for all $h < h_0$, which satisfies*

$$\|u_h - u\|_\alpha \leq C \inf_{\chi_h \in S_h} \|\chi_h - u\|_\alpha. \tag{4.10}$$

We have seen that the Galerkin method has a strong theoretical foundation, and is widely applicable. The only problem with it, as we remarked in Section 1, is that its matrix is expensive to calculate.

Even faster convergence than given by (4.9) is often possible, provided we observe the error in Sobolev norms with $s < \alpha$. We will see this in relation to a simple problem in the next section.

5 Assorted methods for a planar problem

5.1 Introduction

In this last section we widen the class of methods studied, but specialize the problem: we consider only the single-layer equation for a plane region with boundary Γ. For most of the section we take Γ to be smooth, but at the end we turn briefly to problems with corners. We shall assume throughout that the transfinite diameter is different from 1, so that then the single-layer operator is one-to-one.

As in Section 3.2, we parameterize the curve by $t = \nu(x)$, with ν 1-periodic, so that the equation becomes

$$Lu = g, \tag{5.1}$$

where

$$Lu(x) = -2\int_0^1 \log|\nu(x) - \nu(y)|u(y)dy. \tag{5.2}$$

We recall that $L = A + B$, where A is the operator L for the special case of a circle of radius $a = e^{-1/2}$, given in Fourier series form by

$$Au(x) = \hat{u}(0) + \sum_{k\neq 0} \frac{1}{|k|}\hat{u}(k)e^{2\pi ikx}, \tag{5.3}$$

and B is the operator given by (3.11), which has a smooth kernel if Γ is smooth.

For each $n \geq 1$ we introduce a **uniform** mesh on $[0, 1]$, defined by

$$s_i = ih, \ i = 0, \ldots, n, \ \text{with } h = 1/n.$$

(This is not as limiting as it first seems: if Γ has corners we can concentrate points near the corners through our choice of the parameterization function ν.)

On this mesh let S_h denote the space of 1-periodic smoothest splines of order $r \geq 1$. That is to say, $v_h \in S_h$ if v_h is 1-periodic and has $r-2$ continuous derivatives, and if its restriction to $(jh, (j+1)h)$ is a polynomial of degree $\leq r-1$.

In this section we consider the Galerkin method, discussed already in a more general setting in Section 4.3; the collocation method, outlined in Section 1.4; and the qualocation method.

5.2 Galerkin method

The Galerkin method sets the standard. In Section 4 we showed that the Galerkin method gives optimal convergence in the $H^{-1/2}$ norm: from Theorem 4.5, the error in the Galerkin method satisfies

$$\|u_h - u\|_{-1/2} \leq C \inf_{\chi_h \in S_h} \|\chi_h - u\|_{-1/2}.$$

For our splines of order r it is well known (see, for example, Arnold and Wendland [3]) that for $s \leq r$

$$\inf_{\chi_h \in S_h} \|\chi_h - u\|_s \leq C h^{r-s} \|u\|_r, \tag{5.4}$$

assuming (as we shall always do, for simplicity) that $u \in H^r$. Thus the Galerkin method in the $H^{-1/2}$ norm gives

$$\|u_h - u\|_{-1/2} \leq C h^{r+1/2} \|u\|_r. \tag{5.5}$$

It is natural to ask also about the order of convergence in the L_2 norm. To deduce this from (5.5) the standard trick is to use the **inverse estimate** associated with a quasi-uniform spline space (again see Arnold and Wendland [3]): if $t < s < r - 1/2$ then there exists $C > 0$ such that

$$\|v\|_s \leq C h^{t-s} \|v\|_t \ \forall \ v \in S_h. \tag{5.6}$$

Note carefully that the exponent of h here is negative. Note too that this result holds only for v in the finite-dimensional space S_h, and does not hold if we replace S_h by a Sobolev space.

In the present context we may use the inverse estimate this way: for any $\chi_h \in S_h$ we have

$$\begin{aligned} \|u_h - u\|_0 &\leq \|u_h - \chi_h\|_0 + \|\chi_h - u\|_0 \\ &\leq Ch^{-1/2}\|u_h - \chi_h\|_{-1/2} + \|\chi_h - u\|_0 \\ &\leq Ch^{-1/2}\|u_h - u\|_{-1/2} + Ch^{-1/2}\|\chi_h - u\|_{-1/2} + \|\chi_h - u\|_0. \end{aligned}$$

It is possible, it turns out, to choose χ_h to be simultaneously optimal in both the $H^{-1/2}$ and L_2 norms (once more see Arnold and Wendland [3]), from which it follows, with (5.5), that

$$\|u_h - u\|_0 \leq Ch^r\|u\|_r. \tag{5.7}$$

The order of convergence of the Galerkin method can be even better than that in (5.5) if we observe the error in a more negative norm. If the curve is sufficiently smooth, the best result is obtained in the H^{-r-1} norm,

$$\|u_h - u\|_{-r-1} \leq Ch^{2r+1}\|u\|_r. \tag{5.8}$$

The proof, which makes use of duality and 'Nitsche's trick', may be instructive. Given $v \in H^{r+1}$, let z be the weak solution of $Lz = v$, and suppose the curve is regular enough to ensure $z \in H^r$. Then

$$(u_h - u, v) = (u_h - u, Lz) = (L(u_h - u),\ z) = (L(u_h - u),\ z - z_h),$$

where the last step holds for any $z_h \in S_h$ by the definition of the Galerkin approximation u_h given in eqn (4.8). Choosing $z_h \in S_h$ to be the Galerkin approximation to z, we obtain

$$\begin{aligned} |(u_h - u, v)| &\leq C\|u_h - u\|_{-1/2}\|z_h - z\|_{-1/2} \\ &\leq (Ch^{r+1/2}\|u\|_r)(Ch^{r+1/2}\|z\|_r) \\ &\leq Ch^{2r+1}\|u\|_r\|v\|_{r+1}, \end{aligned}$$

which we see is equivalent to (5.8) by the dual definition of $\|u_h - u\|_{-r-1}$, i.e. by using (3.9) with $s = 0$. Apparently Hsiao and Wendland [20] were the first to introduce Nitsche's trick in the boundary integral equation context.

The result we just proved is important enough to deserve some emphasis: we see here an extraordinary property of the Galerkin solution, namely that if we multiply $u_h - u$ by a smooth function and then integrate, then the error appears much, much smaller than the error as measured in the L_2 or $H^{-1/2}$ norms. This is the foundation for 'superconvergence' effects. For example, if we use a piecewise-linear space, i.e. $r = 2$, then the error in the potential φ (given by (1.3)) at a point in Ω has the excellent order $O(h^5)$, if u and Γ are sufficiently smooth. This is because

$$\varphi(t) = -2\int_0^1 \log|t - \nu(y)|u(y)dy,\ t \in \Omega, \tag{5.9}$$

which has the form (u, v) with v smooth.

5.3 The collocation method

In the collocation method we again look for an approximate solution $u_h \in S_h$, but this time we fix u_h by **collocating** the equation at selected points $t_0, \ldots, t_{n-1}$:

$$Lu_h(t_k) = g(t_k), \;\; k = 0, \ldots, n-1. \tag{5.10}$$

How should the points be chosen? For our present case of smoothest splines on a uniform mesh the dimension of the space is n, so we need one point for each subinterval. It is by now well known that the simple rule to follow is this: if r is even (e.g. for the piecewise-linear and cubic spline cases) the collocation points should be the **breakpoints** $t_k = k/n$, whereas if r is odd (e.g. for the piecewise-constant case) the collocation points should be the **midpoints** of the subintervals. (This was shown for r even by Arnold and Wendland, [4] and for r odd by Saranen and Wendland [31].) We shall see that there is more freedom than this might suggest, but whatever happens the advice should not be reversed: for example, midpoint collocation in the piecewise-linear case will certainly lead to disaster! We can easily see that this is so, at least if n is even, by the following single argument. Let v_h be a piecewise-linear function with the values ± 1 at alternating vertices. Then v_h is odd about each midpoint, so Lv_h must vanish at these points. Thus we have constructed a nontrivial piecewise-linear function that satisfies the homogeneous collocation equation exactly if we collocate at the midpoints, making the collocation matrix in that situation **not** one-to-one.

Allowing greater freedom, for fixed ϵ satisfying $0 \leq \epsilon < 1$, let us take the collocation points to be

$$t_k = \frac{k+\epsilon}{n}, \;\; k = 0, \; \ldots, n-1. \tag{5.11}$$

(The collocation method which uses this choice of collocation points is called **ϵ-collocation.**) It is by now well known (Schmidt [32]) that ϵ-collocation is stable for the present problem, provided that $\epsilon \neq 1/2$ if r is even, and $\epsilon \neq 0$ if r is odd. With these exceptions, the convergence has the optimal order $O(h^r)$ in the L_2 norm, and we even have

$$\|u_h - u\|_{-1} \leq Ch^{r+1}\|u\|_r. \tag{5.12}$$

The method of proof will be indicated, in the context of a more general method, in Section 5.4.

Note that in this negative norm sense the collocation method is not as good as the Galerkin method: for example, in the piecewise-linear case the collocation method gives only $O(h^3)$ for the order of convergence of $(u_h - u, v)$ with v smooth, whereas the Galerkin method gives $O(h^5)$. That gives us an incentive to do better, bringing us to the **qualocation method.**

Before leaving the collocation method, however, there are two more things we ought to note. The first is that if r is odd (as in the piecewise-constant case) Saranen [29] has shown that the choice $\epsilon = 1/2$ (i.e. midpoint collocation) makes the order of convergence better by 1 than the general ϵ-collocation result above: now the best result is

$$\|u_h - u\|_{-2} \leq Ch^{r+2}\|u\|_{r+1}. \tag{5.13}$$

This can be viewed as a special case of the forthcoming qualocation method. The second is that for r odd and $\epsilon = 0$ (i.e. breakpoint collocation) there exists a quite different theory, due to Arnold and Wendland [3], that dispenses altogether with the uniform mesh requirement, and still obtains the order of convergence (5.12). That very nice theory relies on repeated integration by parts to convert the collocation method to a Galerkin method with a non-standard inner product. Its limitation is that for this to work the collocation points **must** be the breakpoints of the splines. (The argument makes essential use of the fact that the rth derivative of a smoothest spline of order r consists of δ functions at the breakpoints.) We have no time to go any further in the latter direction in these lectures.

5.4 The qualocation method

The qualocation (='quadrature-modified collocation') method is a quadrature based generalization of the collocation method, that attempts to capture some of that higher-order negative-norms convergence we found with the Galerkin method, while being like the collocation method in that it is relatively easy to implement.

In this section we establish the method for smooth curves, before at the finish describing a very recent application to curves with corners.

Like the **Petrov–Galerkin method**, the qualocation method has associated with it not only the trial space S_h, within which the approximate solution u_h is sought, but also a **test space** S'_h. We take the test space to be the space of smoothest splines of order $r' \geq 1$. Often, but not always, we take $r' = r$, implying $S'_h = S_h$. The Petrov–Galerkin method associated with this pair of spaces is: find $u_h \in S_h$ such that

$$(Lu_h, \chi_h) = (g, \chi_h) \;\forall\; \chi_h \in S'_h. \tag{5.14}$$

Clearly it reduces to the Galerkin method (4.8) if $S'_h = S_h$.

In the qualocation method the inner product $(v, w) = \int v\overline{w}$ is replaced by a discrete equivalent $(v, w)_h$: we seek $u_h \in S_h$ such that

$$(Lu_h, \chi_h)_h = (g, \chi_h)_h \;\forall\; \chi_h \in S'_h, \tag{5.15}$$

where

$$(v, w)_h = Q_h(v\overline{w}), \tag{5.16}$$

and

$$Q_h(f) = h\sum_{i=0}^{n-1}\sum_{j=1}^{J} w_j f(s_i + h\xi_j), \tag{5.17}$$

with

$$0 \le \xi_1 < \xi_2 < \ldots < \xi_J < 1, \quad \sum_{j=1}^{J} w_j = 1, \; w_j > 0. \tag{5.18}$$

Note that Q_h is just the composition of the simple J-point rule

$$Q(f) = \sum_{j=1}^{J} w_j f(\xi_j) \approx \int_0^1 f(x)dx. \tag{5.19}$$

The richness of the qualocation method lies, as we shall see, in our freedom in the choice of this J-point rule: in general we shall **not** choose any of the familiar rules (Gauss, Simpson, etc.), but instead design new rules with magical properties.

Once we choose bases $\{\varphi_1, \ldots, \varphi_n\}$ of S_h and $\{\varphi'_1, \ldots, \varphi'_n\}$ of S'_h, we may write

$$u_h = \sum_{j=1}^{n} a_j \varphi_j,$$

and solve in practice

$$\sum_{j=1}^{n}(L\varphi_j, \varphi'_k)_h a_j = (g, \varphi'_k)_h, \quad k = 1, \ldots, n. \tag{5.20}$$

This is certainly easier to implement than the Galerkin method, since only the inner integral (i.e. the one within $L\varphi_j$) needs to be computed 'exactly', while the outer integral (i.e. the inner product) is now discrete.

What results can we obtain for the qualocation method? And related to this, how should we choose the quadrature rule Q? Interesting rules start with $J = 2$, but let's first look at the case $J = 1$. The general 1-point rule satisfying (5.18) is just

$$Qf = f(\epsilon),$$

where ϵ is a fixed number satisfying $0 \le \epsilon < 1$. **In this case the qualocation method is equivalent to the ϵ-collocation method.** This is because (5.15) is in this case equivalent to

$$h\sum_{i=0}^{n-1} Lu_h(s_i + \frac{\epsilon}{n})\varphi'_k(s_i + \frac{\epsilon}{n}) = h\sum_{i=0}^{n-1} g(s_i + \frac{\epsilon}{n})\varphi'_k(s_i + \frac{\epsilon}{n}), \; k = 0, \ldots, n-1,$$

and hence to

$$Lu_h(s_i + \frac{\epsilon}{n}) = g(s_i + \frac{\epsilon}{n}),\ i = 0, \ldots, n-1,$$

provided the $n \times n$ matrix $\{\varphi'_k(s_i + \frac{\epsilon}{n})\}$ is non-singular.

Guided by the known stability results for the Petrov–Galerkin method, we will require our qualocation methods to satisfy:

Assumption A *We assume r and r' are either both even or both odd.*

Moreover, from our knowledge of the ϵ-collocation method (and in particular the fact that it breaks down for midpoint collocation if r is even, and for breakpoint collocation if r is odd), we make a second assumption that excludes the known 'bad' cases:

Assumption B *If the rule Q is a 1-point rule (i.e. $J = 1$), then the single point ξ_1 is not equal to $1/2$ if r is even, and not equal to 0 if r is odd.*

It is easiest to state the known theoretical results for the qualocation method on smooth curves in two parts. First, we establish that the method is stable, and has at least the basic order of convergence stated in (5.9) for the ϵ-convergence method. And only then (in Section 5.6) do we explore the possibilities for higher-order convergence through special choices of the rule Q. These results are extracted from more general results obtained by Chandler and Sloan [9]. (That paper considers also other operators, not just the single-layer operator (5.2).)

Theorem 5.1 *Assume that A and B hold. Assume also that Γ is smooth and has transfinite diameter different from 1, and that the parameterization function ν in (5.2) is smooth and satisfies $|\nu'(x)| \neq 0$. Then (5.15) has a unique solution $u_h \in S_h$ for all h sufficiently small, which satisfies*

$$\|u_h - u\|_{-1} \leq Ch^{r+1}\|u\|_r. \tag{5.21}$$

5.5 A sketch of the proof of Theorem 5.1

Here is the briefest possible sketch of the proof of Theorem 5.1, taken from Chandler and Sloan [9]. The proof extends that of Arnold and Wendland [4] for the collocation method. The ideas in it trace back to De Hoog [19].

1. The basic strategy is to prove the result for $L = A$, and then to extend the result to the general case by taking $B = L - A$ as a compact perturbation.
2. To prove the result for $L = A$, express u and u_h as Fourier series, and use the Fourier representation (5.3) for A.
3. Because u_h is a 1-periodic smoothest spline of order r on a mesh with uniform spacing $h = 1/n$, its Fourier coefficients satisfy the recurrence relation (Arnold [2])

$$\mu^r \hat{u}_h(\mu) = m^r \hat{u}_h(m) \text{ if } m \equiv \mu (\mathrm{mod}\, n).$$

4. For the theoretical analysis we use a special basis for the spline space S_h: the basis functions (n of them) are defined, for integer values of μ satisfying $-n/2 < \mu \leq n/2$, by

$$\psi_\mu(x) = \begin{cases} 1 & \text{if } \mu = 0, \\ \sum_{m\equiv\mu}(\frac{\mu}{m})^r e^{2\pi imx} & \text{if } -\frac{n}{2} < \mu \leq \frac{n}{2}, \mu \neq 0. \end{cases}$$

It is useful to think of ψ_μ as a spline approximation to $\varphi_\mu := e^{2\pi i\mu x}$; note that $\psi_\mu = \varphi_\mu +$ the higher Fourier modes needed to satisfy the recurrence relation. We can write, since $\psi_\mu(\nu) = \delta_{\mu\nu}$ for $-n/2 < \mu, \nu \leq n/2$,

$$u_h = \sum_{-n/2<\mu\leq n/2} \hat{u}_h(\mu)\psi_\mu.$$

5. The spline basis function ψ_μ has the property

$$\psi_\mu(x+h) = e^{2\pi i\mu h}\psi_\mu(x)\ \forall\, x \in \mathbb{R}.$$

For that reason the basis $\{\psi_\mu\}$ diagonalizes the qualocation matrix: for $-n/2 < \mu \leq n/2$ we easily find

$$(A\psi_\nu, \psi'_\mu)_h = \begin{cases} 1 & \text{if } \mu = \nu = 0, \\ \frac{1}{|\mu|}D(\mu h) & \text{if } \mu = \nu \neq 0, \\ 0 & \text{otherwise.} \end{cases}$$

That means that we can study the qualocation problem with $L = A$ for each Fourier mode separately. Thus we see that: **the qualocation method is stable if and only if the function D is bounded away from zero.**

6. Explicitly, D is given by

$$D(y) = \sum_{j=1}^{J} w_j[1 + \Omega(\xi_j, y)][1 + \overline{\Delta'(\xi_j, y)}], \tag{5.22}$$

with

$$\Delta'(\xi, y) = y^{r'} \sum_{\ell\neq 0} \frac{1}{(\ell + y)^{r'}} e^{2\pi i\ell\xi}, \tag{5.23}$$

and with

$$\Omega(\xi, y) = |y|^{r+1} \sum_{\ell\neq 0} \frac{1}{|\ell + y|^{r+1}} e^{2\pi i\ell\xi} \tag{5.24}$$

if r is even, or

$$\Omega(\xi, y) = \text{ sign } y|y|^{r+1} \sum_{\ell\neq 0} \frac{\text{sign } \ell}{|\ell + y|^{r+1}} e^{2\pi i\ell\xi} \tag{5.25}$$

if r is odd.

7. It can be shown, by appeal to known properties of trigonometric sums (see Brown, Chandler, Sloan and Wilson [8]) that $D(\xi, y)$ given by (5.22) is continuous in y. Moreover, the real part of each term of (5.22) is non negative,

$$\mathrm{Re}\,[1 + \Omega(\xi, y)][1 + \overline{\Delta'(\xi, y)}] \geq 0,$$

with equality holding if and only if $y = \frac{1}{2}$ and $\xi = \frac{1}{2}$ for the case r and r' even, and if and only if $y = \frac{1}{2}$ and $\xi = 0$ for the case r and r' odd. So $\mathrm{Re}D(y) \geq 0$, with equality if and only if $J = 1$ together with $\xi_1 = \frac{1}{2}$ for r and r' even or $\xi_1 = 0$ for r and r' odd. Since these cases are excluded by assumption B, the stability part of the theorem is established.

8. Now suppose that the method is stable, i.e. that D is bounded away from zero. Let's solve the qualocation equation

$$(Au_h, \psi'_\mu)_h = (g, \psi'_\mu)_h \text{ for } -\frac{n}{2} < \mu \leq \frac{n}{2}.$$

For $\mu \neq 0$ this is equivalent to

$$\frac{1}{|\mu|} D(\mu h) \hat{u}_h(\mu) = (Au, \psi'_\mu)_h,$$

which has the obvious solution

$$\begin{aligned} \hat{u}_h(\mu) &= \frac{|\mu|}{D(\mu h)} (Au, \psi'_\mu)_h, \\ &= \frac{1}{D(\mu h)} \sum_j w_j [1 + \overline{\Delta'(\xi_j, \mu h)}] \hat{u}(\mu) + R_n(\mu), \end{aligned}$$

where the last term involves only high Fourier coefficients $\hat{u}(m)$ of u, specifically, those for which $|m| \geq n/2$.

9. For the moment we neglect those terms involving high Fourier coefficients of u (their influence can be made arbitrarily small by insisting that u belong to a high enough Sobolev space), and write (for $-n/2 < \mu \leq n/2,\ \mu \neq 0$)

$$\hat{u}_h(\mu) \approx \frac{1}{D(\mu h)} \sum_j w_j [1 + \overline{\Delta'(\xi_j, \mu h)}] \hat{u}(\mu),$$

giving, to the same approximation,

$$\hat{u}_h(\mu) - \hat{u}(\mu) \approx -\frac{E(\mu h)}{D(\mu h)}, \tag{5.26}$$

where

$$E(y) = \sum_j w_j \Omega(\xi_j, y)[1 + \overline{\Delta'(\xi_j, y)}]. \tag{5.27}$$

10. The $\mu = 0$ component is easier: in the same sense we find

$$\hat{u}_h(0) - \hat{u}(0) \approx 0. \tag{5.28}$$

11. So what rate of convergence can we expect? For fixed $\mu \neq 0$, we see from (5.26) that everything depends on the behavior of the function E as $y \to 0$: indeed, if

$$E(y) = O(y^\rho), \tag{5.29}$$

then $\hat{u}_h(\mu) - \hat{u}(\mu) = O(h^\rho)$. But we see from the definition of Ω given in (5.24) or (5.25) that ρ is at least $r+1$. This is where the $O(h^{r+1})$ order of convergence in Theorem 5.1 comes from.

12. Now write down the appropriate Sobolev norm of $u_h - u$, using the explicit expression for $\hat{u}_h(\mu)$, and the recurrence relation above for the higher Fourier coefficients. From here on the proof of Theorem 5.1 is just a matter of working out detailed bounds on the Sobolev norm of the error. We omit the details. □

5.6 Higher-order convergence

The essence of the qualocation method is to design special quadrature rules so as to obtain a higher order of convergence (of course in a more negative Sobolev norm) than shown in Theorem 5.1. Fortunately, all that hard work we have just done tells us how to do this: **all we need to do is to choose the rule Q so as to eliminate leading terms in the expansion of the function**

$$E(y) = \sum_j w_j \Omega(\xi_j, y)[1 + \overline{\Delta'(\xi_j, y)}]. \tag{5.30}$$

We say that the qualocation rule is of **order** ρ, or of **additional order** b, if $\rho = r + 1 + b$ and

$$E(y) = O(y^\rho) = O(y^{r+1+b}) \text{ as } y \to 0. \tag{5.31}$$

Every stable qualocation method has $\rho \geq r + 1$, or $b \geq 0$. The interesting rules have order higher than $r + 1$. This is the main result:

Theorem 5.2 *If a qualocation method for $Lu = g$, where L is the single-layer operator (5.2), satisfies the condition of Theorem 5.1 and is of order $r + 1 + b$ with $b \geq 0$, then (in addition to (5.21)) we have*

$$\|u_h - u\|_{-1-b} \leq Ch^{r+1+b}\|u\|_{r+b}, \tag{5.32}$$

provided $u \in H^{r+b}$.

For the proof, see Chandler and Sloan [9]. Notice that if $b \geq r$ then the

Table 1 Qualocation rules for the single-layer equation.

r	r'	ξ_1	w_1	ξ_2	w_2	b	ρ
1	1	1/2	1	–	–	1	3
1	1	0	3/7	1/2	4/7	3	5
1	3	0.2308 2965 03	1/2	$1-\xi_1$	1/2	3	5
2	2	0	3/7	1/2	4/7	2	5
2	2	0.2308 2965 03	1/2	$1-\xi_1$	1/2	2	5

order of convergence in the H^{-1-b} norm is at least as good as that attained by the Galerkin method (see (5.8)). However, there is a price: the smoothness condition on u in (5.32) is more demanding than the corresponding condition in (5.8). It is believed that this additional smoothness requirement cannot be weakened. (It comes, essentially, from the quadrature approximation on the right hand side of the qualocation approximation.)

Devising higher-order qualocation methods, then, boils down to expanding the functions $\Omega(\xi, y)$ and $\Delta'(\xi, y)$ in powers of y, and then arranging matters so that the leading terms of the expansion of $E(y)$ disappear.

Table 1 lists some of the qualocation rules found by Chandler and Sloan [9]. All are 2-point rules, except for the first, which is the first of the 1-point (midpoint collocation) results obtained by Saranen [29].

The second and third entries in the table for the piecewise-constant case have some curiosity value, in that they achieve a rate of convergence even higher than the Galerkin method: for this piecewise-constant case the best result for the Galerkin method is $O(h^3)$, compared to $O(h^5)$ here. In reality, however, the fourth and fifth rules – which like the Galerkin method both achieve $O(h^5)$ for the piecewise-linear case – are probably the ones of greater interest, in that in this case the higher order of convergence seems to extend to more complicated situations, such as that which follows in the next (and last) section.

To show that the methods do indeed behave as predicted here, we consider the following simple example from Chandler and Sloan [9]. Here Γ is a circle of radius $\frac{1}{2}$ centred at (0,0), and $f(t) = f(t_1, t_2) = t_1$ in (1.2). The computed quality is the potential φ_h computed at a point $\tau \in \Omega$ by the

Table 2 Errors and computed convergence rates

k	collocation		qualocation $\frac{3}{7}, \frac{4}{7}$ rule		qualocation symmetric 2 pt		discrete Galerkin	
4	6.42(−3)		2.20(−3)		1.90(−3)		2.42(−3)	
8	5.06(−4)	3.66	1.31(−5)	7.39	4.54(−6)	8.71	5.55(−5)	5.44
16	5.98(−5)	3.08	3.71(−7)	5.14	1.12(−7)	5.33	6.05(−6)	3.19
32	7.37(−6)	3.02	1.12(−8)	5.04	3.27(−9)	5.10	7.33(−7)	3.04
64	9.18(−7)	3.00	3.50(−10)	5.00	1.00(−10)	5.02	9.09(−8)	3.01
88	3.53(−7)	3.00	7.11(−11)	5.00	2.03(−11)	5.00	3.49(−8)	3.00

analogue of (5.9),

$$\varphi_h(\tau) = -2 \int_0^1 \log |t - \nu(y)| u_h(y) dy, \;\; \tau \in \Omega.$$

The exact value is

$$\varphi(\tau) = \varphi(\tau_1, \tau_2) = \tau_1.$$

Numerical results with $r = r' = 2$ (i.e. piecewise-linear test and trial spaces) are shown in Table 2. The number k of quadrature points is listed in the first column. Beside the errors are the apparent rates of convergence, each calculated from the ratios of two successive errors. The collocation method uses the recommended breakpoints as collocation points. The two qualocation methods are the fourth and fifth entries in Table 1. They clearly display the predicted $O(h^5)$ convergence, whereas the collocation method, with equal clarity, has $O(h^3)$ convergence.

It is worth saying that there have been many extensions of the qualocation idea, including a number of fully discrete variants. Some references are Sloan and Burn [34], Saranen and Sloan [30], Jeon [22], McLean and Sloan [27], and McLean [26].

5.7 The qualocation method for a region with corners

In very recent work (Elschner, Prössdorf and Sloan [13]), the qualocation method has been applied successfully to a polygon. The key to this application is to choose the parameterization function ν in (5.2) in such a way as to concentrate the breakpoints into the corners. To be specific, let the vertices of the polygon be $x_1, \ldots, x_m$, $x_{m+1} = x_1$ and let $S_1, \ldots, S_m$, where

$$0 < S_1 < S_2 < \ldots < S_m < 1, \; S_{m+1} = S_1 + 1,$$

be the pre-images of $x_1, \ldots, x_m$ under the mapping ν, i.e. $x_j = \nu(S_j)$ for $j = 1, \ldots, m$. The degree of concentration of points near the points is determined by a "grading parameter" $q \geq 1$; we shall indicate below how to determine a suitable value for q. On the jth edge of the polygon the parameterization is defined by

$$\nu(s) = x_j + \frac{(s - S_j)^q}{(s - S_j)^q + (S_{j+1} - s)^q}(x_{j+1} - x_j) \text{ for } S_j \leq s \leq S_{j+1}, \quad (5.33)$$

together with the usual perodicity assumption $\nu(s) = \nu(s + 1)$.

As before, the mesh is to be chosen uniformly spaced with respect to s. With that in mind, the formula (5.33) shows clearly the mesh grading role of the parameter q. If $q = 1$ the points of the mesh are uniformly spaced on an edge, while if $q > 1$ they are concentrated towards the corners, with the degree of concentration increasing as q increases.

For technical reasons the spacing of $S_1, \ldots, S_m$ should also depend on the value of q : the requirement is that

$$S_{j+1} - S_j = \frac{|x_{j+1} - x_j|^{1/q}}{\sum_{k=1}^m |x_{k+1} - x_k|^{1/q}}, \quad j = 1, \ldots, m. \qquad (5.34)$$

Note that if $q = 1$ this choice makes the mesh spacing the same on each edge of the polygon.

The recommended size of q depends, not unreasonably, on the extent to which Γ departs from a smooth curve. Let the interior angle at the corner x_j be denoted by $(1 - \chi_j)\pi$, with $0 < |\chi_j| < 1$. The exterior angle at x_j is therefore $(1 + \chi_j)\pi$. It is well known that the solution of the single-layer equation (1.5) is generally singular at a corner, with the nature of the singularity being determined by whichever is the larger of the internal and external angles.

Before stating the theorem, there is one further thing we need to know. It is that, as with many other boundary element problems involving corners, we are only able to prove stability of the approximation method if the approximation is allowed to be modified near the vertices: the modification consists of replacing the qualocation approximation u_h by zero in some fixed number i^* of intervals adjacent to each vertex. In practice it seems we usually need no modification, i.e. we can set $i^* = 0$, and the following theorem is only applicable if this is the case; but we certainly have no proof that it must be so.

Theorem 5.3 *Assume that the qualocation method employs smoothest* *splines of order r and is of additional order $b \geq 0$. The method is applied to the single-layer equation (1.5) for a polygon, parameterized as in (5.33), with f smooth. Let*

$$\ell := \min(2r, r+b),$$

and assume that

$$q > (\ell + \frac{1}{2}) \max_j (1 + |\chi_j|). \tag{5.35}$$

If the qualocation method is stable without modification, then for v sufficiently smooth,

$$(u_h - u, v) = O(h^{\ell+1}).$$

The theorem asserts that the qualocation method can achieve the same $O(h^{2r+1})$ superconvergent order as the Galerkin method, even for a region with corners, provided the mesh grading parameter is sufficiently large, and the 'additional order of convergence' of the qualocation method is at least r.

To take an example, for the piecewise-linear case the last two qualocation methods in Table 1 should achieve $O(h^5)$ convergence for $(u_h - u, v)$ if v is smooth, provided q is large enough. But the theorem requires q to be very large: for example, if Γ is a rectangle then in the piecewise-linear case with $b = 2$, making $\ell = 4$, we need $q > 27/4$. Probably this is larger than is really needed.

Obviously there is no time here to give the proof. All that can be said is that the analysis rests on writing L, as before, as

$$L = A + B = A(I + M),$$

where $M = A^{-1}B$. But now M is not a compact operator: rather, it is in essence a **Mellin convolution**. The analysis of such operators and their approximations plays a significant role in the modern study of boundary element methods; but that discussion will have to wait for another day.

Bibliography

1. Agranovich, M.S. (1979). Spectral properties of elliptic pseudodifferential operators on a closed curve. *Functional Anal. Appl.*, **13**, 279–281.
2. Arnold, D.N. (1983). A spline-trigonometric Galerkin method and an exponentially convergent boundary integral method. *Math. Comput.*, **41**, 383–397.
3. Arnold, D.N. and Wendland, W.L. (1983). On the asymptotic convergence of collocation methods. *Math. Comput.*, **41**, 349–381.
4. Arnold, D.N. and Wendland, W.L. (1985). The convergence of spline collocation for strongly elliptic equations on curves. *Numer. Math.*, **47**, 317–341.

5. Atkinson, K.E.(1976). *A Survey of Numerical Methods for the Solution of Fredholm Integral Equations of the Second Kind.* SIAM, Philadelphia.
6. Atkinson, K.E. and De Hoog, F.R. (1984). The numerical solution of Laplace's equation on a wedge. *IMA J. Numer. Anal.*, **4**, 19–41.
7. Brebbia, C.A., Telles, J.C.F. and Wrobel, L.C. (1984). *Boundary Element Techniques.* Springer-Verlag, Berlin.
8. Brown, G., Chandler, G.A., Sloan, I.H. and Wilson, D. (1991). Properties of certain trigonometric series arising in numerical analysis. *J. Math. Anal. Appl.*, **2**, 371–380.
9. Chandler, G.A. and Sloan, I.H. (1990). Spline qualocation methods for boundary integral equations. *Numer. Math.*, **58**, 537–567.
10. Ciarlet, P.G. (1978). *The Finite Element Method for Elliptic Problems.* North Holland, Amsterdam.
11. Costabel, M. (1988). Boundary integral operators on Lipschitz domains: elementary results. *SIAM J. Math. Anal.*, **19**, 613–626.
12. Costabel, M. and Wendland, W.L. (1986). Strong ellipticity of boundary integral operators. *J. Reine Angew. Math.*, **372**, 34–63.
13. Elschner, J., Prössdorf, S. and Sloan, I.H. (1994). The qualocation method for Symm's integral equation on a polygon. In preparation.
14. Gaier, D. (1976). Integralgleichungen erster Art und konforme Abbildungen. *Math. Zeitschr.* **147**, 113–129.
15. Gilbarg, D. and Trudinger, N.S. (1983). *Elliptic Partial Differential Equations of Second Order.* 2nd edition, Springer-Verlag, Berlin.
16. Hackbusch, W. (1989). *Integralgleichungen: Theorie und Numerik.* Teubner Studienbücher, Stuttgart.
17. Hildebrandt, S. and Wienholtz, E. (1964). Constructive proofs of representation theorems in separable Hilbert space. *Comm. Pure Appl. Math.* **17**, 369–373.
18. Hille, E. (1962). *Analytic Function Theory.* Vol. II. Ginn and Company, Boston.
19. De Hoog, F.R. (1974). *Product integration techniques for the numerical solution of integral equations.* Ph.D. Thesis, Australian National University, Canberra.
20. Hsiao, G.C. and Wendland, W.L. (1981). The Aubin–Nitsche lemma for integral equations. *J. Int. Eqns*, **3**, 299–315.
21. Jaswon, M.A. and Symm, G. (1977). *Integral Equation Methods in Potential Theory and Elastostatics.* Academic Press, London.
22. Jeon, Y. (1994). An unconventional method for logarithmic-kernel integral equations on closed curves. Submitted for publication.
23. Kress, R. (1989). *Linear Integral Equations.* Springer-Verlag, Berlin.

24. Lions, J.L. and Magenes, E. (1974). *Non-homogeneous Boundary Value Problems and Applications.* Vol.I. Springer-Verlag, Berlin.

25. McLean, W. (1991). Local and global descriptions of periodic pseudodifferential operators. *Math. Nachr.*, **150**, 151–161.

26. McLean, W. (1994). *Discrete Petrov–Galerkin methods for periodic pseudodifferential equations.* Preprint.

27. McLean, W. and Sloan, I.H. (1994). A fully-discrete and symmetric boundary integral method. To appear in *IMA J. Numerical Anal.*

28. Mikhlin, S.G. (1970). *Mathematical Physics, an Advanced Course.* North-Holland, Amsterdam.

29. Saranen, J. (1988). The convergence of even degree spline collocation solution for potential problems in smooth domains of the plane. *Numer. Math.* **53**, 499–512.

30. Saranen, J. and Sloan, I.H. (1992). Quadrature methods for logarithmic-kernel integral equations on closed curves. *IMA J. Numerical Anal.*, **12**, 167–187.

31. Saranen, J. and Wendland, W.L. (1985). On the asymptotic convergence of collocation methods with spline functions of even degree. *Math. Comput.*, **45**, 91–108.

32. Schmidt, G. (1985). On spline collocation methods for boundary integral equations in the plane. *Math. Meth. Appl. Sci.*, **7**, 74–89.

33. Sloan, I.H. (1992). Error analysis of boundary integral methods. *Acta Numerica* **1**, 287–339.

34. Sloan, I.H. and Burn, B.J. (1992). An unconventional quadrature method for, logarithmic – kernel integral equations and closed curves. *J. Int. Eqns. and Applics*, **4**, 117–151.

35. Sloan, I.H. and Spence, A. (1988). The Galerkin method for integral equations of the first kind with logarithmic kernel: theory. *IMA J. Numer. Anal.* **8**, 105–122.

36. Stephan, E.P. and Wendland, W.L. (1976) Remarks on Galerkin and least squares methods with finite elements for general elliptic problems. In *Partial Differential Equations (Lecture Notes in Mathematics 564)*, Springer-Verlag, Berlin, and in *Manuscripta Geodaetica* **1**, 93–123.

37. Wendland, W.L. (1983). Boundary element methods and their asymptotic convergence. In *Theoretical Acoustics and Numerical Techniques, CISM Courses* **277**, edited by P. Filippi. Springer-Verlag, Berlin.

38. Wendland, W.L. (1987). Strongly elliptic boundary integral equations. In *The State of the Art in Numerical Analysis*, edited by A. Iserles and M.J.D. Powell. Clarendon Press, Oxford.

39. Wendland, W.L. (1990). Boundary element methods for elliptic problems. In *Mathematical Theory of Finite and Boundary Element*

Methods, edited by A.H. Schatz, V. Thomée and W.L. Wendland. Birkhäuser, Basel.

Perturbation Theory for Infinite Dimensional Dynamical Systems

Andrew Stuart

Program in Scientific Computing and Computational Mathematics,
Division of Applied Mechanics,
Durand 252,
Stanford University,
California 94305-4040,
USA [1]

Abstract

When considering the effect of perturbations on initial value problems over long time intervals it is not possible, in general, to uniformly approximate individual trajectories. This is because well-posed initial value problems allow exponential divergence of trajectories and this fact is reflected in the error bound relating trajectories of the perturbed and unperturbed problems. In order to interpret data obtained from numerical simulations over long time intervals, and from other forms of perturbations, it is hence often necessary to ask different questions concerning the behavior as the approximation is refined. One possibility, which we concentrate on in this review, is to study the effect of perturbation on sets which are invariant under the evolution equation. Such sets include equilibria, periodic solutions, stable and unstable manifolds, phase portraits, inertial manifolds and attractors; they are crucial to the understanding of long-time dynamics.

An abstract semilinear evolution equation in a Hilbert space X is considered, yielding a semigroup $S(t)$ acting on a subspace V of X. A general class of perturbed semigroups $S^h(t)$ are considered which are C^1 close to $S(t)$ uniformly on bounded subsets of V and time intervals $[t_1, t_2]$ with $0 < t_1 < t_2 < \infty$. A variety of perturbed problems are shown to satisfy these approximation properties. Examples include a Galerkin method based on the eigenfunctions of the linear part of the abstract sectorial evolution equation, a backward Euler approximation of the same equation and a singular perturbation of the Cahn–Hilliard equation arising from the phase-field model of phase transitions. The invariant sets of $S(t)$ and $S^h(t)$ are compared and convergence properties established.

[1]This work was supported by the Office of Naval Research, contract number N00014-92-J-1876 and by the National Science Foundation, contract number DMS-9201727. I am greatly indebted to Don Jones, Stig Larsson and Tony Shardlow for help in the proof-reading of, and suggestions of improvements to, the material in this article.

1 Introduction

The accumulation of errors caused by perturbation of a well-posed evolution equation is governed by the properties of the evolution equation itself. In regions where nearby solutions of the equation tend to diverge exponentially in time, the error introduced by perturbation will tend to grow exponentially in time. Consider, for example, the perturbation caused by numerical approximation of an evolution equation. Because of the exponential divergence of trajectories, typical a priori error estimates between true and numerical solutions contain error constants which grow as the exponential of the time interval under consideration. For many nonlinear problems this exponential divergence may not persist for all time and it is possible to obtain greatly reduced error constants by means of a posteriori error analysis (see Eriksson *et al.* [33]) whereby the error constant is estimated as the computation proceeds rather than being majorized by an exponential in time. It is nonetheless a fact that, for most problems and for both a priori and a posteriori error estimates, *it is not possible to approximate a true trajectory with a numerical trajectory starting from the same point, uniformly on an infinite time interval.* The only exception is the case where trajectories are contracting which, for autonomous problems, implies that the solution being approximated is asymptotic to a stable steady state. It is thus natural to ask how numerical data from long time simulations involving more complicated behavior than convergence to a steady state should be interpreted. Such long time simulations are of great importance in science and engineering arising, for example, in the simulation of turbulent fluid flow, in phase transition calculations and in interplanetary interactions. In all these examples it is commonplace to integrate past the time at which standard a priori or a posteriori estimates guarantee closeness of trajectories. Although we have talked primarily about the effect of perturbations introduced through numerical approximation, similar considerations are relevant, for example, in studying the effect of singular perturbations of the coefficients in evolution equations.

The properties of an evolution equation on a Hilbert space V are captured by a semigroup $S(t)$ mapping an initial data point $v \in V$ to the solution of the equation at time t, namely $S(t)v$. The semigroup can also be extended to act on subsets of V in the natural way. Of crucial importance to understanding the long time behavior of an evolution equation are the sets which remain invariant under $S(t)$. Such sets may include simple objects such as equilibria or periodic solutions and also more complicated sets such as those arising in chaotic systems; also of importance are invariant manifolds such as the stable and unstable manifolds of equilibrium points and the inertial manifold, a set on which the dynamics of certain partial differential equations is governed by ordinary differential equations. Here we study the effect of small perturbations to $S(t)$, such as those arising from numerical approximation or from singular perturbations to partial differential equations, on these invariant sets. As we shall see the analysis is considerably streamlined by assuming that the perturbation is C^1 close to $S(t)$, in a sense to be made precise. Such C^1 closeness results can be proved for many numerical methods and for singular perturbations of partial differential equations. They are are established in this article in a variety of contexts.

In Section 2 we introduce the class of evolution equations which we use to illustrate the analysis. These equations are formulated as ordinary differential equations in a Hilbert space and take the form of a linear differential operator (whose solutions decay in time) with a lower-order nonlinear perturbation. Examples which are considered include an abstract sectorial evolution equation, reaction-diffusion equations, some important fourth-order pattern formation problems and the Navier–Stokes equation. Existence, uniqueness and regularity results are described for the abstract sectorial evolution equation, using a variation of constants formula for the problem. Appropriate references are given to the analogous results for other classes of problems. Section 2 concludes with a summary of the important definitions and results from the theory of dynamical systems that will be useful to us here.

In Section 3 we state our basic assumptions concerning the perturbed semigroup $S^h(t)$: it is C^1 close to the semigroup $S(t)$ generated by the underlying equation itself, uniformly in bounded subsets of V and on compact time-intervals disjoint from the origin. We then describe various examples where this is satisfied. We initially consider the abstract sectorial evolution equation and introduce a spectral method based on the eigenfunctions of the linear differential operator. We prove standard error estimates for trajectories together with results concerning the effect of numerical approximation of the Fréchet derivative of the semigroup with respect to initial data. Throughout, the variation of constants formula is used to unify the presentation. Similar results are described for a singular perturbation of the Cahn–Hilliard equation, known as the viscous Cahn–Hilliard equation.

In Section 4 we consider the neighborhood of an equilibrium point. We first show that, provided $\bar{u}$ is a hyperbolic equilibrium point of $S(t)$, then there is a nearby equilibrium point $\bar{u}^h$ of the approximate semigroup $S^h(t)$. We then show that, if the solution being approximated approaches an exponentially attracting equilibrium point as $t \to \infty$, then it *can be uniformly approximated over infinite time intervals.* This situation is the only case where uniform-in-time approximation of individual trajectories is possible for autonomous problems.

Following this we construct a local phase portrait comprising the union of solutions of $u(t) = S(t)u(0)$ in the neighborhood of an equilibrium point $\bar{u}$ of saddle type. This local phase portrait is shown to persist under the basic approximation assumptions. Since these solutions are defined over arbitrarily long time intervals, a standard error estimate cannot be used directly in proving this result; instead we show that each solution of $S(t)$ may be approximated by a solution of $S^h(t)$ *starting from a different initial condition.* Understanding the effect of numerical perturbation on phase portraits is of use for the interpretation of data found when directly computing phase portraits close to equilibria. It is also useful as a building block for the proof of other results such as the uniform in time, piecewise continuous, error estimates for gradient systems which are described in Section 8.

In Section 5 we look at unstable manifolds of a hyperbolic equilibrium point $\bar{u}$. Locally in the neighborhood of $\bar{u}$ these may be constructed as part of the phase portrait discussed in Section 4; however, we describe an alternative presentation based on a technique known as a graph transform. Again, persistence of a nearby local unstable manifold under the assumptions on the perturbation is proved. Every point on the true local unstable manifold is close to a point on the approximate local unstable manifold and vice versa. Having studied local unstable manifolds the results are extended to the global unstable manifold by a compactness argument. There are two main reasons for studying unstable manifolds: they are of interest in their own right and, furthermore, are an important building block in the study of attractors — something pursued further in Section 7.

In Section 6 we study inertial manifolds. These objects are of theoretical importance in the study of a partial differential equation since they show that certain infinite-dimensional systems are governed by finite-dimensional systems for large time. Indeed, on its inertial manifold, the equation reduces to an ordinary differential equation known as an inertial form. It is thus of some interest to understand the effect of perturbation on the inertial manifold. The techniques we use are very similar to the graph transform techniques used in Section 5 for the construction of unstable manifolds. Under appropriate conditions we prove that $S(t)$ and $S^h(t)$ have inertial manifolds $\mathcal{M}$ and $\mathcal{M}^h$ and that every point of $\mathcal{M}$ is close to a point in $\mathcal{M}^h$ and vice versa.

In Section 7 we study attractors — these are sets which attract an open neighborhood of themselves under the evolution equation. Simple examples are stable equilibrium points and periodic solutions; however, the reason for abstracting to this general object is that in very complex problems (for example those exhibiting chaos) the most basic description of what we observe after a long time is contained in the notion of attractor. We assume that $S(t)$ and $S^h(t)$ have attractors $\mathcal{A}$ and $\mathcal{A}^h$ and then show that every point on $\mathcal{A}^h$ is close to a point on $\mathcal{A}$ for h small. Unfortunately it is not possible to prove the converse in general; simple counter-examples exist to illustrate why. Thus parts of an attractor may disappear under small perturbations. However, we know by considering the simple example of an exponentially attracting stable equilibrium point that, for some problems, the whole attractor *will* be well approximated. The remainder of the chapter is devoted to studying two classes of attractor for which the whole object perturbs by a small amount under the basic assumptions about the perturbing semigroup. The first class is the class of attractors $\mathcal{A}$ and $\mathcal{A}^h$ which are uniformly exponentially attracting; the second class is the class of attractors made up of the union of unstable manifolds and in this connection we exploit the theory of Section 5.

In Section 8 we turn our attention to error estimates for gradient systems. These systems are characterized by a globally decreasing Lyapunov functional which drives solutions towards an equilibrium point for large time. Using this property and the results of Sections 3 and 4, it is possible to approximate uniformly almost all trajectories of gradient systems by solutions of the perturbed system. However, in so doing, the error constant behaves badly with respect to initial data. In particular, the error constant is not uniformly bounded as the initial data is varied in a bounded set. For this reason the value of the error estimate is severely diminished. To obviate this problem we introduce a weaker concept of piecewise approximation of trajectories: we seek approximation of true solutions on $(0, \infty)$ by a piecewise continuous trajectory of the approximate problem with a finite number of discontinuities. In this context we derive uniform in time error estimates with error constants bounded uniformly in a bounded set of initial data.

Note that the analysis outlined so far is primarily concerned with proving persistence of invariant sets, and their convergence, under a basic assumption about the perturbed semigroup which includes, but is not restricted to, a variety of numerical approximation techniques. As such no distinction is made between the practical value of different numerical methods for dynamical systems other than in their rate of convergence. In order to assess the practical value of various schemes for dynamical systems it is of value to generalize the idea of *practical numerical stability* to classes of nonlinear problems arising in practice. In Section 9 we briefly turn our attention to this question.

Certain themes persist throughout the article and important results, frequently used and not necessarily proved, have been summarized in the Appendices. The first recurrent theme is the use of a *variation of constants* approach to the evolution equations: we use it in Sections 2 and 3 to prove existence, regularity and error estimates for the abstract sectorial evolution equation and its spectral approximation. We also use it in Sections 4 and 5 to enable us to replace the differential equation in the neighborhood of an equilibrium point by a mapping which retains the "linear plus small nonlinear" structure inherent in such problems; similar considerations apply for the construction of inertial manifolds in Section 6. The use of mappings is motivated by a desire to prove results which apply to time-discrete as well as time-continuous perturbations. The essential material for understanding the variation of constants approach to evolution equations in a Hilbert space is given in Appendix A. The second recurrent theme is the use of *uniform contraction principles* to construct objects of interest (such as equilibria, phase portraits and invariant manifolds) and at the same time to incorporate the effect of perturbation in the analysis. Contraction principles of various types used throughout the article are described in Appendix B; the *Taylor expansion in a Hilbert space* is also given in that Appendix and used several times in the main body of the text. The third recurrent theme is *attractive invariant manifolds.* Simple examples are unstable manifolds and inertial manifolds and both of these are constructed by use of an abstract theorem concerning attractive invariant manifolds given in Appendix C.

Some of the material described here is also presented in the context of time-discretization of ordinary differential equations in $\mathbb{R}^m$ in Stuart [88]. However there are certain technically challenging difficulties which arise in the consideration of partial differential equations which mean that different or modified techniques need to be applied.

2 Evolution equations in a Hilbert space

2.1 Introduction

This section commences with an introduction to the basic properties of sectorial operators in a Hilbert space. The important material, together with some examples, is sketched. More details and references are given in Appendix A. This material is followed by a brief summary of some of the important notation used throughout the article. In Section 2.4 the nonlinear evolution equation which we study is introduced and the assumptions about its solution given. These basically amount to asking that the solution depend on time and initial data in a C^1 sense.

We then describe a variety of situations in which the assumptions hold. We concentrate on an abstract evolution equation in a Hilbert space, governed by a sectorial evolution equation. The primary result of importance here is Theorem 2.6 which describes the existence and regularity for the equation. The **Important remark** following the theorem introduces the use of a variation of constants approach to the nonlinear partial differential equation, a theme which recurs throughout. We also mention briefly the Navier–Stokes equation, the Cahn–Hilliard equation and ordinary differential equations, all of which fit into our framework.

The section concludes with a summary of some basic results in the theory of dynamical systems.

2.2 Sectorial operators

We consider the background theory of certain abstract evolution equations in a Hilbert space. For this we need the idea of a sectorial operator. For details concerning sectorial operators together with definitions of their fractional powers and exponentials see Appendix A. We will refer freely to results in this Appendix throughout the article, and the reader is encouraged to study it at this point.

The easiest context in which to understand sectorial operators is the theory of self-adjoint operators. Consider a separable Hilbert space X with inner product $\langle \bullet, \bullet \rangle$ and norm defined by $| \bullet |^2 = \langle \bullet, \bullet \rangle$. We assume that A is a closed, densely defined, self-adjoint, positive operator with compact inverse; we denote the associated eigenfunctions by $\{\varphi_i\}$ and the positive real eigenvalues by $\{\lambda_i\}$ with the ordering chosen so that

$$0 < \lambda_1 \leq \lambda_2 \leq \dots \quad . \tag{2.1}$$

In the case primarily considered here, where X is infinite-dimensional, we have $\lambda_n \to \infty$ as $n \to \infty$.

The properties of A ensure that it is a sectorial operator so that we may define fractional powers of A, A^α, and also the resulting Hilbert spaces $X^\alpha = D(A^\alpha)$ with norms $| \bullet |_\alpha = |A^\alpha \bullet |$. For each $\alpha > \beta \geq 0$ it follows that the inclusion $X^\alpha \subset X^\beta$ is compact. Furthermore, we may define the operator e^{-At} for $t > 0$. Of particular interest to us is the space X^β where $\beta \in [0, 1)$ appears in Assumption 2.3 and in (2.11) below. We employ the notation

$$V \equiv X^\beta \quad \text{and} \quad \| \bullet \| \equiv | \bullet |_\beta .$$

Initial data for our basic equation (2.9) will be specified in V.

Example 2.1 Let $X = L_2(\Omega)$, $A = -\frac{d^2}{dx^2}$, $D(A) = H^2(\Omega) \cap H_0^1(\Omega)$ and $\Omega = (0, 1)$. Thus

$$(v, w) = \int_0^1 v(x) w(x) dx,$$

and $| \bullet |$ denotes the L_2 norm. Also $\varphi_j = \sqrt{2} \sin(j\pi x)$ and $\lambda_j = j^2 \pi^2$ for $j = 1, 2, \dots, \infty$. It is well known that any function $v \in X$ can be written as

$$v = \sum_{j=1}^{\infty} v_j \varphi_j, \quad v_j = (v, \varphi_j) \tag{2.2}$$

where the $v_j \in \mathbb{R}, j = 1, 2, \dots, \infty$ satisfy

$$\sum_{j=1}^{\infty} v_j^2 < \infty.$$

The fractional powers of A are generated by defining

$$A^\alpha v = \sum_{j=1}^{\infty} \lambda_j^\alpha v_j \varphi_j .$$

Thus

$$|v|_\alpha^2 = \sum_{j=1}^{\infty} \lambda_j^{2\alpha} v_j^2.$$

Roughly speaking $D(A^\alpha)$ may be thought of as the set of functions v such that

$$\sum_{j=1}^{\infty} \lambda_j^{2\alpha} v_j^2 < \infty,$$

with v_j given by (2.2). For example, if $\alpha = \frac{1}{2}$, we find that $X^{\frac{1}{2}} \equiv H_0^1(\Omega)$. This may be seen by noting that, formally,

$$|v|_{\frac{1}{2}}^2 = \sum_{j=1}^{\infty} j^2\pi^2 v_j^2 = \langle \sum_{j=1}^{\infty} j\pi v_j \sqrt{2}\cos(j\pi x), \sum_{j=1}^{\infty} j\pi v_j \sqrt{2}\cos(j\pi x)\rangle$$

$$= |\sum_{j=1}^{\infty} j\pi v_j \sqrt{2}\cos(j\pi x)|^2 = |v_x|^2.$$

Recalling that $|v_x|$ is a norm on $H_0^1(\Omega)$ the result follows.

For future use we also note here the following basic norm inequalities:

$$\begin{aligned} |v| &\leq \tfrac{1}{\pi}|v|_{\frac{1}{2}} \quad \forall v \in H_0^1(\Omega), \\ \|v\|_\infty := \sup_{x\in(0,1)} |v(x)| &\leq |v|_{\frac{1}{2}} \quad \forall v \in H_0^1(\Omega). \end{aligned} \tag{2.3}$$

Note that the heat equation

$$u_t + Au = 0, \; u(0) = u_0 := \sum_{j=1}^{\infty} u_j \varphi_j$$

has solution

$$u(t) = e^{-At}u_0 = \sum_{j=1}^{\infty} e^{-\lambda_j t} u_j \varphi_j;$$

this gives the appropriate way to think about exponentials of sectorial operators in the self-adjoint case.

It cannot be over-emphasized that the preceding example, and the straightforward calculations given therein, are very specific to the self-adjoint problem. To understand fully the concepts of fractional powers and exponentials of sectorial operators, the reader should pursue in detail the references given in Appendix A. In any case it will be beneficial to the reader to study Appendix A before proceeding.

2.3 Notation

In subsequent sections we will need an appropriate definition of the distance between sets in V. Thus we introduce the following notation:

$$\left\{\begin{aligned} \operatorname{dist}(u, A) &= \inf_{v \in A} \|u - v\|, \\ \operatorname{dist}(B, A) &= \sup_{u \in B} \operatorname{dist}(u, A), \\ \mathcal{N}(A, \varepsilon) &= \{u \in V : \operatorname{dist}(u, A) < \varepsilon\}, \\ \partial\mathcal{N}(A, \varepsilon) &= \{u \in V : \operatorname{dist}(u, A) = \varepsilon\}. \end{aligned}\right\} \tag{2.4}$$

It follows from (2.4) that, if $\operatorname{dist}(B, A) < \varepsilon$, then $\bar{B} \subseteq \mathcal{N}(\bar{A}, \varepsilon)$. Hence

$$\operatorname{dist}(B, A) = 0 \Leftrightarrow \bar{B} \subseteq \bar{A}.$$

Thus "dist" only defines a semidistance — the asymmetric *Hausdorff semi-distance*. The *Hausdorff distance* between two sets A and B is defined by

$$d_{\mathrm{H}}(B, A) = \max\{\operatorname{dist}(A, B), \operatorname{dist}(B, A)\}. \tag{2.5}$$

Thus

$$d_{\mathrm{H}}(A, B) = 0 \Leftrightarrow \bar{B} \equiv \bar{A}.$$

We also employ the notation

$$\begin{aligned} B(v, \varepsilon) &:= \{u \in V : \|u - v\| < \varepsilon\}, \\ \partial B(v, \varepsilon) &:= \{u \in V : \|u - v\| = \varepsilon\}. \end{aligned} \tag{2.6}$$

Thus $B(v, \varepsilon) = \mathcal{N}(v, \varepsilon)$ and $\partial B(v, \varepsilon) = \partial\mathcal{N}(v, \varepsilon)$. With this notation we make a definition.

Definition 2.2 *A family of sets $\mathcal{A}^h$, $h \in [0, h_c]$, is termed upper semi-continuous at $h = 0$ if dist($\mathcal{A}^h, \mathcal{A}_0$) $\to 0$ as $h \to 0$. The family is termed lower semi-continuous at $h = 0$ if dist($\mathcal{A}_0, \mathcal{A}^h$) $\to 0$ as $h \to 0$. The family is continuous at $h = 0$ if it is both upper semi-continuous and lower semi-continuous at $h = 0$ so that $d_{\mathrm{H}}(\mathcal{A}_0, \mathcal{A}^h) \to 0$ as $h \to 0$.*

In the remainder of the article we use the standard induced operator norm on linear mappings $L \in \mathcal{L}(V, V)$ namely

$$\|L\| := \sup_{\|\xi\|=1} \|L\xi\|. \tag{2.7}$$

We define the operator norm of $L \in \mathcal{L}(X, X)$ by

$$|L| := \sup_{|\xi|=1} |L\xi|. \tag{2.8}$$

2.4 The evolution equation

Throughout this article we study the behavior of the abstract evolution equation

$$\frac{du}{dt} + Au = F(u), t > 0, \quad u(0) = u_0. \tag{2.9}$$

The operator A is assumed to satisfy the properties given in Section 2.1: it is a linear, closed, densely defined, self-adjoint, positive operator with compact inverse. The function F is a continuous nonlinear operator from X^{ζ} to X, for some $\zeta \in [0,1)$, whose properties are specified below. Its Fréchet derivative is denoted by $dF(\bullet)$. The precise definition of a solution to eqn (2.9) may be found in Lemma 10.8. Throughout this article we make the following assumptions concerning solutions of this equation.

Assumption 2.3 *There is $\beta \in [0,1)$ such that, for every $u_0 \in V \equiv X^{\beta}$, eqn (2.9) has a unique solution $u(t;u_0) \in V$ defined for $t \in [0,\infty)$. We denote by $S(t) : V \mapsto V$ the operator defined by*

$$S(t)v := u(t;v).$$

There exists $\eta > \beta$ such that, for every $t > 0$ and every $v \in V$, $S(t)v \in X^{\eta}$. Furthermore, we assume that $S(\bullet)\bullet \in C^1(\mathbb{R}^+ \times V, V)$.

Briefly, this assumption implies existence and uniqueness of a solution, continuous dependence upon the data, and a compactness property for the solution operator. *This assumption is made throughout without being explicitly stated in the results in the remainder of the article.*

We shall use the notation $dS(v;t)$ to denote the Fréchet derivative of $S(t)u$ with respect to u, evaluated at a point v. Note that, since the solution operator $S(t)\bullet$ is C^1 we deduce the following result.

Lemma 2.4 *For any $R > 0$ there is a positive and increasing function $C(t)$, defined on $\mathbb{R}^+$, such that*

$$\|S(t)u - S(t)v\| \leq C(t)\|u - v\| \quad \forall u, v \in B(0,R). \tag{2.10}$$

In Sections 2.5, 2.6, 2.7 and 2.8 we consider various examples illustrating that Assumption 2.3 is satisfied for a wide variety of problems.

2.5 Sectorial evolution equations

As our first example of a class of problems satisfying Assumptions 2.3 we consider (2.9) as an ordinary differential equation in the Hilbert space X. We make the following assumption concerning the function $F(u)$ appearing in (2.9). There exists a constant $K > 0$ and $\beta \in [0,1)$ such that, for all

$u, v, w, \in V$:

$$\begin{aligned} F &\in C^1(V, X); \\ |F(u)| &\leq K; \\ |F(u) - F(v)| &\leq K|u - v|_\beta; \\ |dF(u)v| &\leq K|v|_\beta; \\ |dF(u)w - dF(v)w| &\leq K|u - v|_\beta |w|_\beta. \end{aligned} \tag{2.11}$$

(Recall that $V \equiv X^\beta$ and that $\|\bullet\| \equiv |\bullet|_\beta$.) Under this assumption, and the standing assumptions on the operator A, we will prove that Assumption 2.3 holds. We start with an example illustrating (2.11).

Example 2.5 As a first example consider the scalar reaction-diffusion equation

$$\begin{gathered} u_t = u_{xx} + f(u), \quad (x,t) \in (0,1) \times (0,\infty), \\ u(0,t) = u(1,t) = 0, \quad t > 0, \\ u(x,0) = u_0(x). \end{gathered} \tag{2.12}$$

We establish that (2.11) holds for this problem, under the assumption that $f \in C^\infty(\mathbb{R}, \mathbb{R})$ and that there exists a constant $C > 0$ such that

$$|f(u)| + |f'(u)| + |f''(u)| \leq C \quad \forall u \in \mathbb{R}. \tag{2.13}$$

Here A and $D(A)$ may be defined as in Example 2.1; recall that $D(A^{\frac{1}{2}}) = H_0^1(\Omega)$. We define $F(u)$ by

$$F(u)(x) := f(u(x))$$

and then

$$(dF(u)v)(x) = f'(u(x))v(x).$$

That F is C^2 follows from the smoothness of f. First note that

$$|F(u)|^2 = \int_0^1 f(u(x))^2 dx \leq C^2.$$

Secondly note that, if $u, v \in H_0^1(\Omega)$, then u and v are bounded pointwise by (2.3); thus $f'(\xi(x))$ is bounded pointwise if $\xi(x) = su(x) + (1-s)v(x)$ for some $s = s(x) \in [0,1]$. By the mean value theorem and (2.3) it follows that

$$|F(u) - F(v)|^2 = \int_0^1 [f(u(x)) - f(v(x))]^2 dx$$

$$\begin{aligned} &\leq \int_0^1 [f'(\xi(x))]^2(u(x)-v(x))^2dx \\ &\leq C^2|u-v|^2 \leq C^2|u-v|_{\frac{1}{2}}^2/\pi^2. \end{aligned}$$

Thirdly we have that, by (2.3),

$$|dF(u)v|^2 = \int_0^1 [f'(u(x))]^2 v(x)^2 dx \leq C^2|v|^2 \leq C^2|v|_{\frac{1}{2}}^2/\pi^2.$$

Finally note that, by arguments similar to those used in bounding $|F(u) - F(v)|$, we have

$$\begin{aligned} |dF(u)w - dF(v)w|^2 &= \int_0^1 [f'(u(x)) - f'(v(x))]^2 w^2(x)dx \\ &\leq \int_0^1 f''(\eta(x))^2(u(x)-v(x))^2 w(x)^2 dx \\ &\leq C^2\|u-v\|_\infty^2 |w|^2 \\ &\leq C^2|u-v|_{\frac{1}{2}}^2 |w|_{\frac{1}{2}}^2/\pi^2. \end{aligned}$$

Thus (2.11) holds with $\beta = 1/2$. The same result may be established in dimensions 2 and 3 by a slightly more subtle analysis.

As a second example consider the equation

$$\begin{gathered} u_t = -u_{xxxx} + f(u), \quad (x,t) \in (0,1)\times(0,\infty), \\ u(0,t) = u(1,t) = u_{xx}(0,t) = u_{xx}(1,t) = 0, \quad t > 0, \\ u(x,0) = u_0(x). \end{gathered}$$

This is sometimes known as the Swift–Hohenberg equation. Let A_0 denote the operator denoted by A in Example 2.1 and now set $A = A_0^2$; the eigenvalues of A are $j^4\pi^4$ and now $|v_x|^2 = |v|_{\frac{1}{4}}^2$. We make the same assumptions about f as for the reaction-diffusion equation and define F in the same way. Consequently, by following the analysis of the previous example, we have that (2.11) holds with $\beta = 1/4$.

We now prove the following result.

Theorem 2.6 *Let (2.11) hold. Then, for every $u_0 \in V$, there exists a unique solution $u(t)$ of (2.9). Furthermore, there exist constants $C_1 = C_1(T,R)$ and $C_2 = C_2(\alpha,T,R) > 0$, such that, for all $t \in (0,T)$ and $u_0 \in B(0,R)$:*

$$\begin{aligned} |u(t)|_\alpha &\leq \frac{C_1}{t^{\alpha-\beta}}, \quad \forall\alpha\in[\beta,1]; \\ \left|\frac{du}{dt}(t)\right|_\alpha &\leq \frac{C_2}{t^{\alpha-\beta+1}}, \quad \forall\alpha\in[\beta-1,1). \end{aligned} \tag{2.14}$$

Finally the solution operator $S(t)u_0 := u(t)$ *satisfies Assumptions* 2.3.

Proof The existence of a solution globally defined in $t > 0$ follows from Theorem 10.9, using (2.11) to establish (10.6) and (10.9). The bound on du/dt follows directly from the estimate in Theorem 10.9. Now let $\alpha \in [\beta, 1]$. Note that from (2.9),

$$|u|_\alpha = |A^{\alpha-1}Au| \leq \left|\frac{du}{dt}\right|_{\alpha-1} + |F(u)|_{\alpha-1}.$$

By Lemma 10.6, $A^{\alpha-1}$ is bounded on X for $\alpha \leq 1$ and so, by (2.11) and the bound on du/dt we obtain

$$|u|_\alpha \leq \frac{K_1}{t^{\alpha-\beta}} + K_2|F(u)| \leq \frac{K_1}{t^{\alpha-\beta}} + K_2 K \leq \frac{C(T)}{t^{\alpha-\beta}},$$

the required result on $|u|_\alpha$ for $\alpha \in [\beta, 1]$. It follows that $S(\bullet)\bullet$ is in $C^1(\mathbb{R}^+ \times V, V)$ by Theorem 10.10, since $F \in C^1(V, X)$ by (2.11). This completes the proof. □

Important remark Since the proof of the bounds in Theorem 2.6 follow directly from the stated bound on du/dt in Theorem 10.9 and are hence somewhat obscure to a reader unfamiliar with Henry [54] or Pazy [81], we sketch a direct proof of the bounds on u which is valid for $\alpha \in [\beta, 1)$. This introduces an approach to the analysis of (2.9) and its applications, based on the variation of constants formula, that will be useful to us in a variety of contexts. By formally using e^{-At} as an integrating factor and noting that it is the solution operator for the linear problem (10.7), it follows that $u(t)$ satisfies

$$u(t) = e^{-At}u_0 + \int_0^t e^{-A(t-s)}F(u(s))ds. \tag{2.15}$$

Applying A^α to (2.15) we obtain

$$A^\alpha u(t) = A^\alpha e^{-At}u_0 + \int_0^t A^\alpha e^{-A(t-s)}F(u(s))ds.$$

Noting that fractional powers of A commute with e^{-At} we obtain

$$|u(t)|_\alpha \leq |A^{\alpha-\beta}e^{-At}A^\beta u_0| + \int_0^t |A^\alpha e^{-A(t-s)}|\,|F(u(s))|ds.$$

Applying Lemma 10.6, (2.11) and (10.4), we obtain

$$|u(t)|_\alpha \leq \frac{C_1}{t^{\alpha-\beta}}\|u_0\| + \int_0^1 \frac{C_2}{(t-s)^\alpha}ds.$$

The second term is integrable and proportional to $(1-\alpha)^{-1}$ if $\alpha < 1$ and the bound on $|u(t)|_\alpha$ follows for $\alpha \in [\beta, 1)$.

It is also of interest to prove the Lipschitz property (2.10) without appealing to the abstract Theorem 10.10. In fact, under (2.11), we have a global Lipschitz property. From (2.15), Lemma 10.6 and (2.11) we see that

$$\|S(t)u - S(t)v\| \le \|u - v\| + \int_0^t \frac{CK}{(t-s)^\beta}\|S(s)u - S(s)v\|ds.$$

Applying the Gronwall Lemma 10.11 we obtain

$$\exists C = C(t) > 0 : \|S(t)u - S(t)v\| \le C\|u - v\| \quad \forall u, v \in V. \tag{2.16}$$

Without loss of generality we may assume that $C(t)$ is bounded as $t \to 0$ and that it is monotonically increasing in t.

Note that by Theorem 2.6 the solution of (2.9) subject to (2.11) is a C^1 function of time t and the initial data u_0. This allows us to consider the concept of the derivative of the solution with respect to initial data. We can now try and find the equation which the derivative satisfies. This can be calculated by linearizing eqn (2.9) about a given solution with initial data u_0. Setting

$$w(t) = u(t) + \eta v(t),$$

where $u(t)$ and $w(t)$ both satisfy (2.9) and $\eta \in \mathbb{R}$, we find that

$$\frac{du}{dt} + Au = F(u), \quad u(0) = u_0,$$
$$\frac{dw}{dt} + Aw = F(w), \quad w(0) = w_0.$$

Hence $v(t)$ satisfies

$$\eta[\frac{dv}{dt} + Av] = F(u + \eta v) - F(u), \quad v(0) = \xi.$$

Linearizing so that $F(u+\eta v) \approx F(u) + \eta dF(u)v + \mathcal{O}(\eta^2)$ and letting $\eta \to 0$ gives the equation

$$\frac{dv}{dt} + Av = dF(u)v, \;\; t > 0, \quad v(0) = \xi, \tag{2.17}$$

where $u = u(t)$ solves (2.9). Thus $v(t)$ is the function found by calculating the derivative of the solution $u(t)$ with respect to initial data u_0 and applying it to ξ. For the following theorem, concerning the existence and regularity of the solution to (2.17), we need the concept of mild solution given in Lemma 10.7.

Theorem 2.7 *Let (2.11) hold. For every $\xi \in V$ there exists a mild solution $v(t)$ of (2.17). Furthermore, there exists $C = C(T) > 0$ such that, for* *all $t \in (0,T)$, $u_0 \in V$:*

$$\|v(t)\| \leq C\|\xi\|;$$
$$|v(t)|_\alpha \leq \frac{C\|\xi\|}{(1-\alpha)t^{\alpha-\beta}}, \quad \forall \alpha \in (\beta, 1). \tag{2.18}$$

Finally, if $v(t) = dS(u_0,t)\xi$ then $v(t)$ satisfies (2.17).

Proof We employ Theorem 10.10 to deduce that (2.17) has a mild solution and that it represents the action of the operator $dS(u_0,t)$ on ξ. Thus the solution satisfies an integral equation obtained by the variation of constants formula

$$v(t) = e^{-At}\xi + \int_0^t e^{-A(t-s)} dF(u(s))v(s)ds. \tag{2.19}$$

Hence

$$A^\gamma v(t) = A^{\gamma-\beta} e^{-At} A^\beta \xi + \int_0^t A^\gamma e^{-A(t-s)} dF(u(s))v(s)ds.$$

Taking norms and applying Lemma 10.6 we obtain

$$|v(t)|_\gamma \leq \frac{C_1}{t^{\gamma-\beta}}\|\xi\| + \int_0^t \frac{C_2}{(t-s)^\gamma}\|v(s)\|ds. \tag{2.20}$$

Letting $\gamma = \beta$ and applying the Gronwall Lemma 10.11 we obtain the first result. Having obtained this we return to (2.20) with $\gamma = \alpha \in (\beta, 1)$ and integrate to obtain the second result:

$$\begin{aligned} |v(t)|_\alpha &\leq \frac{C_1\|\xi\|}{t^{\alpha-\beta}} + \int_0^t \frac{C_2 C\|\xi\|}{(t-s)^\alpha} ds \\ &= \frac{C_1\|\xi\|}{t^{\alpha-\beta}} + \frac{t^{1-\alpha} C_2 C\|\xi\|}{(1-\alpha)} \\ &\leq \frac{C_1 + C_2 C T^{1-\beta}}{(1-\alpha)t^{\alpha-\beta}}\|\xi\|. \end{aligned}$$

This completes the proof. □

Important remark At the expense of some unwieldy calculation it is possible to obtain greater regularity on $v(t)$ than that given here. However, in the context in which we are interested when considering the spectral approximation of (2.9), for example, the fact that the numerical solution converges in a C^1 sense is all that we need and the rate of convergence is not required. The rate of convergence of the spectral method is governed by the regularity of the solution being approximated and hence, for our purposes, it is not necessary to derive greater regularity on $v(t)$.

2.6 The Navier–Stokes equations

These equations may be written as

$$u_t + u \cdot \nabla u = \tfrac{1}{R}\Delta u - \nabla p + h(x), \quad x \in \Omega,$$

$$\nabla \cdot u = 0, \quad x \in \Omega,$$

$$u = 0, \quad x \in \partial\Omega.$$

Let X be the Hilbert space $\mathcal{H}$ comprising divergence free velocity fields contained in the space $L_2(\Omega)^2$ — see, for example, Temam [91]. This can be used to formulate an abstract evolution equation of the form (2.9) where A denotes the Stokes operator and $F(u)$ comprises the effect of convection and the body forcing h. Then in two dimensions it may be shown that

$$|F(u) - F(v)| \leq K(R)|u - v|_\beta, \quad \forall u, v \in B(0, R)$$

provided that $\beta > \frac{1}{2}$. See, for example, Hale [50]. Thus (2.11) is not satisfied directly. However, under suitable conditions on Ω, a priori bounds on the solution enable the construction of a modified F which satisfies (2.11) and yields a problem which is equivalent to (2.21) for sufficiently large time. See Temam [92].

An alternative approach to the existence theory is to use the Faedo-Galerkin approach as described in Temam [91] and in Constantin and Foias [19]. This approach can be used to deduce that Assumptions 2.3 are satisfied with $X = V = \mathcal{H}$.

2.7 The Cahn–Hilliard equation

The Cahn–Hilliard equation (see Elliott [28] and the references therein) subject to Dirichlet boundary conditions may be written in the form

$$u_t = \Delta w, \quad x \in \Omega,$$

$$0 = \Delta u + f(u) + w, \quad x \in \Omega,$$

$$u = w = 0, \quad x \in \partial\Omega,$$

$$u = u_0(x), \quad t = 0.$$

Here a typical choice for $f(u)$ arising in applications is $f(u) = \beta(u - u^3)$ for some $\beta > 0$. Equation (2.21) may be formulated as an abstract evolution equation in the form of (2.9) by setting

$$u_t + A_0^2 u = A_0 F(u), \quad u(0) = u_0, \tag{2.21}$$

where A_0 is the operator denoted by A in Example 2.1, and letting

$$F(u)(x) := f(u(x)).$$

The existence theory described in Section 2.5 and in Appendix A does not apply directly to this equation. However a similar theory can be developed and the Assumptions 2.3 shown to hold — see Elliott and Larsson [30] and Elliott and Stuart [32] for details.

2.8 Ordinary differential equations

As a final example of a system satisfying Assumptions 2.3, consider the system of ordinary differential equations of the form

$$u_t = f(u), \quad u(0) = u_0. \tag{2.22}$$

Assume that $f \in C^1(\mathbb{R}^m, \mathbb{R}^m)$ and that structure is imposed on the function $f(u)$ ensuring global existence in time for all initial data in $\mathbb{R}^m$. Then solutions of eqn (2.22) are readily seen to generate a semigroup $S(t) : \mathbb{R}^m \mapsto \mathbb{R}^m$ such that Assumptions 2.3 hold with $V = \mathbb{R}^m$.

2.9 Semigroups

By Assumption 2.3 we know that a unique solution of (2.9) exists for all $t \geq 0$ and any $u_0 \in V$ and we have encountered several examples of classes of equations which do satisfy these assumptions. We have defined a semigroup $S(t) : V \to V$ in such a way that the solution $u(t)$ of (2.9) is given by

$$u(t) = S(t)u_0.$$

The one parameter mapping $S(t)$ satisfies the usual semigroup properties

(1) $S(0) = I$, the identity on V;

(2) $S(t+s) = S(t)S(s) \quad \forall t, s \in \mathbb{R}^+$.

For certain u_0 the operator $S(t)$ may be defined for $t < 0$; in such instances we will freely use $S(t)$ with negative arguments. A simple example is when u_0 is an equilibrium point. We now give some basic definitions and results concerning dynamical systems.

Definition 2.8 *The action of $S(t)$ on a set $E \subset V$ is defined by*

$$S(t)E = \bigcup_{x \in E} S(t)x. \tag{2.23}$$

A set E is said to be invariant (resp. positively invariant, negatively invariant) if, for any $t \geq 0$, $S(t)E \equiv E$ (resp. $S(t)E \subseteq E$, $S(t)E \supseteq E$).

Example 2.9 The simplest example of an invariant set is an equilibrium point $\bar{u} \in D(A) \subset V$ satisfying $A\bar{u} = F(\bar{u})$ (see (4.13)). Since $du/dt \equiv 0$ if $u_0 = \bar{u}$ in (2.9) it is clear that $\bar{u}$ is invariant. Another simple example of an invariant set is a periodic solution of the equation.

To construct a simple positive invariant set assume that, for some $r > 0$,

$$\frac{d}{dt}\|u(t)\|^2|_{t=0} < 0 \quad \forall u_0 \in \partial B(0,r),$$

where $B(0,r)$ and $\partial B(0,r)$ are defined in (2.6).

Then no solutions starting in the set $B(0,r)$ can leave and hence it is positively invariant.

When comparing $S(t)$ and the perturbed semigroup $S^h(t)$ it is thus natural to compare the effect of approximation on the invariant sets (or positively invariant sets or negatively invariant sets) of $S(t)$. This is the approach we take in the remainder of the article. In general, given an initial data point u_0, we define the *forward orbit* to be $\{S(t)u_0, t \geq 0\}$. If there is $\varphi : (-\infty, 0] \to V$ with $\varphi(0) = u_0$ and $S(t)\varphi(s) = \varphi(t+s)$ for $0 \leq t \leq -s$ then a *negative orbit* of u_0 is $\{\varphi(t), t \leq 0\}$. This orbit will not exist for general u_0; when it does exist it may not be unique. If a negative orbit does exist then a *complete orbit* is the union of the positive and a negative orbits. The notion of backward orbits is very useful in the study of unstable manifolds (see Chapter 5). The notion of complete orbits will also be particularly useful in the study of attractors (see Chapter 7). A forward orbit is a simple example of a positively invariant set; a negative orbit is a simple example of a backward invariant set; a complete orbit is a simple example of an invariant set. Indeed it is a simple exercise to show that every point in an invariant set lies on a complete orbit.

It is important to be able to capture all the possible behavior of a dynamical system for large time. This is the motivation behind the following definition which leads to the identification of some further important invariant sets:

Definition 2.10 *The ω-limit set of a point u_0 is defined by*

$$\omega(u_0) = \{x \in V | \exists \{t_i\}, t_i \to \infty : S(t_i)u_0 \to x \text{ as } t_i \to \infty\}.$$

An equivalent definition is

$$\omega(u_0) = \bigcap_{s \geq 0} \overline{\bigcup_{t \geq s} S(t)u_0}. \tag{2.24}$$

Similary we may define the ω-limit set of a set $E \subset V$ by

$$\omega(E) = \{x \in V | \exists \{t_i\}, \{u_i\}, t_i \to \infty, u_i \in E : S(t_i)u_i \to x \text{ as } t_i \to \infty\}.$$

An equivalent definition is

$$\omega(E) = \bigcap_{s \geq 0} \overline{\bigcup_{t \geq s} S(t)E}. \tag{2.25}$$

Given a particular negative orbit through u_0, say $\{\varphi(t)|t \leq 0\}$, we define its α-limit set by

$$\alpha(u_0) = \{x \in V | \exists \{t_i\}, t_i \to -\infty : \varphi(t_i) \to x \text{ as } t_i \to -\infty\}.$$

Examples of ω-limit and α-limit sets of individual points are equilibrium points, periodic solutions, quasi-periodic solutions and more complicated objects such as the strange attractors observed in chaotic systems like the Lorenz equations. Note that, in general, we only have

$$\bigcup_{x \in E} \omega(x) \subset \omega(E).$$

Thus the ω-limit sets of sets may be more complicated than simply the union of limit sets of individual trajectories — they also contain *heteroclinic and homoclinic orbits* connecting individual limit sets of trajectories. These are complete orbits with the same α- and ω-limit sets in the homoclinic case and differing ones in the heteroclinic case.

The following property of limit sets is fundamental.

Theorem 2.11 *Assume that $E \subset V$ is non-empty and that there exists $t_0 \geq 0$ such that $\bigcup_{t \geq t_0} S(t)E$ is relatively compact. Then $\omega(E)$ is non-empty, compact and invariant. Furthermore, for any point $u_0 \in V$, $\omega(u_0)$ is connected.*

For any negative orbit $\{\varphi(t), t \leq 0\}$ through u_0 for which there exists $t_1 \leq 0$ such that $\bigcup_{t \leq t_1} \varphi(t)$ is relatively compact, $\alpha(u_0)$ is non-empty, compact, invariant and connected.

Proof Note that

$$\omega(E) = \bigcap_{s \geq 0} \overline{\bigcup_{t \geq s} S(t)E} = \bigcap_{s \geq t_0} \overline{\bigcup_{t \geq s} S(t)E}.$$

By the assumptions of the theorem, this is an intersection of nested non-empty compact sets; it is therefore non-empty and compact.

Now we show positive invariance: assume that $x \in \omega(E)$. If $S(t_i)v_i \to x$ then by continuity of $S(t)\bullet$,

$$S(t + t_i)v_i = S(t)S(t_i)v_i \to S(t)x \quad \forall t \geq 0.$$

Thus $S(t + t_i)v_i \to S(t)x$ and, hence, $S(t)x \in \omega(E)$ by Definition 2.10. Thus we deduce positive invariance of $\omega(E)$.

Now we establish negative invariance: assume that $x \in \omega(E)$. We wish to show that, for any $t > 0$, $\exists y$: $S(t)y = x$ and $y \in \omega(E)$. Let $S(t_i)v_i \to x$, where, without loss of generality, we may choose $t_1 \geq 1 + t_0 + t$. Now consider the sequence $S(t_i - t)v_i$. Since $\bigcup_{t \geq t_0} S(t)E$ is relatively compact it follows that there exists a convergent subsequence

$$S(t_{i_j} - t)v_{i_j} \to y.$$

Now

$$x = \lim_{j\to\infty} S(t_{i_j})v_{i_j} = \lim_{j\to\infty} S(t)S(t_{i_j} - t)v_{ij}$$

$$= S(t) \lim_{j\to\infty} S(t_{i_j} - t)v_{i_j} = S(t)y.$$

Hence the result is proved.

Finally we show that $\omega(u_0)$ is connected. Assume for contradiction that $\omega(u_0)$ has two disjoint components P and Q with $\mathcal{N}(P,\varepsilon) \cap \mathcal{N}(Q,\varepsilon) = \emptyset$, for some $\varepsilon > 0$. Then there exist sequences $t_i \to \infty$ and $\tau_i \to \infty$ such that $S(t_i)u_0 \to x \in P$ and $S(\tau_i)u_0 \to y \in Q$. Without loss of generality we may assume that $t_i < \tau_i$ and that $S(t_i)u_0 \in \mathcal{N}(P,\varepsilon)$, $S(\tau_i)u_0 \in \mathcal{N}(Q,\varepsilon)$ $\forall i \geq 1$. By continuity of $S(\bullet)u_0$ it follows that there exists $T_i \in (t_i, \tau_i)$ such that $S(T_i)u_0 \in \partial\mathcal{N}(P,\varepsilon)$. But the set $\partial\mathcal{N}(P,\varepsilon)$ is compact, since P is compact. Thus there exists a convergent subsequence $S(T_{i_j})u_0 \to z$, where $z \in \partial\mathcal{N}(P,\varepsilon)$. But this is a contradiction since then, by definition, $z \in \omega(u_0)$ but $z \notin P \cup Q$. This completes the proof. The statements about α-limit sets follow similarly. □

We conclude with three categories of dynamical systems which will be useful to us throughout this article.

Definition 2.12 *The semigroup $S(t) : V \mapsto V$ is said to be contractive in the neighborhood of an equilibrium point $\bar{u}$ if there exist constants $\alpha, \sigma > 0$ such that, if $u_1, u_2 \in B(\bar{u}, \sigma)$, then*

$$\|S(t)u_1 - S(t)u_2\| \leq e^{-\alpha t}\|u_1 - u_2\| \quad \forall t \geq \tau.$$

In Section 4.3 we will prove that hyperbolic, stable equilibrium points yield contractive semigroups under certain natural conditions on the nonlinearity.

Definition 2.13 *The semigroup $S(t) : V \mapsto V$ is said to be dissipative if there is a bounded set $B \subset V$, known as an absorbing set, such that for any bounded set $E \subset V$ there is $T = T(E, B)$ such that $S(t)E \subseteq B$ for all $t \geq T$.*

Example 2.14 Consider eqn (2.12) under (2.13). By using the techniques outlined in Temam [92] it is possible to show that the equation generates a semigroup $S(t) : X \mapsto X$, where $X = L_2(0, 1)$. Let $\langle \bullet, \bullet \rangle$ and $| \bullet |$ denote the inner product and norm on X. To see that the equation is dissipative on X note that, by (2.3), for any $\epsilon > 0$ we have

$$\begin{aligned} \frac{1}{2}\frac{d}{dt}|u|^2 &= \langle u, u_t\rangle \\ &= \langle u, u_{xx}\rangle + \langle u, f(u)\rangle \\ &\le -|u_x|^2 + \frac{\varepsilon^2}{2}|u|^2 + \frac{1}{2\varepsilon^2}|f(u)|^2 \\ &\le -\pi^2|u|^2 + \frac{\varepsilon^2}{2}|u|^2 + \frac{C^2}{2\varepsilon^2}. \end{aligned}$$

By choosing $\epsilon = \pi$ we have

$$\frac{d}{dt}|u|^2 \le \frac{C^2}{\pi^2} - \pi^2|u|^2.$$

Integration shows that the equation is dissipative on X with absorbing set $B = B(0, \rho)$ for any $\rho : \rho^2 > C^2/\pi^4$.

Definition 2.15 *The C^1 semigroup $S(t)$ generated by (2.9) is said to define a gradient system if there exists $\mathcal{V} \in C(V, \mathbb{R})$, called a Lyapunov function, satisfying*

(i) $\mathcal{V}(u) \ge 0$ for all $u \in V$;
(ii) $\mathcal{V}(u) \to \infty$ as $\|u\| \to \infty$;
(iii) $\mathcal{V}(S(t)u)$ is nonincreasing in t for each $u \in V$;
(iv) if u is such that $S(t)u$ is defined for all $t \in \mathbb{R}$ and $\mathcal{V}(S(t)u) = \mathcal{V}(u)$ for $t \in \mathbb{R}$ then u is an equilibrium point satisfying $Au = F(u)$.

Example 2.16 Consider the reaction-diffusion equation (2.12) subject to (2.13). By Theorem 2.6 this forms a dynamical system on $V \equiv H_0^1(\Omega)$. (Note that, in contrast, we considered the same equation as a dynamical system in $L_2((0,1))$ in Example 2.14.) If we define $h(u)$ to be a primitive of $f(u)$ so that $h'(u) = f(u)$ and set

$$\mathcal{V}(\varphi) := \int_0^1 \{\frac{1}{2}\varphi_x^2 - h(\varphi)\}dx$$

then $\mathcal{V}(\bullet)$ is a Lyapunov functional for (2.12) and hence we have a gradient system. To see this note that

$$h(u) = c + \int_0^u f(w)dw \tag{2.26}$$

for some arbitrarily chosen $c \in \mathbb{R}$. Thus, by (2.13), for any $\varepsilon > 0$,

$$\int_0^1 h(\varphi)d\varphi \le c + C\varphi \le c + \frac{C}{2\varepsilon^2} + \frac{C\varepsilon^2}{2}|\varphi|^2.$$

Thus, by (2.3),

$$\mathcal{V}(\varphi) \geq \frac{1}{2}\|\varphi\|^2 - \frac{C\varepsilon^2}{2\pi^2}\|\varphi\|^2 - c - \frac{C}{2\varepsilon^2}.$$

Choosing ε such that $4C\varepsilon^2 = 2\pi^2$ and c such that $2\varepsilon^2 c = -C$, we obtain $\mathcal{V}(\varphi) \geq \frac{1}{4}\|\varphi\|^2$, as required for (i), (ii). Also, multiplying (2.12) by u_t and integrating by parts, we find that

$$\frac{d}{dt}\{\mathcal{V}(u(t))\} = -|u_t(t)|^2$$

so that (iii), (iv) follow.

In this article we will start by studying equilibrium points and their neighborhoods in Sections 4 and 5. We will also consider ω-limit sets of sets, leading to the study of *global attractors* — see Section 7. There are currently certain limiting factors in the study of convergence of attractors and, in Section 6, we study an object which contains the global attractor, namely an *inertial manifold.* Because of its stronger attractivity, it is possible to prove stronger results about the affect of perturbation on the inertial manifold than on the global attractor. Section 8 uses our analysis of Section 4 to derive piecewise continuous error bounds for gradient systems. Finally, in Section 9, we briefly describe numerical methods which preserve the dissipative or gradient structure of Defintions 2.13 and 2.15.

2.10 Bibliography

The theory of ordinary differential equations in a Hilbert space that we exploit here may be found in Henry [54] and Pazy [81]; the majority of the results in Appendix A are taken from these two sources.

The basic theory and definitions for dynamical systems in finite dimensions is presented very clearly in Bhatia and Szego [11]. Generalizations to partial differential equations include Babin and Vishik [4], Hale [50], Ladyzhenskaya [72] and Temam [92]. See also Chueshov [18] for a review of the subject. General references concerning the numerical analysis of dynamical systems can be found in Beyn [9], Broomhead and Iserles [14], Kloeden and Palmer [69] and Stuart [88].

3 Basic approximation of trajectories

3.1 Introduction

Section 3.2 contains our basic assumptions about the approximating semigroup. Roughly this states that the error is small in the C^1 sense, uniformly on compact time intervals disjoint from the origin and on bounded sets in V. In the remaining sections a variety of perturbations to equations of the form (2.9) are described and shown to satisfy the assumptions of Section 3.2.

In Section 3.3 a spectral method is defined for the abstract evolution equation introduced in Section 2.5 and its existence and regularity properties stated. We then derive error estimates for the spectral method. Theorem 3.6 does this by use of a variation of constants approach to the error analysis. (For an introduction to the use of variation of constants, refer back to the **Important remark** following Theorem 2.6). Theorem 3.7 is an extension to obtain C^1 error estimates, namely estimates for the difference between the Fréchet derivatives of the true and spectral solution operators with respect to initial data. Such C^1 error estimates can be derived for many methods and are particularly useful in the context of dynamical systems. As mentioned they form the basis of our assumptions outlined in Section 3.2. Their importance follows from the fact that C^1 closeness implies closeness of the Lipschitz constants of certain nonlinear operators used in the construction of objects of interest in the context of dynamical systems. We emphasize that the spectral method is *not* being recommended for practical computation; it is simply a good example with which to illustrate the abstract approximation theory that follows in remaining sections.

In Section 3.4 we consider the viscous Cahn–Hilliard equation which can be derived from a simplification of the phase-field model of phase transitions and contains the Cahn–Hilliard equation as a singular limit. This singular limit is studied and the assumptions of Section 3.2 shown to hold for the singular perturbation. Section 3.5 contains a discussion of ordinary differential equations.

Those readers not interested in the error analysis for particular perturbations can jump straight to Section 4 after reading Section 3.2. Thus the important point in Section 3 is to understand the basic Assumptions 3.2 which will be used throughout the remainder of the article.

It is our aim in this article to study the effect of various perturbations over long time intervals. In this context it is clear that the error estimates made in Assumptions 3.2 are of no direct use since they contain constants which typically grow with the time interval under consideration. (Indeed the growth is typically exponential). Sections 4–9 deal with a variety of results enabling us to interpret the relationship between the underlying unperturbed dynamical systems and the perturbed dynamical system over long time intervals.

3.2 Approximation assumptions

In the rest of the article we will consider a whole class of approximations to the semigroup $S(t)$ given in Assumptions 2.3 yielding approximate semigroups $S^h(t) : V \mapsto V$ satisfying certain natural approximation properties. In some applications, such as time discretization, t may not take on values in the whole of $\mathbb{R}^+$ but in a subset $g(h)\mathbb{N}^+ = \{0, g(h), 2g(h), \ldots\}$. For example, in backward Euler approximation with time step h we will have $g(h) = h$. To enable us to make statements about time-discrete and time-continuous semigroups $S^h(t)$ we use $\mathbf{S}$ to denote $\mathbb{R}^+$ or $g(h)\mathbb{N}^+$ as appropriate.

We denote the Fréchet derivative of $S^h(t)u$ with respect to u evaluated at a point $v \in V$ by $dS^h(v,t)$. Before stating the basic assumptions concerning $S^h(t)$ we make the following definition.

Definition 3.1 *The approximation error for the semigroup $S^h(t) : V \mapsto V$ as an approximation to the semigroup $S(t) : V \mapsto V$ generated by the partial differential equation (2.9) at a point $u_0 \in V$ is defined by*

$$E(u_0; t) := S(t)u_0 - S^h(t)u_0.$$

The Fréchet derivative of $E(u_0; t)$ with respect to $u_0 \in V$ evaluated at a point $v \in V$ is denoted by

$$dE(v; t) := dS(v; t) - dS^h(v; t).$$

Throughout the remainder of this article we make the following assumption concerning the relationship between the semigroup $S(t)$ and its perturbation $S^h(t)$.

Assumption 3.2 *For all $u \in B(0, R)$ and all $t \in \mathbf{S}$, $t > 0$, there exist constants $C_i = C_i(t, R) < \infty, i = 1, 2$, and a function $\kappa : \mathbb{R}^+ \mapsto \mathbb{R}^+$ such that the semigroups $S(\bullet)\bullet$ and $S^h(\bullet)\bullet$ satisfy*

$$\|E(u; t)\| \leq C_1 h,$$

$$\|dE(u; t)\| \leq C_2 \kappa(h),$$

where $\kappa(h) \to 0$ as $h \to 0_+$.

We make this assumption throughout the remainder of the paper. We will not state it explicitly in the results. What makes the analysis particularly challenging, in comparison with analogous theories for ordinary differential equations, is that C_1, C_2 are not assumed to be bounded as $t \to 0$. Many perturbations, such as those arising from numerical approximation or singular perturbations of the terms in (2.9), have the property that the C_i are unbounded as $t \to 0$, and hence it is important to incorporate it in our assumptions.

In the remainder of this section we detail a variety of situations in which this assumption can be shown to hold. Examples include spectral approximation of (2.9) based on the eigenfunctions of A, time approximation of (2.9) based on the backward Euler method and singular perturbations of the Cahn–Hilliard equation arising by considering (2.21) as singular limit of the phase-field equations for phase transitions.

3.3 Spectral method for the sectorial equation

We now introduce a spectral method for the approximation of the abstract sectorial evolution equation (2.9) under (2.11). Let $\mathbb{P}$ denote the projection of X into $\text{span}\{\varphi_j\}_{j=1}^N$; thus, given v expressed as

$$v = \sum_{j=1}^{\infty} v_j \varphi_j,$$

we set

$$\mathbb{P}v = \sum_{j=1}^{N} v_j \varphi_j.$$

We also let $\mathbb{Q}$ denote the orthogonal complement of $\mathbb{P}$ so that $\mathbb{Q} = I - \mathbb{P}$. We use the notation $V^{\mathrm{N}} = \mathbb{P}X$. The Galerkin approximation to (2.9) which we consider is to find $u^{\mathrm{N}} \in V^{\mathrm{N}}$ satisfying

$$\frac{du^{\mathrm{N}}}{dt} + Au^{\mathrm{N}} = \mathbb{P}F(u^{\mathrm{N}}), \;\; t > 0, \quad u^{\mathrm{N}}(0) = \mathbb{P}u_0^{\mathrm{N}}. \tag{3.1}$$

As for (2.9) it will also be important to consider the linearized problem

$$\frac{dv^{\mathrm{N}}}{dt} + Av^{\mathrm{N}} = \mathbb{P}dF(u^{\mathrm{N}})v^{\mathrm{N}}, t > 0, \quad v^{\mathrm{N}}(0) = \mathbb{P}\xi^{\mathrm{N}}. \tag{3.2}$$

By methods analogous to those used to prove Theorems 2.6, 2.7 we may prove the following two results about (3.1) and (3.2). The proofs are identical after noting that $|\mathbb{P}| = 1$.

Theorem 3.3 *Let (2.11) hold. Then, for every $u_0^{\mathrm{N}} \in V$ there exists a unique solution $u^{\mathrm{N}}(t)$ of (3.1). Furthermore, there exists $C_1 = C_1(T, R) > 0$ and $C_2 = C_2(\alpha, T, R)$ such that, for all $t \in (0, T)$ and $u_0^{\mathrm{N}} \in B(0, R)$*

$$\begin{aligned} |u^{\mathrm{N}}(t)|_\alpha &\le \frac{C_1}{t^{\alpha-\beta}}, \forall \alpha \in [\beta, 1]; \\ \left|\frac{du^{\mathrm{N}}}{dt}(t)\right|_\alpha &\le \frac{C_2}{t^{\alpha-\beta+1}}, \forall \alpha \in [\beta - 1, 1). \end{aligned} \tag{3.3}$$

Finally, the semigroup $S^{\mathrm{N}}(t)(t) : V \mapsto V$ defined by $S^{\mathrm{N}}(t)(t)u_0^{\mathrm{N}} = u^{\mathrm{N}}(t)$ satisfies Assumptions 2.3.

Theorem 3.4 *Let (2.11) hold. Then, for every $\xi^{\mathrm{N}} \in V$ there exists a mild solution $v^{\mathrm{N}}(t)$ of (3.2). Furthermore, there exists $C = C(T, \|u_0^{\mathrm{N}}\|) > 0$ such that, for all $t \in (0, T)$:*

$$\begin{aligned} \|v^{\mathrm{N}}(t)\| &\le C\|\xi^{\mathrm{N}}\|; \\ |v^{\mathrm{N}}(t)|_\alpha &\le \frac{C\|\xi^{\mathrm{N}}\|}{(1-\alpha)t^{\alpha-\beta}}, \quad \forall \alpha \in (\beta, 1). \end{aligned} \tag{3.4}$$

Finally, if $v^N(t) = dS^N(t)(u_0{}^N; t)\xi^N$, *then* $v^N(t)$ *satisfies (3.2).*

Thus the numerical method (3.1) generates a semigroup $S^N(t) : V \to V$ — denoted by $S^h(t)$ in the foregoing theorems — in such a way that the solution u^N of (3.1) is given by

$$u^N(t) = S^N(t)u_0{}^N.$$

Note also that $S^N(t)$ may be viewed as a mapping from V^N to V^N since it is V^N valued for $t > 0$ and $V^N \subset V$. Here the superscript N is used simply to emphasize the dependence of the numerical method on the dimension of the projection $\mathbb{P}$. This semigroup satisfies properties analogous to those for $S(t)$:

(1) $S^N(0) = \mathbb{P}$, the projection $V \mapsto V^N$;
(2) $S^N(t+s) = S^N(t)S^N(s) \quad \forall t, s \in \mathbb{R}^+$.

(Actually a true semigroup must satisfy $S^N(0) = I$ but having $S^N(0) = \mathbb{P}$ does not affect the analysis given here in any way. We can view $S^N(t)$ as a true semigroup satisfying $S^N(0) = I$ if we restrict its domain to V^N.)

We denote the Fréchet derivative of $S^N(t)u_0$ with respect to $u_0 \in V$, evaluated at a point $v \in V$, by $dS^N(v;t)$.

Example 3.5 Consider A given by Example 2.1 and the heat equation

$$u_t + Au = 0, \quad u(0) = u_0 \in H_0^1(\Omega).$$

For $v \in H_0^1(\Omega) \subset L_2(\Omega)$ we may write v as a series as in (2.2). Similarly any $\xi \in H_0^1(\Omega)$ can be written as

$$\xi = \sum_{j=1}^{\infty} \xi_j \varphi_j.$$

It follows that

$$S(t)v = \sum_{j=1}^{\infty} e^{-\lambda_j t} v_j \varphi_j.$$

Furthemore we have that

$$dS(v;t)\xi = \sum_{j=1}^{\infty} e^{-\lambda_j t} \xi_j \varphi_j.$$

Note that $dS(v;t)$ is independent of v because of the linearity of the problem.

The spectral approximation generates semigroup $S^N(t)$ satisfying

$$S^N(t)v = \sum_{j=1}^{N} e^{-\lambda_j t} v_j \varphi_j$$

and

$$dS^{\mathrm{N}}(v;t)\xi = \sum_{j=1}^{N} e^{-\lambda_j t}\xi_j\varphi_j.$$

We now prove that the approximation error for (3.1) as an approximation of (2.9) is small in both a C^0 and a C^1 sense. The closeness of the approximation depends on the regularity of the solution being approximated. For many particular equations it is possible to obtain greater regularity than we prove under the assumptions here; this yields stronger approximation results — see Example 3.8.

Theorem 3.6 **(C^0 Error Estimates)** *Let (2.11) hold. There exists a constant $C = C(T,R)$ such that the solutions of (2.9) and (3.1) with $u_0 \in B(0,R)$ satisfy*

$$\|u(t) - u^{\mathrm{N}}(t)\| \le \frac{C}{t^{1-\beta}}[\|\mathbb{P}(u_0 - u_0^{\mathrm{N}})\| + \lambda_{N+1}^{\beta-1}] \quad \forall t \in (0,T].$$

Proof Let

$$e(t) = \mathbb{P}u(t) - u^{\mathrm{N}}(t), \quad E(t) = u(t) - u^{\mathrm{N}}(t), \quad q(t) = \mathbb{Q}u(t).$$

(Recall that $\mathbb{Q} = I - \mathbb{P}$.) Note that, if

$$u(t) = \sum_{j=1}^{\infty} u_j\varphi_j$$

then

$$\|q(t)\|^2 = \sum_{j=N+1}^{\infty} \lambda_j^{2\beta}u_j^2 \le \lambda_{N+1}^{2(\beta-1)} \sum_{j=N+1}^{\infty} \lambda_j^2 u_j^2 \le \lambda_{N+1}^{2(\beta-1)}|u(t)|_1^2.$$

By applying Theorem 2.6 we obtain, for $C_1 = C_1(T,R)$

$$\|q(t)\| \le \frac{C_1}{(t\lambda_{N+1})^{1-\beta}}. \tag{3.5}$$

Now note that $e(t)$ satisfies the equation

$$e_t + Ae = \mathbb{P}[F(u) - F(u^{\mathrm{N}})].$$

Use of variation of constants gives

$$e(t) = e^{-At}e(0) + \int_0^t e^{-A(t-s)}\mathbb{P}[F(u(s)) - F(u^{\mathrm{N}}(s))]ds.$$

Taking norms, using (2.11), (10.4), $|\mathbb{P}| = 1$ and Lemma 10.6, we obtain

$$\|e(t)\| \le \|e(0)\| + \int_0^t \frac{K}{(t-s)^{\beta}}\|E(s)\|ds.$$

Hence, since $E(t) = e(t) + q(t)$, we have from (3.5),

$$\|E(t)\| \le \|\mathbb{P}(u_0 - u_0^{\mathrm{N}})\| + \frac{C_1}{(t\lambda_{N+1})^{1-\beta}} + \int_0^t \frac{K}{(t-s)^\beta}\|E(s)\|ds.$$

Thus

$$\|E(t)\| \le \frac{1}{t^{1-\beta}}\left\{T^{1-\beta}\|\mathbb{P}(u_0 - u_0^{\mathrm{N}})\| + C_1\lambda_{N+1}^{\beta-1}\right\} + \int_0^t \frac{K}{(t-s)^\beta}\|E(s)\|ds.$$

Application of the Gronwall Lemma 10.11 gives the desired result for $E(t) = u(t) - u^{\mathrm{N}}(t)$. □

Important remark Note that the error estimate blows up as $t \to 0$. To understand this consider the set of functions $u_0 \in B(0,R)$ and their spectral approximations u_0^{N}. Let $u_0 = w^{\mathrm{N}}$ where

$$w^{\mathrm{N}} = \sum_{j=1}^{\infty} w^{\mathrm{N}}{}_j \varphi_j$$

and

$$w^{\mathrm{N}}{}_j = 0,\ j \ne N+1, \quad w^{\mathrm{N}}{}_{N+1} = R/\lambda_j^\beta.$$

Then $\|w^{\mathrm{N}}\|^2 = R^2$ so that $w^{\mathrm{N}} \in B(0,R)$. But $\mathbb{P}w^{\mathrm{N}} = 0$ and so

$$\|w^{\mathrm{N}} - \mathbb{P}w^{\mathrm{N}}\| = R \quad \forall N \ge 0.$$

Hence there are sequences of functions in $B(0,R)$ for which the spectral approximation of the initial data in X^β yields a constant error R for each $N \ge 0$. Thus the estimate $\mathcal{O}(\lambda_{N+1}^{\beta-1})$ for the error cannot hold uniformly for all solutions with initial data in $B(0,R)$ as $t \to 0$. This observation is particular to partial differential equations and does not arise in the approximation of ordinary differential equations. It means that certain proofs employed in ordinary differential equations which rely on small time behavior need to be modified. To understand this further the reader should compare the proofs of results in this article with similar results concerning the convergence under perturbation of invariant sets of dynamical systems in ordinary differential equations given in Stuart [88]. Note also that on any compact time interval $[t_1, t_2]$ disjoint from the origin the error estimate is uniform within a ball $B(0,R)$ of initial data. This is not true of the error estimate in Corollary 4.10 for uniform-in-time approximation of trajectories asymptotic to a stable equilibrium point — see the remark following that corollary.

Recall the standard induced operator norm on linear mappings $L \in \mathcal{L}(V,V)$ given in (2.7).

Theorem 3.7 **(C^1 Error Estimates)** *Let* 2.11 *hold. There exists a constant $C = C(T, \|u_0\|)$ such that, for any $\alpha \in (\beta, 1)$, the Fréchet derivative of* *the approximation error generated by solutions of (2.17) and (3.2) satisfies*

$$\|dS(u_0;t) - dS^{\mathrm{N}}(u_0;t)\| \leq \frac{C}{(1-\alpha)(t\lambda_{N+1})^{\alpha-\beta}} \quad \forall t \in (0,T].$$

Proof We let

$$d(t) = \mathbb{P}v(t) - v^{\mathrm{N}}(t), \quad D(t) = v(t) - v^{\mathrm{N}}(t), \quad r(t) = \mathbb{Q}v(t)$$

where $v(t)$ and $v^{\mathrm{N}}(t)$ satisfy (2.17), (3.2) with $\xi^{\mathrm{N}} = \xi$ and with $u(t)$ and $u^{\mathrm{N}}(t)$ generated from (2.9), (3.1) with $u_0^{\mathrm{N}} = u_0$. Using the regularity established for $v(t)$ in Theorem 2.7, it follows that

$$\|r(t)\| \leq \frac{C_1\|\xi\|}{(1-\alpha)(t\lambda_{N+1})^{\alpha-\beta}}; \tag{3.6}$$

the derivation is similar to that for (3.5). Now $d(t)$ satisfies the equation

$$d_t + Ad = \mathbb{P}[dF(u)v - dF(u^{\mathrm{N}})v] + \mathbb{P}[dF(u^{\mathrm{N}})(v - v^{\mathrm{N}})].$$

Applying the variation of constants formula we obtain

$$\begin{aligned} d(t) = e^{-At}d(0) &+ \int_0^t e^{-A(t-s)}\mathbb{P}[dF(u)v - dF(u^{\mathrm{N}})v]ds \\ &+ \int_0^t e^{-A(t-s)}\mathbb{P}[dF(u^{\mathrm{N}})(v - v^{\mathrm{N}})]ds. \end{aligned}$$

Applying A^β, taking norms and using (2.11), (10.4) and Lemma 10.6, we find that

$$\|d(t)\| \leq \|d(0)\| + \int_0^t \frac{K\|u(s) - u^{\mathrm{N}}(s)\|\,\|v(s)\|}{(t-s)^\beta}ds + \int_0^t \frac{K\|D(t)\|}{(t-s)^\beta}ds.$$

Since $\xi^{\mathrm{N}} = \xi$ we have $d(0) = 0$. Also, by Theorem 3.6, we have that since $u_0^{\mathrm{N}} = u_0$,

$$\|u(t) - u^{\mathrm{N}}(t)\| \leq \frac{C}{(\lambda_{N+1}t)^{1-\beta}}. \tag{3.7}$$

By Theorem 2.7 we have $\|v(t)\| \leq C\|\xi\|$ and hence, putting all this together, we find that

$$\|d(t)\| \leq \int_0^t \frac{K\|\xi\|}{(t-s)^\beta s^{1-\beta}\lambda_{N+1}^{1-\beta}}ds + \int_0^t \frac{K\|D(s)\|}{(t-s)^\beta}ds. \tag{3.8}$$

It may be shown that, if $\beta > 0$, then

$$\int_0^t \frac{ds}{(t-s)^\beta s^{1-\beta}} \leq \frac{1}{\beta(1-\beta)};$$

we return to the case $\beta = 0$ below. Hence, using (3.6) and (3.8), we find that

$$\begin{aligned}\|D(t)\| &\leq \|d(t)\| + \|r(t)\| \\ &\leq \frac{C_2\|\xi\|}{\beta(1-\beta)\lambda_{N+1}^{1-\beta}} + \frac{C_1\|\xi\|}{(1-\alpha)(\lambda_{N+1}t)^{\alpha-\beta}} + \int_0^t \frac{K\|D(t)\|}{(t-s)^\beta} ds.\end{aligned}$$

Hence

$$\|D(t)\| \leq \frac{C_3\|\xi\|}{(1-\alpha)(\lambda_{N+1}t)^{\alpha-\beta}} + \int_0^t \frac{K\|D(t)\|}{(t-s)^\beta} ds.$$

Applying the Gronwall Lemma 10.11 yields

$$\|v(t) - v^N(t)\| \leq \frac{C\|\xi\|}{(1-\alpha)(t\lambda_{N+1})^{\alpha-\beta}}, \quad \forall t \in (0,T].$$

Since $v(t) = dS(u_0;t)\xi$ and $v^N(t) = dS^N(t)(u_0;t)\xi$ the required bound follows from (2.7).

The case $\beta \ll 1$ can be handled similarly by using the (weaker) error bound

$$\|u(t) - u^N(t)\| \leq \frac{C}{(\lambda_{N+1}t)^{\alpha-\beta}}$$

in place of (3.7); this bound may be derived by modifying the proof of Theorem 3.6. □

Remark Recall the remark following Theorem 2.7 which indicates why it is not necessary for us to obtain convergence in the C^1 norm at the same rate as in the C^0 norm. □.

The following example shows, however, that for many particular problems results far stronger than those proved here will hold.

Example 3.8 For the heat equation Example 3.5 we have, for $\beta = \frac{1}{2}$ so that $\| \bullet \|$ is the norm on $H_0^1(\Omega)$,

$$E(v;t) = \sum_{j=N+1}^{\infty} e^{-\lambda_j t} v_j \varphi_j.$$

Thus

$$\|E(v;t)\|^2 = \sum_{j=N+1}^{\infty} e^{-2\lambda_j t} \lambda_j v_j^2 \leq e^{-2\lambda_{N+1}t}\|v\|^2.$$

This shows the exponential rate of convergence of the spectral method with respect to N, for any fixed $t > 0$. The same rate of convergence holds for $\|dE(v;t)\|$. This rate of convergence is a consequence of the high degree of regularity of the solution for each $t > 0$ given by the exponential decay of the Fourier coefficients. In general the addition of nonlinear terms will destroy this smoothness and hence the weaker error bounds proved in Theorems 3.6, 3.7 are typical. However, many nonlinear problems do possess a form of regularity known as *Gevrey class* which implies rapid decay of the co-efficients in eigenfunction expansions; such regularity can be exploited in the error estimates for spectral methods. Ferrari and Titi [35] give general conditions under which (2.9) yields solutions of Gevrey class regularity.

Note that Theorems 3.6 and 3.7 show that, at any positive finite time, both the semigroup and its derivative are well-approximated. The constants appearing in the error bounds depend only on the norm of the initial data u_0 and the time interval under consideration. Thus with the definition

$$S^h(t) := S^N(t) \quad \forall h \in [\lambda_{N+1}^{(1-\beta)}, \lambda_N^{(1-\beta)}), \tag{3.9}$$

Theorems 3.6, 3.7 show that Assumptions 3.2 are satisfied for the spectral approximation.

3.4 Backward Euler method for sectorial equations

As another example of a method satisfying the approximation Assumptions 3.2, consider the backward Euler method

$$U^{n+1} - U^n + \Delta t A U^{n+1} = \Delta t F(U^{n+1}), \quad U^0 = u_0.$$

This equation generates an approximation $U^n \approx u(n\Delta t)$. The approximation assumptions used in this paper are established for the method applied to a reaction-diffusion equation and to the Kuramtoto-Sivashinsky equation respectively in Hale *et al.* [51] and Alouges and Debussche [1].

3.5 The phase-field and viscous Cahn–Hilliard equations

Now we introduce an example where the effect of perturbation is not from numerical approximation but from a singular perturbation to the partial differential equation. The phase-field equations are

$$c\theta_t + \frac{l}{2}u_t = k\Delta\theta, \quad x \in \Omega, \quad t > 0,$$
$$\alpha u_t = \Delta u + f(u) + \delta\theta, \quad x \in \Omega, \quad t > 0.$$

We consider these equations subject to the Dirichlet boundary conditions

$$u = \theta = 0 \qquad x \in \partial\Omega, \quad t > 0,$$

and initial conditions on u and θ. These equations model phase transitions such as that between ice and water; see Caginalp [15].

If we set $c = 0$, re-scale time and the function f and introduce a scaled version of θ, namely w, then we obtain equations in the form

$$
\begin{aligned}
u_t &= \Delta w, \quad x \in \Omega, \\
\epsilon u_t &= \Delta u + f(u) + w, \quad x \in \Omega, \\
u &= w = 0, \quad x \in \partial\Omega, \\
u &= u_0(x), \quad t = 0.
\end{aligned}
\tag{3.10}
$$

These are a form of the viscous Cahn–Hilliard equation. We use A_0 to denote the same operator as in (2.21) and then (3.10) may be written as

$$u_t + (\varepsilon + A_0^{-1})^{-1}A_0 u = (\varepsilon + A_0^{-1})^{-1}F(u).$$

If ϵ is small this may be viewed as a non-local singular perturbation of the Cahn–Hilliard equation (2.21). In Elliott and Stuart [32] it is shown that for any $\varepsilon \geq 0$ equations (3.10) generate a dynamical system with semigroup $S(\bullet)\bullet \in C^2(\mathbb{R}^+ \times V, V)$ where $V = H_0^1(\Omega)$ so that Assumptions 2.3 hold. Furthermore the semigroup is shown to be C^1 in ϵ, uniformly on compact time intervals disjoint from the origin and on bounded sets in V. Thus Assumptions 3.2 hold with $h = \kappa(h) = \epsilon$.

3.6 Ordinary differential equations

The system of ordinary differential equations

$$u_t = f(u; \epsilon), \quad u(0) = U,$$

where f is smooth in u and ϵ, generates a semigroup satisfying Assumptions 2.3 and 3.2 with $h = \kappa(h) = \varepsilon$. Thus the theory in this article applies.

3.7 Bibliography

The approach to C^0 error estimates given here is similar to that developed for finite element methods applied to reaction-diffusion equations in Larsson [74] and generalized to the Cahn–Hilliard equation in Elliott and Larsson [30]; that work in turn builds on the work of Thomee [96] and of Johnson *et al.* [61]. These works employ the semigroup approach to facilitate the error analysis. The use of C^1 error estimates is particularly important in the context of dynamical systems and examples of such results may be found in Alouges and Debussche [1] and Hale *et al.* [51] for specific approximations of reaction-diffusion equations, in Stuart [88] for arbitrary one-step methods applied to ordinary differential equations and in Jones and Titi [64] and Jones [62] for spectral and finite difference approximations of certain partial differential equations of the form (2.9).

The spectral method based on the eigenfunctions of A is typically only useful as a numerical computational technique for problems with simple geometry and simple differential operators such as those appearing in Examples 2.1, 2.5. An early use of such a computational method for (2.9) is described in Orszag [80]; a more recent reference, directly relevant to the computation of invariant manifolds in partial differential equations, is Bai *et al.* [5]. The spectral method (3.1) is also of theoretical importance as a tool to prove existence results about (2.9); indeed, by use of a spectral method, such results can be proved under weaker hypotheses concerning the initial data u_0 than those given here – see Friedman [40], Lions [76], Constantin and Foias [19], and Temam [91] [92].

4 Equilibria and phase portraits

4.1 Introduction

This section starts with a consideration of the effect of the approximation error on general, not necessarily stable, equilibrium points. The approach is to use the implicit function theorem. The core of the analysis is contained in Lemma 4.2 but the main result is stated as Theorem 4.3.

In Section 4.3 we study the one case where error estimates for (2.9) may be obtained which are independent of the time-interval, namely for solutions approaching an exponentially stable equilibrium point. This introduces the important idea, which we use throughout the article, of combining the standard finite time error estimates together with some underlying property concerning the behavior of the differential equation. The most important result in this context is Corollary 4.10. An important stepping-stone along the way is Lemma 4.7 which shows that, near a stable hyperbolic equilibrium point, the semigroup is contractive.

In Section 4.4 we move from stable equilibria to local phase portraits near saddle points — the union of all solutions of (2.9) in a small ball around the equilibrium point. Since such solutions can spend an arbitrarily long time near the equilibrium point standard error analysis does not apply. The key is to compare true and numerical solutions with different initial data and the important results are Theorems 4.18 and 4.19. The method of analysis used in this section is to construct the solutions of interest by use of a contraction mapping argument (see Theorem 4.13) and then use the uniform contraction principle (see Appendix B) to incorporate the effect of perturbation.

4.2 Equilibria

In this section we are concerned with the behavior of steady solutions of (2.9) and solutions in their neighborhood.

Definition 4.1 *A point $\bar{u} \in V$ is a fixed point of $S(\tau)$ for some $\tau \in \mathbb{R}$ if $S(\tau)\bar{u} = \bar{u}$. A point $\bar{u} \in V$ is an equilibrium point if it is a fixed point of $S(t)$ for all $t \in \mathbb{R}$. A fixed point $\bar{u}$ of a semigroup $S(t)$ is said to be hyperbolic if $dS(\bar{u}, t)$ has no eigenvalues on the unit circle. An equilibrium point $\bar{u}$ of a semigroup $S(t)$ is said to be hyperbolic if $dS(\bar{u}, t)$ has no eigenvalues on* *the unit circle for all $t \neq 0$.* □

Throughout we will use the following notation for the set of fixed points of $S(t)$ and $S^h(t)$; recall the notation $\mathbf{S}$, given before Definition 3.1, enabling us to consider time-continuous and time-discrete semigroups together.

$$\begin{aligned} \mathcal{E} &= \{v \in V : S(t)v = v \ \ \forall t \in \mathbb{R}\} \\ \mathcal{E}^h &= \{v \in V : S^h(t)v = v \ \ \forall t \in \mathbf{S}\} \end{aligned} \tag{4.1}$$

Let us assume that $S(t)$ has an equilibrium point $\bar{u}$. We introduce the new variable

$$v(t) = u(t) - \bar{u}$$

and change variables in (2.9). Then $v(t)$ satisfies the equation

$$\begin{aligned} &v_t + Cv = g(v), \quad v(0) = v_0 := u_0 - \bar{u}, \\ &C = A - dF(\bar{u}), \quad g(v) = [F(v + \bar{u}) - F(\bar{u}) - dF(\bar{u})v]. \end{aligned} \tag{4.2}$$

Recall the operator norm of $L \in \mathcal{L}(X, X)$ given by (2.8). Note that $|(C - A)A^{-\beta}| = |dF(\bar{u})A^{-\beta}| \leq K$ by (2.11) and hence, by Lemma 10.13, the operator C is sectorial so that e^{-Ct} may be defined. Thus we see that $dS(\bar{u}, t) = e^{-Ct}$.

The hyperbolicity of $\bar{u}$ is equivalent to the operator C having no eigenvalues with zero real part. When $\bar{u}$ is hyperbolic, by Theorem 10.14 we can split the space X into $X = Y' \oplus Z'$ where Y' (resp. Z') is the subspace of X spanned by the generalized eigenspace of C corresponding to eigenvalues with negative (resp. positive) real parts. We denote by $\mathcal{P}$ and $\mathcal{Q}$ the spectral projections $\mathcal{P} : X \to Y'$ and $\mathcal{Q} : X \to Z'$ and then denote $Y = \mathcal{P}V, Z = \mathcal{Q}V$ so that $V = Y \oplus Z$. Using Theorem 10.15 it follows that, for any $a < 1$, there exists $T^* > 0$ such that

$$\begin{aligned} &\|e^{Ct}v\| \leq a\|v\| \quad \forall t \geq T^*, \ \forall v \in Y \\ &\|e^{-Ct}v\| \leq a\|v\| \quad \forall t \geq T^*, \ \forall v \in Z. \end{aligned} \tag{4.3}$$

Our aim is to show that if $\bar{u}$ is hyperbolic then, under Assumptions 3.2, $S^h(t)$ will also have a fixed point and hence an equilibrium point. We start by showing in the next lemma that for any given fixed time t the semigroup $S^h(t)$ has a fixed point $\bar{u}^h(t)$. In the theorem following the lemma we establish that the fixed point of $S^h(t)$ is actually an equilibrium point for $S^h(t)$.

Lemma 4.2 *Assume that $S(\bullet)\bullet \in C^2(\mathbb{R}^+ \times V, V)$. Let $\bar{u}$ be a hyperbolic equilibrium point of (2.9) and assume that $t \geq T^*$ given by (4.3) and that $t \in \mathbf{S}$. Then there exists $h^*, C > 0$ such that $S^h(t)$ has a fixed point, $\bar{u}^h = \bar{u}^h(t)$ unique in $B(\bar{u}; Ch)$, for all $h \in (0, h^*]$.*

Proof The proof is similar to the proof of the implicit function theorem. Consider the mapping

$$\begin{aligned} W^{k+1} &= F(W^k;t) \\ F(W;t) &:= W - D[W - S^h(t)W] \end{aligned} \tag{4.4}$$

where $D = D(t) = [I - dS(\bar{u},t)]^{-1}$ and $t \geq T^*$, where T^* is given in (4.3). As a preliminary to the proof we prove that $\|D\|$ is bounded. Consider the equation

$$(I - e^{-Ct})\mathbf{a} = \mathbf{b}$$

and note that, in the induced operator norm (2.7) we have

$$\|D\| = \sup_{\|\mathbf{b}\|=1} \|\mathbf{a}\|,$$

since $dS(\bar{u},t) = e^{-Ct}$. We write $\mathbf{a} = a_p + a_q$ and $\mathbf{b} = b_p + b_q$ where $a_p, b_p \in Y$ and $a_q, b_q \in Z$; here Y and Z are the invariant subspaces of C. Thus

$$\begin{aligned} (I - e^{-Ct})a_p &= b_p \\ (I - e^{-Ct})a_q &= b_q. \end{aligned}$$

Using the hyperbolicity property we deduce that (4.3) holds and hence that

$$\begin{aligned} \|a_p\| &\leq a(1-a)^{-1}\|b_p\| \\ \|a_q\| &\leq (1-a)^{-1}\|b_q\|. \end{aligned} \tag{4.5}$$

Note that $\| \bullet \|$ is equivalent to the norm $\| \bullet \|_V$ given by

$$\| \bullet \|_V = \max\{\|\mathcal{P} \bullet \|, \|\mathcal{Q} \bullet \|\}.$$

Hence it follows from (4.5) that there exists a constant $K > 0$ such that

$$\|\mathbf{a}\| \leq K(1-a)^{-1}\|\mathbf{b}\|$$

so that $\|D\| \leq K(1-a)^{-1}$.

Fix $\varepsilon > 0$. Recall the constants $C_1 = C_1(t, \|\bar{u}\| + \varepsilon)$ and $C_2 = C_2(t, \|\bar{u}\| + \varepsilon)$ given by Assumptions 3.2 and evaluated here at time t and within a ball of radius $\|\bar{u}\| + \epsilon$. To prove existence of a fixed point of (4.4) we show that the iteration (4.4) maps $B(\bar{u}; Ch)$ into itself for $C = (1+\alpha)\|D\|C_1$ and that it is a contraction on that set. The constant α has been introduced to facilitate the proof of Theorem 4.3 following this lemma. We assume that h is sufficiently small that

$$(1+\alpha)\|D\|C_1 h \leq \varepsilon \tag{4.6}$$

so that $u \in B(\bar{u}; Ch)$ implies that $\|u\| \leq \|\bar{u}\| + \varepsilon$; thus, since all our analysis takes place in $B(\bar{u}; Ch)$, the constants C_1 and C_2 as defined are the appropriate constants in the following argument. Clearly a fixed point of the mapping (4.4) is necessarily a fixed point of $S^h(t)$. To show that the mapping (4.4) is into, note that by Definition 3.1, (4.4) may be written as

$$W^{k+1} = W^k - D[W^k - S(t)W^k + E(W^k;t)]. \tag{4.7}$$

Also, since $\bar{u}$ is a fixed point of $S(t)$ it follows that

$$\bar{u} = \bar{u} - D[\bar{u} - S(t)\bar{u}]. \tag{4.8}$$

Let $W^k \in B(\bar{u}; Ch)$ and set $e^k = W^k - \bar{u}$. Then (4.7), (4.8) yield, upon appplication of Taylor's Theorem 11.4,

$$\|e^{k+1}\| = \|e^k - D[(I - dS(\bar{u},t))e^k + Q_1 + E(W^k;t)]\|$$

where

$$\|Q_1\| \leq K_1\|e^k\|^2,$$

for some $K_1 > 0$. Thus, using the definition of D,

$$\|e^{k+1}\| \leq \|D\|\|Q_1\| + \|D\|\|E(W^k;t)\|.$$

Hence, by Assumption 3.2,

$$\|e^{k+1}\| \leq (1+\alpha)^2\|D\|^3K_1C_1^2h^2 + \|D\|C_1h.$$

Choosing h sufficiently small so that

$$(1+\alpha)^2K_1\|D\|^2C_1h \leq \alpha, \tag{4.9}$$

we deduce that the mapping (4.4) takes $B(\bar{u}, Ch)$ into itself.

To show that the mapping (4.4) is a contraction, let V^k satisfy (4.7) with $W^k \to V^k$ and define $d^k = W^k - V^k$; assume that $W^0, V^0 \in B(\bar{u}, Ch)$. Then

$$\|d^k\| \leq 2(1+\alpha)\|D\|C_1h. \tag{4.10}$$

A similar manipulation to that used in showing that the mapping (4.4) is "into" yields

$$\|d^{k+1}\| \leq \|d^k - D[(I - dS(\bar{u},t))d^k + Q_2 + E(W^k;t) - E(V^k;t)]\|$$

where

$$\|Q_2\| \leq K_2\|d^k\|^2,$$

for some constant $K_2 > 0$. Hence, by Assumption 3.2 and Taylor's Theorem 11.4, (4.10) gives

$$\begin{aligned} \|d^{k+1}\| &\leq \|D\|\|Q_2\| + \|D\|\|E(W^k;t) - E(V^k;t)\| \\ &\leq 2(1+\alpha)\|D\|^2K_2C_1h\|d_k\| + \|D\|C_2\kappa(h)\|d_k\|. \end{aligned} \tag{4.11}$$

Thus the mapping (4.4) is a contraction for h sufficiently small so that

$$2(1+\alpha)K_2\|D\|^2C_1h + \|D\|C_2\kappa(h) \leq \frac{1}{2}. \tag{4.12}$$

The existence of a fixed point $\bar{u}^h$ of S^h follows for $h \leq h^* = h^*(\|\bar{u}\|, \varepsilon, t, \alpha)$ (given by (4.6), (4.9), (4.12) and $t \geq T^*$.) □

Now we deduce that the fixed point $\bar{u}^h(t)$ from the previous lemma is actually independent of t and hence an equilibrium point of $S^h(t)$.

Theorem 4.3 (Equilibrium Points Under Approximation) *Assume that $S(\bullet)\bullet \in C^2(\mathbb{R}^+ \times V, V)$ and assume that $S^h(t)$ is a time-continuous semigroup. Let $\bar{u}$ be a hyperbolic equilibrium point of (2.9). Then there exist $h_c > 0$ and $C > 0$ such that $S^h(t)$ has an equilibrium point $\bar{u}^h$, unique in $B(\bar{u}; Ch)$, for all $h \in (0, h_c]$.*

Proof Consider the fixed point $\bar{u}^h$ found in Lemma 4.2 by setting $t = T \geq T^*$. First we show that the point $\bar{u}^h$ is a fixed point of $S^h(t)$ for all t. Note that h^* and C in the previous theorem both depend upon α; since this is the only dependency of interest to us here we denote these quantities by $h^*(\alpha)$ and $C(\alpha)$; recall that α was introduced to parameterize the radius of the ball in which the contraction argument of Lemma 4.2 was performed and note that $C(1) < C(2)$. By Lemma 4.2 we deduce that, if $h \leq \max\{h^*(1), h^*(2)\}$, then there is a fixed point of $S^h(t)$ in $B(\bar{u}, C(1)h)$ which is unique in $B(\bar{u}, C(2)h)$. Clearly $\bar{u}^h$ must lie on a periodic solution with minimum period τ where, if $\tau \neq 0$, then T is an integer multiple of τ, say $T = m\tau$. We wish to show that $\tau = 0$. Denote the set of points on the periodic solution by

$$\mathcal{S} = \{S^h(t)\bar{u}^h : 0 \leq t \leq \tau\}.$$

If $\tau > 0$ then, by continuity, there exists $v \in \mathcal{S}$ with $v \neq \bar{u}^h$ and $v \in B(\bar{u}, C(2)h)$ and satisfying $S^h(m\tau)v = v$. This contradicts the uniqueness of $\bar{u}^h$ in $B(\bar{u}, C(2)h)$ and it follows that $\tau = 0$ as required. Thus $\bar{u}^h(t)$ is actually an equilibrium point of $S^h(t)$.

To see that this equilibrium point is unique in a ball of sufficiently small radius note that, if it is not, then Lemma 4.2 is contradicted since any other equilibrium point would also be a fixed point of $S^h(t)$. □

A similar argument proves the following, an analog of Theorem 4.3 when $S^h(t)$ is time-discrete.

Theorem 4.4 (Equilibrium Points Under Approximation) *Assume that $S(\bullet)\bullet \in C^2(\mathbb{R}^+ \times V, V)$ and assume that $S^h(t)$ is a time-discrete semigroup so that $S^h(g(h))$ is a one-step map from V into V. Let $\bar{u}$ be a hyperbolic equilibrium point of (2.9). Then there exists $h_c, C > 0$ such that $S^h(g(h))$ has a fixed point $\bar{u}^h$, unique in $B(\bar{u}; Ch)$, for all $h \in (0, h_c]$.*

4.3 Trajectories asymptotic to a stable steady state

We continue our analysis of the approximation of dynamical systems by generalizing Example 3.8, which shows uniform convergence for $t \in (0, \infty)$ for a solution approaching an exponentially stable equilibrium point of the linear heat equation, to the general nonlinear problem. In the following section we consider the neighborhood of an equilibrium of saddle type and derive uniform-in-time approximation results for trajectories near the saddle. In Section 5 we study unstable manifolds near a saddle point. Thus for both this section, the following section and Section 5, we assume that there exists $\bar{u} \in D(A)$ such that

$$A\bar{u} = F(\bar{u}). \tag{4.13}$$

We introduce the notation $\bar{S} : V \to V$ to denote the semigroup constructed so that $v(t) = \bar{S}(t)v_0$ solves (4.2). Hence

$$\bar{S}(t)v = S(t)(\bar{u} + v) - \bar{u}. \tag{4.14}$$

For partial differential equations of the form (2.9) it follows from (4.2) that the nonlinear function $g(\bullet)$ is quadratic in v. Thus we make the following assumption:

Assumption 4.5 *There exists $\zeta < 1 - \beta$ such that*

$$\|g(v) - g(w)\|_{-\zeta} \le k(\rho)\|v - w\|_{\beta} \quad \forall v, w \in B(0, \rho). \tag{4.15}$$

Here $k : \mathbb{R}^+ \mapsto \mathbb{R}^+$ is nondecreasing and satisfies

$$k(\rho) \to 0 \ \ as \ \ \rho \to 0_+. \tag{4.16}$$

The following example illustrates the assumption:

Example 4.6 Consider the reaction-diffusion equation

$$u_t = \Delta u + f(u), \quad x \in \Omega$$

$$u = 0, \quad x \in \partial\Omega$$

$$u = u_0, \quad t = 0.$$

If Ω is a sufficiently smooth domain in $\mathbb{R}^d$ and $f(u)$ satisfies

$$|f^{(j)}(u)| \le C(1 + |u|^{\delta - j}), \quad j = 0, 1, 2, \ u \in \mathbb{R},$$

for some $C > 0$ where $\delta = 3$ if $d = 3$ and $\delta < \infty$ if $d < 3$, then (4.15) and (4.16) hold for g given by (4.2). In that case $\beta = \frac{1}{2}$ and $\zeta = 0$. See Larsson and Sanz-Serna [75] for details.

Consider the Cahn–Hilliard equation (2.21) which can be written as

$$u_t = -\Delta\{\Delta u + f(u)\}, \quad x \in \Omega$$

$$u = \Delta u + f(u) = 0, \quad x \in \partial\Omega$$

$$u = u_0, \quad t = 0.$$

If f satisfies the conditions described for the preceding reaction-diffusion equation then (4.15), (4.16) hold with $\beta = \frac{1}{4}$ and $\zeta = \frac{1}{2}$. See Elliott and Stuart [32] for details.

This assumption about g enables us to prove the following result about the semigroup in the neighborhood of stable equilibria. Recall Definition 2.12.

Lemma 4.7 *Let Assumption* 4.5 *hold. If the equilibrium point* $\bar{u}$ *is hyperbolic and stable (so that* $Z \equiv V$ *and* $Y \equiv \emptyset$*) then the semigroup generated by (2.9) is contractive in a neighborhood of* $\bar{u}$.

Proof We work in the v variable so that $\bar{u}$ translates to the origin. We may use the variation of constants formula to write the solution of (4.2) as

$$v(t) = e^{-Ct}v(0) + \int_0^t e^{-C(t-s)}g(v(s))ds. \tag{4.17}$$

Thus a second solution $w(t)$ satisfies

$$w(t) = e^{-Ct}w(0) + \int_0^t e^{-C(t-s)}g(w(s))ds.$$

Letting $e(t) = v(t) - w(t)$, applying Theorem 10.15 and using Assumption 4.5 we find that

$$\|e(t)\| \le C_1 e^{-\gamma t}\|e(0)\| + \int_0^t \frac{k(\rho)C_1 e^{-\gamma(t-s)}\|e(s)\|}{(t-s)^{\beta+\zeta}}ds, \tag{4.18}$$

provided $v(t)$, $w(t) \in B(0,\rho)$.

By Lemma 10.12 there exists $K > 0$ such that, for $\nu = 1 - \beta - \zeta > 0$,

$$\|e(t)\| \le 2C_1\|e(0)\| \exp[\{K(C_1 k(\rho))^{1/\nu} - \gamma\}t]$$

whilst $v(t), w(t) \in B(0,\rho)$. Choose ρ sufficiently small that

$$K(C_1 k(\rho))^{1/\nu} \le \frac{\gamma}{2} \tag{4.19}$$

and let $v(0) = 0$ so that $v(t) = 0$ for all $t \ge 0$. If $\|w(0)\| \le \sigma := \rho/2C_1$ then, arguing by contradiction, we have that $w(t) \in B(0,\rho)$ for all $t \ge 0$

and, furthermore, that

$$\|w(t)\| \le \rho e^{-\gamma t/2} \quad \forall t \ge 0. \tag{4.20}$$

Hence any two solutions with $v(0), w(0) \in B(0,\sigma)$ satisfy

$$\|v(t) - w(t)\| \le 2C_1 e^{-\gamma t/2} \|v(0) - w(0)\|.$$

If $\tau = 4\log(2C_1)/\gamma$ then

$$\|v(t) - w(t)\| \le e^{-\gamma t/4} \|v(0) - w(0)\| \quad \forall t \ge \tau. \tag{4.21}$$

Thus, by Definition 2.12, contractivity holds at the origin. Converting back to the u variable gives the desired result with

$$\alpha = \gamma/4, \quad \tau = 4\log(2C_1)/\gamma, \quad \sigma = \rho/2C_1 \tag{4.22}$$

and ρ sufficiently small that (4.19) holds. □

Theorem 4.8 (Time Uniform Approximation of Trajectories) *Let Assumption 4.5 hold and let $\bar{u}$ be hyperbolic and stable. Then there exists $\alpha > 0$ such that, for any $u_0 \in B(0,\sigma)$, the solution of (2.9) satisfies*

$$\|S(t)u_0 - \bar{u}\| \le e^{-\alpha t} \|u_0 - \bar{u}\| \quad \forall t \ge \tau.$$

Furthermore, there exists a constant $K = K(\tau, \|\bar{u}\|, \sigma) > 0$ such that, for any $u_0 \in B(0, \sigma/2)$ the semigroup $S(t)$ generated by the solution of (2.9) and the approximate semigroup $S^h(t)$ satisfy

$$\|S(t)u_0 - S^h(t)u_0\| \le Kh \quad \forall t \ge 2\tau,$$

if h is sufficiently small.

Proof The first result follows by taking $u_1 = u_0, u_2 = \bar{u}$ in Definition 2.12 and noting that $S(t)\bar{u} = \bar{u}\ \forall t \ge 0$. Note that, if $\xi \ge 1$ and $u_0 \in B(0, \sigma/\xi)$, then $S(t)u_0 \in B(0, \sigma/\alpha)$ for all $t \ge \tau$ by (4.21). Let $r = \|\bar{u}\| + \sigma$ and note that

$$\|S(t)u - S^h(t)v\| \le C(t,r)[\|u - v\| + h] \quad \forall u, v \in B(0,\sigma), \tag{4.23}$$

where we may assume that $C(t,r)$ is non-decreasing in t for $t \ge \tau$ without loss of generality — this follows from Assumption 3.2 and Lemma 2.4. Also $C(t,r) < \infty$ for any $t \in (0,\infty)$. Let

$$E(t) = \|S(t)u_0 - S^h(t)u_0\|, \quad E_m := E(m\tau).$$

Assume for induction that

$$\|E_m\| \le \left[\frac{1 - e^{-m\alpha\tau}}{1 - e^{-\alpha\tau}}\right] C(\tau, r)h, \tag{4.24}$$

noting that this holds for $m = 1$ by (4.23) with $u = v = u_0$ and $t = \tau$. Assume that h is sufficiently small so that

$$\|E_m\| \leq \frac{C(\tau, r)h}{1 - e^{-\alpha\tau}} \leq \sigma/2.$$

Since $S(m\tau)u_0 \in B(0, \sigma/2)$ it follows that $S^h(m\tau)u_0 \in B(0, \sigma)$. Now, by Definition 2.12, (4.21), (4.22) and (4.23),

$$\begin{aligned}
\|E_{m+1}\| &= \|S(\tau)S(m\tau)u_0 - S^h(\tau)S^h(m\tau)u_0\| \\
&\leq \|S(\tau)S(m\tau)u_0 - S(\tau)S^h(m\tau)u_0\| \\
&\qquad + \|S(\tau)S^h(m\tau)u_0 - S^h(\tau)S^h(m\tau)u_0\| \\
&\leq e^{-\alpha\tau}\|E_m\| + C(\tau, r)h.
\end{aligned}$$

Hence, by (4.24), we have

$$\begin{aligned}
\|E_{m+1}\| &= e^{-\alpha\tau}\left(\frac{1 - e^{-m\alpha\tau}}{1 - e^{-\alpha\tau}}\right) C(\tau, r)h + C(\tau, r)h \\
&\leq \left(\frac{1 - e^{-(m+1)\alpha\tau}}{1 - e^{-\alpha\tau}}\right) C(\tau, r)h.
\end{aligned}$$

Thus (4.24) holds for all $m \geq 1$ by induction so that

$$\|E_m\| \leq \frac{C(\tau, r)h}{1 - e^{-\alpha\tau}} \quad \forall m \geq 1.$$

It remains to fill in the error between times $t = m\tau$ for $m \geq 2$. By (4.23), if $t = m\tau + T$ with $m \geq 1$ and $T \in [\tau, 2\tau)$, then

$$\begin{aligned}
\|S(t)u_0 - S^h(t)u_0\| &\leq \|S(T)S(m\tau)u_0 - S^h(T)S^h(mT)u_0\| \\
&\leq C(T, r)[\|E_m\| + h] \\
&\leq C(2\tau, r)[\|E_m\| + h].
\end{aligned}$$

The result follows. □

The following is a straightforward corollary of Theorem 4.8.

Corollary 4.9 (Time Uniform Approximation of Trajectories) *Let Assumption 4.5 hold and let $\bar{u}$ be a hyperbolic, stable equilibrium point. Then, for any $u_0, u_0^h \in B(0, \sigma/2)$, there exists $K = K(\tau, \|\bar{u}\|, \sigma)$ such that*

$$\|S(t)u_0 - S^h(t)u_0^h\| \leq e^{-\alpha t}\|u_0 - u_0^h\| + Kh \quad \forall t \geq 2\tau,$$

for all h sufficiently small.

Proof Note that, by Theorem 4.8 and Definition 2.12,

$$\begin{aligned}\|S(t)u_0 - S^h(t)u_0^h\| &\leq \|S(t)u_0 - S(t)u_0^h\| + \|S(t)u_0^h - S^h(t)u_0^h\| \\ &\leq e^{-\alpha t}\|u_0 - u_0^h\| + Kh\end{aligned}$$

for all $t \geq 2\tau$. □

Corollary 4.10 (Time Uniform Approximation of Trajectories) *Let Assumption 4.5 hold and let $\bar{u}$ be hyperbolic and stable. Assume also that the solution of (2.9) satisfies $u(t) \to \bar{u}$ as $t \to \infty$. Then there exists a constant $K_1 = K_1(\tau, \|\bar{u}\|, u_0) > 0$ such that the semigroup $S(t)$ generated by the solution of (2.9) and the approximate semigroup $S^h(t)$ satisfy*

$$\|S(t)u_0 - S^h(t)u_0\| \leq K_1 h \quad \forall t \geq 2\tau,$$

for all h sufficiently small.

Proof If $u_0 \in B(0, \sigma/2)$ the result follows directly from Theorem 4.8. If $u_0 \notin B(0, \sigma/2)$ then there exists $T = T(u_0)$ such that $S(t)u_0 \in B(0, \sigma/4)$ for all $t \geq T$, since $u(t) \to \bar{u}$. Without loss of generality we may assume that $T \geq 2\tau$. By Assumption 3.2 there is $C = C(T + 2\tau, \|u_0\|)$ such that

$$\|S(t)u_0 - S^h(t)u_0\| \leq Ch \quad \forall t \in [2\tau, T + 2\tau].$$

By choosing h sufficiently small we have $S^h(t)u_0 \in B(0, \sigma/2)$. Applying Corollary 4.9 gives $K = K(\tau, \|\bar{u}\|, \sigma)$ such that

$$\begin{aligned}\|S(t)u_0 - S^h(t)u_0\| &= \|S(t-T)S(T)u_0 - S^h(t-T)S^h(T)u_0\| \\ &\leq e^{-\alpha(t-T)}\|S(T)u_0 - S^h(t)u_0\| + Kh \\ &\leq (C + K)h,\end{aligned}$$

for all $t \geq T + 2\tau$. This completes the proof by defining $K_1 = C + K$. □

Important remark Note that K, the error constant, depends explicitly on u_0. Furthermore it is not necessarily uniform in $u_0 \in B(0, \rho)$. This is since it is possible for there to be a sequence ${u_0}^{(i)} \to {u_0}^{(\infty)}$ for which $S(t){u_0}^{(i)} \to \bar{u}$ for each i but $S(t){u_0}^{(\infty)}$ does not tend to $\bar{u}$. By continuity it follows that $T({u_0}^{(i)}) \to \infty$ as $i \to \infty$ and hence $K(\tau, \|\bar{u}\|, u_0)$ is not uniformly bounded.

This lack of uniformity is undesirable in some circumstances since $T(u_0)$ is usually not known a priori. By weakening the notion of approximation to piecewise approximation, the non-uniformity with respect to initial data can be overcome in some circumstances. See Section 8. □

4.4 Phase portraits near a saddle

In this section our aim is to construct solutions of (2.9) (or equivalently (4.2)) in the neighborhood of a hyperbolic equilibrium point $\bar{u} \in \mathcal{E}$ (or equivalently $v = 0$) of saddle type. The union of these solutions will form a "local phase portrait" near $\bar{u}$. We will then show that, under Assumptions 3.2, each of these solutions can be approximated by $S^h(t)$ *independently of the time interval* over which the solution is being considered; thus the complete local phase portrait perturbs smoothly with h. Recall that standard error estimates involve constants which grow exponentially with the time interval under consideration.

By virtue of the smallness properties of $g(\bullet)$ given by Assumptions 4.5 it is reasonable to expect that, for hyperbolic equilibria, the properties of the linear equation $w_t + Cw = 0$ describe the dynamics of solutions to (4.2) in the neighborhood of $v = 0$. Our aim is to prove such a result and then show continuity with respect to the perturbations introduced by the approximating semigroup of Assumption 3.2.

We now formulate the solution of (4.2) as a mapping over time interval T. Recalling (4.14), (4.17) we may write

$$v(t) = L(t)v(0) + G(v(0), t) \tag{4.25}$$

where

$$L(t) := e^{-Ct}, \quad G(v,t) := \int_0^t L(t-s)g(\bar{S}(s)v)ds. \tag{4.26}$$

Now we define $t_n = nT$, $v_n = v(t_n)$ and deduce from (4.25) that

$$v_{n+1} = Lv_n + G(v_n) \tag{4.27}$$

where $L := L(T)$ and $G(\bullet) := G(\bullet, T)$. Using the spectral projections $\mathcal{P}$ and $\mathcal{Q}$ we can decompose v_n as $v_n = p_n + q_n$ where $p_n = \mathcal{P}v_n \in Y$ and $q_n = \mathcal{Q}v_n \in Z$ to obtain

$$\begin{aligned} p_{n+1} &= Lp_n + \mathcal{P}G(p_n + q_n), \\ q_{n+1} &= Lq_n + \mathcal{Q}G(p_n + q_n). \end{aligned} \tag{4.28}$$

This follows since $\mathcal{P}$ and $\mathcal{Q}$ commute with L since they commute with C. This decomposition of the variable v will also be useful to us when studying local unstable manifolds in Section 5. We now prove a lemma summarizing the important properties of L and G which we need.

Lemma 4.11 *For any $a \in (0,1)$ there exists $T^* > 0$ such that for all $T \geq T^*$*

$$\begin{aligned} &\|L^{-1}v\| \leq a\|v\| \quad \forall v \in Y, \\ &\|Lv\| \leq a\|v\| \quad \forall v \in Z. \end{aligned} \tag{4.29}$$

Furthermore, if Assumption 4.5 *holds, then there exist* $K_i(t) > 0,\ i = 1, 2$ *such that*

$$\begin{aligned} \|\mathcal{R}[G(v,t) - G(w,t)]\| &\le K_1(t)k(\rho)\|v - w\| \\ \|\mathcal{R}G(v,t)\| &\le K_2(t)k(\rho)\rho \end{aligned} \tag{4.30}$$

for all $v, w \in B(0,\rho)$ *and for* $\mathcal{R} = I, \mathcal{P}, \mathcal{Q}$, *where*

$$K_1(t), K_2(t) \to 0 \ \ as \ \ t \to 0 \tag{4.31}$$

and

$$\sup_{0 \le t \le 2T^*} K_1(t) = K_1 < \infty, \qquad \sup_{0 \le t \le 2T^*} K_2(t) = K_2 < \infty. \tag{4.32}$$

Proof To prove (4.29) apply Theorem 10.15 and choose T^* such that $C_1 e^{-\gamma T^*} = a$ as in the derivation of (4.3). Now let $v, w \in B(0,\rho)$. Note that, by Lemma 2.4, we have that there exists $C = C(t) > 0$ such that

$$\|\bar{S}(t)v - \bar{S}(t)w\| \le C(t)\|v - w\| \quad \forall v, w \in B(0,\rho); \tag{4.33}$$

here $C(0)$ is bounded and, without loss of generality, we may assume $C(t)$ is monotonically increasing in t.

To prove (4.30) note that, by Lemmas 10.6 and 10.13, Assumptions 4.5, (4.26) and the boundedness of the projections $\mathcal{P}, \mathcal{Q}$ there exists $K > 0$ such that

$$\|\mathcal{R}[G(v,t) - G(w,t)]\| \le \int_0^t \frac{Kk(\rho)}{(t-s)^{\beta+\zeta}} \|[\bar{S}(s)v - \bar{S}(s)w]\| ds.$$

Thus we have from (4.33)

$$\|\mathcal{R}[G(v,t) - G(w,t)]\| \le \frac{KC(t)t^{1-\beta-\zeta}k(\rho)}{(1-\beta-\zeta)}\|v - w\|. \tag{4.34}$$

A bound for $\|\mathcal{R}G(v)\|$ then follows in a similar fashion to the calculation of the Lipschitz constant in (4.34); we obtain

$$\|\mathcal{R}[G(v)]\| \le \frac{Kt^{1-\beta-\zeta}k(\rho)\rho}{(1-\beta-\zeta)}. \tag{4.35}$$

The required result follows since $C(t)$ is bounded as $t \to 0_+$ and since $\beta + \zeta < 1$. □

Remark Note that $K_1(t)$ and $K_2(t)$ are $\mathcal{O}(t^{1-\beta-\zeta})$; if $\zeta + \beta > 0$ then this presents certain technical difficulties not present when considering ordinary differential equations — for many of the arguments we present it is not possible to let $t \to 0$. Instead it is neceesary to fix $t > 0$ and develop a separate argument to show that the objects of interest persist as $t \to 0$. □

For the remainder of Section 4, whenever estimating v or its numerical counterpart, we employ the norm on V given by

$$\|v\|_V = \max\{\|\mathcal{P}v\|, \|\mathcal{Q}v\|\}. \tag{4.36}$$

This choice of norm simplifies the exposition considerably since the majority of the estimation takes place either in Y or Z where $\|\bullet\|_V \equiv \|\bullet\|$. Note that

$$\|v\| \leq 2\|v\|_V. \tag{4.37}$$

Thus by (4.30), if $\|v\|_V \leq \varepsilon$ and $T \in [T^*, 2T^*]$, then

$$\begin{aligned} \|\mathcal{R}[G(v,t) - G(w,t)]\| &\leq K_1(t)k(2\varepsilon)\|v - w\| \\ \|\mathcal{R}G(v,t)\| &\leq 2K_2(t)k(2\varepsilon)\varepsilon \end{aligned} \tag{4.38}$$

In the remainder of this section, the ball $B(0,r)$, for any $r > 0$, is always measured in the $\|\bullet\|_V$ norm wherever the v variable is under consideration.

We now seek a solution of (4.28) which satisfies the boundary conditions

$$p_m = \xi \in Y, \quad q_0 = \eta \in Z \tag{4.39}$$

where $\|\xi\|, \|\eta\| \leq \varepsilon/2$. Recalling that $v_n = p_n + q_n$, induction on (4.28) gives

$$\begin{aligned} p_n &= L^{n-m}p_m - \sum_{j=n}^{m-1} L^{n-1-j}\mathcal{P}G(v_j), \\ q_n &= L^n q_0 + \sum_{j=0}^{n-1} L^{n-1-j}\mathcal{Q}G(v_j). \end{aligned} \tag{4.40}$$

We wish to solve (4.40) subject to (4.39) for arbitrary $m > 0$. Such a solution corresponds to solving the eqn (4.2) with boundary conditions conditions specified in Y at $t = mT$ and in Z at $t = 0$; since m is arbitrary, the time of flight between these points can be arbitrarily large. Finding all such solutions with ε small corresponds to constructing the local phase portrait for equation (4.2) near $v = 0$. Figure 1 illustrates the situation pictorially. At time $t = 0$ each solution lies on the dotted line intersecting the Z-axis at η; at time $t = mT$ each solution lies on the dotted line intersecting the Y-axis at ξ. Four solutions are shown corresponding to four values of m: $m_1 \leq m_2 \leq m_3 \leq m_4$. As $m \to \infty$ the solution of (4.39, 4.40) gets closer and closer to the origin.

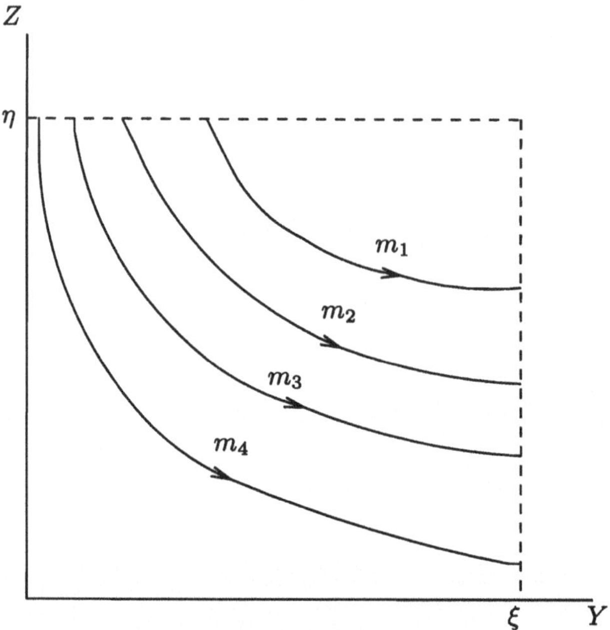

FIG. 1.

Example 4.12 Consider the equation

$$v_t + Av = 2\pi^2 v$$

where A is given by Example 2.1. In this case $C = A - 2\pi^2 I$ and $g(\bullet) \equiv 0$. Thus C has eigenvalues $\{\lambda_m\}_{m=1}^{\infty}$, where $\lambda_m = m^2\pi^2 - 2\pi^2$, and corresponding eigenfunctions $\{\sin(m\pi x)\}_{m=0}^{\infty}$. Hence we have

$$\mathcal{P}V = \text{span}\{\sin(\pi x)\}, \quad \mathcal{Q}V = \text{span}\{\sin(2\pi x), \sin(3\pi x), \ldots\}.$$

The analog of (4.39) for this problem is

$$\mathcal{P}v(\tau) = \xi_1 \sin(\pi x), \quad \mathcal{Q}v(0) = \sum_{j=2}^{\infty} \eta_j \sin(j\pi x),$$

where $\tau = mT$. Separation of variables yields the unique solution

$$v(t) = \xi_1 e^{\pi^2(t-\tau)} \sin(\pi x) + \sum_{j=2}^{\infty} \eta_j e^{(2-j^2)\pi^2 t} \sin(j\pi x).$$

In the following we aim to generalize situations like the one shown in the previous example to incorporate the addition of the small nonlinear term $g(\bullet)$. Our approach to this problem will be to use a contraction mapping argument exploiting the fact that e^{-Ct} is contractive on Z and e^{Ct} is contractive on Y, by virtue of Lemma 10.15 and (4.29). We need an appropriate functional setting for this argument and we let $\mathcal{V} = \{v_n\}_{n=0}^m$ denote an element of the product Hilbert space $\Psi = \{V\}^m$ and define

$$\|\mathcal{V}\|_\infty = \max_{0\leq n\leq m} \|v_n\|_V.$$

We also define the set

$$\Psi_\epsilon = \{\mathcal{V} \in \Psi : \|\mathcal{V}\|_\infty \leq \varepsilon\}.$$

To study (4.40), (4.39) we use the contraction mapping theorem in Ψ_ϵ. We generate iterates $\mathcal{V}^l = \{v_n^l\}_{n=0}^m$ through the definition $\mathcal{V}^{l+1} = M\mathcal{V}^l$ where

$$M\mathcal{V} = \{Mp_n + Mq_n\}_{n=0}^m$$

and where $Mp_n \in Y$ and $Mq_n \in Z$ are defined by

$$\begin{aligned} Mp_n &= L^{n-m}\xi - \textstyle\sum_{j=n}^{m-1} L^{n-1-j}\mathcal{P}G(v_j), \\ Mq_n &= L^n\eta + \textstyle\sum_{j=0}^{n-1} L^{n-1-j}\mathcal{Q}G(v_j). \end{aligned} \tag{4.41}$$

Clearly a fixed point of M is a solution of (4.39), (4.40).

From the bounds (4.29) in Lemma 4.11 it follows that

$$\begin{aligned} \sum_{j=n}^{m-1} \|L^{n-j-1}v_j\| &\leq \sum_{j=n}^{m-1} a^{1+j-n}\|v_j\| \leq \sum_{l=1}^{m-n} a^l\|\mathcal{V}\|_\infty \leq \frac{\|\mathcal{V}\|_\infty}{1-a} \qquad \forall v_j \in Y, \\ \sum_{j=0}^{n-1} \|L^{n-j-1}v_j\| &\leq \sum_{j=0}^{n-1} a^{n-j-1}\|v_j\| \leq \sum_{l=0}^{n-1} a^l\|\mathcal{V}\|_\infty \leq \frac{\|\mathcal{V}\|_\infty}{1-a} \qquad \forall v_j \in Z. \end{aligned} \tag{4.42}$$

We may now prove

Theorem 4.13 *Let Assumption* 3.2 *hold and let a and T^* be as in Lemma 4.11. Assume that $T \in [T^*, 2T^*]$ and that ϵ is chosen so that*

$$4\max\{K_1, 2K_2\}k(2\varepsilon) \leq 1 - a.$$

Then, for any $m > 0$ and any $\xi \in Y$, $\eta \in Z$ with $\|\xi\|, \|\eta\| \leq \frac{\varepsilon}{2}$ there exists a unique solution of (4.27) subject to (4.39) satisfying

$$\max_{0\leq n\leq m} \|\mathbf{v}_n\|_V \leq \varepsilon.$$

Proof Note that the bound on ε in the theorem implies that

$$2\max\{K_1, 2K_2\}k(2\varepsilon) \leq 1 - a. \tag{4.43}$$

The bound has been chosen to be robust under doubling of K_1 and K_2 to allow for an analogous proof for the approximation of (2.9) under Assumptions 3.2.

Since (4.28) is equivalent to (4.40) we examine (4.40), (4.39). To prove the result we show that $M : \Psi_\epsilon \mapsto \Psi_\epsilon$ and is a contraction. From (4.41) we have, using (4.29), (4.38), (4.30) and (4.42), that

$$\begin{aligned} \|Mp_n\| &\leq \|L^{n-m}\xi\| + \sum_{j=n}^{m-1} \|L^{n-j-1}\mathcal{P}G(v_j)\| \\ &\leq \|\xi\| + \frac{1}{1-a}K_2 k(2\varepsilon)2\varepsilon. \end{aligned}$$

Hence, by (4.43) it follows that

$$\|Mp_n\| \leq \varepsilon \quad \forall n : 0 \leq n \leq m.$$

Likewise it may be shown that

$$\|Mq_n\| \leq \varepsilon \quad \forall n : 0 \leq n \leq m$$

so that $M\mathcal{V} \in \Psi_\epsilon$.

To show that the mapping contracts, consider (4.41) with $p_n \mapsto x_n$, $q_n \mapsto y_n, v_n \mapsto w_n$, define $w_n = x_n + y_n$ and set $\Omega = \{w_n\}_{n=0}^m$. Then, using (4.29), (4.30) and (4.42) we obtain from (4.41)

$$\|Mp_n - Mx_n\| \leq \frac{1}{1-a}K_1 k(2\varepsilon)\|\mathcal{V} - \Omega\|_\infty \quad \forall n : 0 \leq n \leq m$$

and

$$\|Mq_n - Mz_n\| \leq \frac{1}{1-a}K_1 k(2\varepsilon)\|\mathcal{V} - \Omega\|_\infty \quad \forall n : 0 \leq n \leq m.$$

Thus it follows from (4.43) that

$$\|\mathcal{V}^{k+1} - \Omega^{k+1}\|_\infty \leq \frac{1}{2}\|\mathcal{V}^k - \Omega^k\|_\infty.$$

Hence $M : \Psi_\varepsilon \mapsto \Psi_\varepsilon$ is a contraction and the result follows. □

We now use the previous lemma to solve eqn (4.2), and hence (2.9), subject to specified boundary conditions. Consider the problem

$$u_t + Au = f(u), \quad \mathcal{P}[u(\tau) - \bar{u}] = \xi, \; \mathcal{Q}[u(0) - \bar{u}] = \eta. \tag{4.44}$$

Corollary 4.14 (Phase Portraits) *Let Assumption* 4.5 *hold. Assume that* ϵ *is chosen so that*

$$4\max\{K_1, 2K_2\}k(2\varepsilon) \leq 1 - a.$$

Then, for any $\tau \geq T^*$ *and any* $\xi \in Y$, $\eta \in Z$ *with* $\|\xi\|, \|\eta\| \leq \frac{\varepsilon}{2}$ *there exists a solution* $u(t)$*, unique in* $\overline{B(\bar{u}, 2\varepsilon)}$*, of (4.44).*

Proof Note that existence and uniqueness for (4.44) is equivalent to the following problem with which we work during the course of the proof:

$$v_t + Cv = g(v), \quad \mathcal{P}v(\tau) = \xi, \ \mathcal{Q}v(0) = \eta. \tag{4.45}$$

The existence of a solution of (4.45) for $\tau = mT$ with $T \in [T^*, 2T^*]$ and any integer m follows from Theorem 4.13. Since any real number larger than T^* can be expressed as an integer multiple of a number in the interval $[T^*, 2T^*]$ the existence result follows. Assume for the purposes of contradiction that this solution of (4.45) is not unique in $B(0, \varepsilon)$. Then any other solution also generates a solution of (4.28) subject to (4.39); by the uniqueness of Theorem 4.13 the second solution must agree with the first at the points $t = nT$ and hence, by uniqueness for the initial value problem, they must agree everywhere. Uniqueness for (4.45) follows. By (4.37) the constant 2 appears when changing norms from v to u. □

We now extend this analysis to the approximate semigroup $S^h(t)$ and prove an approximation result for the trajectories constructed in Corollary 4.14 which is independent of the time τ. We start by converting the approximation to new coordinates and introduce the semigroup $\bar{S}^h(t) : V \mapsto V$ defined by

$$\bar{S}^h(t)v = S^h(t)(\bar{u} + v) - \bar{u}. \tag{4.46}$$

Compare this with (4.14). Here $v = \bar{u}^h - \bar{u}$ is a fixed point of $\bar{S}^h(t)$ for all $t > 0$; it is simply the equilibrium given by Theorem 4.3 in the new coordinates. We define the approximation error in the new coordinates by

$$\bar{E}(v; t) = \bar{S}(t)v - \bar{S}^h(t)v. \tag{4.47}$$

The Fréchet derivative of $\bar{E}(v; t)$ with respect to v exists by Assumptions 3.2. When evaluated at a point $w \in V$, we denote it by $d\bar{E}(w; t)$. The following lemma is a straightforward consequence of Assumptions 3.2.

Lemma 4.15 *For all $v \in B(0, R)$ and all $t \in \mathbf{S}$, $t > 0$, there exist constants $C_i = C_i(t, R) < \infty, i = 1, 2$, and a function $\kappa : \mathbb{R}^+ \mapsto \mathbb{R}^+$ such that*

$$\|\bar{E}(v; t)\| \leq C_1 h$$

$$\|d\bar{E}(v; t)\| \leq C_2 \kappa(h),$$

where $\kappa(h) \to 0$ as $h \to 0_+$.

Now we introduce the notation $V_n = \bar{S}^h(t_n)V_0$ where $t_n = nT$; this is analogous to the notation $v_n = \bar{S}(t_n)v_0$ used in the construction of phase portraits for (4.2). Note that

$$V_{n+1} = \bar{S}^h(T)V_n = \bar{S}(T)V_n - \bar{E}(V_n; T)$$

$$= \quad LV_n + G(V_n) - \bar{E}(V_n;T).$$

Hence we may write

$$V_{n+1} = LV_n + \tilde{G}(V_n) \tag{4.48}$$

where

$$\tilde{G}(v) = G(v) - \bar{E}(v;T). \tag{4.49}$$

Recall T^*, K_1, K_2 from Lemma 4.11. The following properties of $\tilde{G}$ will be needed:

Lemma 4.16 *For any $\rho > 0$ there exists $h^* = h^*(\rho) > 0$ such that for all $h \leq h^*$ and $T \in [T^*, 2T^*]$*

$$\begin{aligned} \|\mathcal{R}[\tilde{G}(v) - \tilde{G}(w)]\| &\leq 2K_1 k(\rho)\|v - w\|, \\ \|\mathcal{R}\tilde{G}(v)\| &\leq 2K_2 k(\rho)\rho \end{aligned} \tag{4.50}$$

for all $v, w \in B(0, \rho)$.

Proof From (4.30), (4.32), (4.49), Lemma 4.15 and Theorem 11.4 we have

$$\begin{aligned} \|\mathcal{R}[\tilde{G}(v) - \tilde{G}(v)]\| &\leq \|\mathcal{R}[G(v) - G(w)]\| + \|\mathcal{R}[\bar{E}(v;T) - \bar{E}(w;T)]\| \\ &\leq K_1 k(\rho)\|v - w\| + \|\mathcal{R}\textstyle\int_0^1 d\bar{E}(sv + (1-s)w;T)[v-w]ds\| \\ &\leq K_1 k(\rho)\|v - w\| + C\kappa(h)\|v - w\|. \end{aligned}$$

By choice of h sufficiently small the first result follows. The second result may be proved similarly. □

By applying appropriate projections to (4.48) we find that

$$\begin{aligned} P_{n+1} &= LP_n + \mathcal{P}\tilde{G}(P_n + Q_n), \\ Q_{n+1} &= LQ_n + \mathcal{Q}\tilde{G}(P_n + Q_n), \end{aligned} \tag{4.51}$$

where $P_n = \mathcal{P}V_n \in Y$ and $Q_n = \mathcal{P}V_n \in Z$. Using Lemma 4.16 we prove existence of solutions to (4.48) subject to

$$P_m = \xi \in Y, \quad Q_0 = \eta \in Z. \tag{4.52}$$

Theorem 4.17 *Let Assumption 4.5 hold. Assume that $T \in [T^*, 2T^*]$ and that ϵ is chosen so that*

$$4\max\{K_1, 2K_2\}k(2\varepsilon) \leq 1 - a.$$

Then if $h \in (0, h^*)$, where $h^* = h^*(\varepsilon)$ is given by Lemma 4.16, there exists $C > 0$ such that, for any $m > 0$ and any $\xi \in Y$, $\eta \in Z$ with $\|\xi\|, \|\eta\| \leq \frac{\varepsilon}{2}$

there exists a unique solution of (4.48) subject to (4.52), satisfying

$$\max_{0\leq n\leq m} \|V_n\| \leq \varepsilon$$

and, furthermore,

$$\max_{0\leq n\leq m} \|v_n - V_n\| \leq Ch,$$

where v_n *is given by Theorem* 4.13.

Proof The problem under consideration is equivalent to finding a solution of (4.51) which satisfies the boundary conditions (4.52), where $\|\xi\|, \|\eta\| \leq \varepsilon/2$ and for any $T \in [T^*, 2T^*]$. As for (4.28), induction on (4.51) gives

$$\begin{aligned} P_n &= L^{n-m}P_m - \textstyle\sum_{j=n}^{m-1} L^{n-1-j}\mathcal{P}\tilde{G}(V_j), \\ Q_n &= L^n Q_0 + \textstyle\sum_{j=0}^{n-1} L^{n-1-j}\mathcal{Q}\tilde{G}(V_j). \end{aligned} \tag{4.53}$$

Thus our objective is to solve (4.53) subject to (4.52).

By analogy with (4.41) we generate iterates $\mathcal{V}^l = \{v_n^l\}_{n=0}^m$ through the definition $\mathcal{V}^{l+1} = M_h\mathcal{V}^l$ where

$$M_h\mathcal{V} = \{M_hP_n + M_hQ_n\}_{n=0}^k$$

and where $M_hP_n \in Y$ and $M_hQ_n \in Z$ are defined by

$$\begin{aligned} M_hP_n &= L^{n-m}\xi - \textstyle\sum_{j=n}^{k-1} L^{n-1-j}\mathcal{P}\tilde{G}(V_j), \\ M_hQ_n &= L^n\eta + \textstyle\sum_{j=0}^{n-1} L^{n-1-j}\mathcal{Q}\tilde{G}(V_j). \end{aligned} \tag{4.54}$$

Clearly a fixed point of M_h is a solution of (4.52), (4.53). The same proof as used in Theorem 4.13 proves the existence of a solution to this problem; note that that proof was constructed to be robust under enlargement of K_1 and K_2 by a factor of 2 and that comparison of Lemmas 4.11, 4.16 shows that this is all that is necessary to extend the proof.

It remains to find the error bound and for this we use the uniform contraction principle. Let $\{v_n\}_{n=0}^m$ denote the fixed point of M found in Theorem 4.13 and let $\{V_n\}_{n=0}^m$ denote the fixed point of M_h. Since the contraction constant is $\frac{1}{2}$ for both M and M_h we obtain from Theorem 11.2

$$\max_{0\leq n\leq m} \|v_n - V_n\|_V \leq 2 \sup_{\|\mathcal{V}\|_\infty \leq \varepsilon} \|M\mathcal{V} - M_h\mathcal{V}\|_\infty.$$

Here we denote the elements of Ψ_ϵ by $\mathcal{V} = \{w_n\}_{n=0}^\infty$ where $w_n \in V$ and we set $w_n = x_n + y_n$ where $x_n \in Y$ and $y_n \in Z$. Then (4.41) and (4.54) show that

$$\|Mx_n - M_hx_n\| \leq \sum_{j=n}^{m-1} \|L^{n-1-j}\mathcal{P}\bar{E}(w_n;T)\|.$$

Thus, by Lemma 4.15 and (4.42), we have

$$\|Mx_n - M_hx_n\| \leq \sum_{j=n}^{m-1} a^{n-1-j}C(T)h \leq \frac{C(T)}{1-a}h.$$

Note that the result is independent of m. A similar result holds for $\|My_n - M_hy_n\|$. Combining these we deduce that, since $C(T)$ may be assumed monotonic increasing for $T \geq T^*$,

$$\max_{0\leq n\leq k} \|v_n - V_n\|_V \leq \frac{C(2T^*)}{1-a}h.$$

□

We now consider the following analog of (4.44): find $u^h(t) \in V$ such that

$$u^h(t) = S^h(t)u^h(0) \quad \mathcal{P}[u^h(\tau) - \bar{u}] = \xi, \; \mathcal{Q}[u^h(0) - \bar{u}] = \eta. \tag{4.55}$$

Theorem 4.18 (Convergence of Phase Portraits) *Let Assumption* 4.5 *hold. Assume also that ϵ is chosen so that*

$$4\max\{K_1, 2K_2\}k(2\varepsilon) \leq 1 - a$$

and $h \in (0, h^*)$, $h^* = h^*(\varepsilon)$ given by Lemma 4.16. *Then there is a constant $C > 0$ such that, for any $\tau > T^*$ and any $\xi \in Y$, $\eta \in Z$ with $\|\xi\|, \|\eta\| \leq \frac{\varepsilon}{2}$ there exists a solution $u^h(t)$, unique in $\overline{B(\bar{u}, 2\varepsilon)}$, of (4.55) satisfying*

$$\sup_{T^*\leq t\leq\tau} \|u(t) - u^h(t)\| \leq Ch,$$

where $u(t)$ is the solution of (4.44) given in Corollary 4.14.

Proof Theorem 4.17 gives the existence of a unique solution in $B(\bar{u}, 2\varepsilon)$ the constant 2 appearing when changing norms by (4.37). The required error estimate at integer multiples of T also follows; the error between these points can be obtained by use of the standard finite time error bound in Assumptions 3.2. □

Instead of (4.39) it is also of interest to study (4.28) subject to

$$\|v_n\|_V \leq \epsilon \; \forall n \geq 0, \quad q_0 = \eta \in Z, \quad \|\eta\| \leq \frac{\varepsilon}{2}. \tag{4.56}$$

This is equivalent to finding a solution of the problem

$$u_t + Au = f(u), \quad \mathcal{Q}[u(0) - \bar{u}] = \eta, \quad \|\eta\| \leq \frac{\varepsilon}{2}, \quad \|u(t) - \bar{u}\|_V \leq \varepsilon \; \forall t \geq 0. \tag{4.57}$$

The analogous problem for the approximation is

$$u^h(t) = S^h(t)u^h(0), \quad Q[u^h(0)-\bar{u}] = \eta, \quad \|\eta\| \le \frac{\varepsilon}{2}, \quad \|u^h(t)-\bar{u}\|_V \le \varepsilon \, \forall t \ge 0. \tag{4.58}$$

It is straightforward to show that solutions of (4.57) (resp. (4.58)) must approach the equilibrium point $\bar{u}$ (resp. $\bar{u}^h$) as $t \to \infty$. Hence by solving these problems we are constructing a set of points known as the *stable set*. This set actually has a manifold structure but we will not use that fact here. By following the method of proofs of Corollary 4.14 and Theorem 4.18 we obtain

Theorem 4.19 (Convergence of Stable Sets) *Let Assumption* 4.5 *hold and let* a *and* T^* *be as in Lemma 4.11. Assume also that* ϵ *is chosen so that*

$$4\max\{K_1, 2K_2\}k(2\varepsilon) \le 1 - a$$

and $h \in (0, h^*)$, $h^* = h^*(\varepsilon)$ *given by Lemma* 4.16. *Then there is a constant* $C > 0$ *such that, for any* $\tau > T^*$ *and any* $\eta \in Z$ *with* $\|\eta\| \le \frac{\varepsilon}{2}$ *there exist solutions of (4.57), (4.58), unique in* $B(\bar{u}, 2\varepsilon)$, *satisfying*

$$\sup_{T^* \le t \le \tau} \|u(t) - u^h(t)\| \le Ch.$$

4.5 Bibliography

The persistence of hyperbolic equilibrium points under appropriate perturbations is a consequence of the implicit function theorem in a Hilbert space — see Chow and Hale [16], for example. Theorem 4.3 simply adapts the proof of the implicit function theorem to our semigroup formulation of the problem, in particular to allow for the fact that the perturbations in Assumption 3.2 are not bounded as $t \to 0$.

Uniformly valid error estimates for $t \in (0, \infty)$ when the true solution approaches an exponentially stable equilibrium point are proved for finite element approximations of reaction-diffusion equations and Navier–Stokes equations in Larsson [73] and Heywood and Rannacher [55] respectively, and for finite difference methods applied to a reaction diffusion equation in Sanz-Serna and Stuart [85]. In the context of ordinary differential equations such results may be found in Stetter [87].

The existence of phase portraits for nonlinear problems can be deduced in finite dimensions from the Hartman–Grobman theorem [53]. Some infinite-dimensional versions of the Hartman–Grobman theorem are also available — see, for example, Bates and Lu [6]. The proof given here is motivated by the construction of stable and unstable sets in Henry [54] and its generalization to the phase portraits of reaction-diffusion equations in Larsson and Sanz-Serna [75]. See also Alouges and Debussche [1]. The idea of using the uniform contraction principle to understand the effect of perturbations is pervasive in the theory of dynamical systems; see Hale [49], Henry [54], Chow and Hale [16], for example. In the context of ordinary differential equations and numerical approximation Beyn [8] was the first to study phase portraits. That work has been taken further by Garay [41] [42] [43] [44] [45] [46] [47] [48].

Although not covered here the properties of periodic solutions under approximation has also been widely studied — see Beyn [7], Braun and Hershenov [13], Doan [25], Eirola [27] and Pugh and Shub [82] for ordinary differential equations, Alouges and Debussche [2] and Titi [95] for partial differential equations. The idea of local phase portrait for periodic solutions, and the effect of perturbation, is also analysed by Alouges and Debussche [2].

5 Unstable manifolds

5.1 Introduction

In this section we study the unstable set of an equilibrium point — that is the set of points from which solutions defined backwards in time converge to the equilibrium point. In Section 5.2 we introduce the definitions and background theory sufficient for the remainder of Section 5. In this regard Lemma 5.2 is crucial since it shows how the unstable set may be broken into two parts, the first a local part (the local unstable manifold) near the equilibrium point and the second being found by evolving the boundary of the first part forward in time. The local part of the unstable set has a manifold structure — it can be represented as a graph. Theorem 5.3 is at the core of the construction of this graph whilst Corollary 5.4 shows that the graph has the desired properties. Again by use of the uniform contraction principle from Appendix B, we incorporate the effect of perturbation into the graph representing the local unstable manifold. In Section 5.4 we return to the second part of the unstable set and use its characterization as the evolution of the boundary of the local unstable manifold to estimate the effect of perturbation. See Theorem 5.7.

5.2 Background theory

We start with a precise definition of the objects of interest to us in Section 5.

Definition 5.1 *The unstable set of an equilibrium point $\bar{u}$ of (2.9) is the set*

$$W^u(\bar{u}) := \{u_0 \in V : \text{ a negative orbit } \{\varphi(t), t \leq 0\} \text{ exists through } u_0 \text{ and } \varphi(t) \to \bar{u} \text{ as } t \to -\infty\}.$$

The local unstable manifold of $\bar{u}$ is the set

$$W^{u,\varepsilon}(\bar{u}) := \{u_0 \in W^u(\bar{u}) : \|u(t) - \bar{u}\| \leq \varepsilon \; \forall t \leq 0\}.$$

Analogous definitions hold for $S^h(t)$ and the notation $(W_h{}^{u,\varepsilon}(\bar{u}^h))$ $W_h{}^u(\bar{u}^h)$ is used to denote the (local) unstable set (manifold) of an equilibrium $\bar{u}^h$ satisfying $S^h(t)\bar{u}^h = \bar{u}^h \;\; \forall t > 0$.

Note that the unstable set is invariant whilst the local unstable manifold is negatively invariant. For brevity, however, we will refer to the local unstable manifold as invariant in the following proofs. The basic idea of Section 5 is first to prove convergence of the local unstable manifold and then to use this as a building block to prove convergence of the global unstable set by means of a compactness argument. The following lemma is essential in this regard.

Lemma 5.2

The unstable set $W^u(\bar{u})$ of (2.9) is invariant and, furthermore,

$$W^u(\bar{u}) = W^{u,\varepsilon}(\bar{u}) \cup \bigcup_{t>0} S(t)\Gamma \tag{5.1}$$

where

$$\Gamma = W^{\bar{u},\varepsilon}(\bar{u}) \bigcap \partial B(\bar{u}, \varepsilon). \tag{5.2}$$

Furthermore, if $W^u(\bar{u})$ is contained in a bounded set $B \subset V$, then the set $\bigcup_{t \geq 0} S(t)\Gamma$ is relatively compact.

Proof It follows from the definition that, if $u \in W^u(\bar{u})$ then, for every $\tau > 0$ there exists $v^\tau \in V$ such that

$$\begin{aligned} S(\tau)v^\tau &= u \\ v^\tau \longrightarrow \bar{u} &\text{ as } \tau \longrightarrow \infty. \end{aligned} \tag{5.3}$$

The converse is also true: if (5.3) holds for every $\tau > 0$ then $u \in W^u(\bar{u})$.

Let $u \in W^u(\bar{u})$. Then

$$S(\tau)S(t)v^\tau = S(\tau + t)v^\tau = S(t)u$$

and, since (5.3) holds, we deduce that $S(t)u \in W^u(\bar{u})$ so that $S(t)W^u(\bar{u}) \subseteq W^u(\bar{u})$. Furthermore, since $S(t)v^t = u$ we have that, for every $t > 0$ and every $\tau > t$, $S(\tau - t)v^\tau = v^t$. Thus, from (5.3), we deduce that $v^t \in W^u(\bar{u})$. Thus $W^u(\bar{u}) \subseteq S(t)W^u(\bar{u})$ and the first part of the proof is complete.

We now establish (5.1) and (5.2). First we show that

$$W^u(\bar{u}) \subseteq W^{u,\varepsilon}(\bar{u}) \cup \bigcup_{t>0} S(t)\Gamma. \tag{5.4}$$

Let $u \in W^u(\bar{u})\backslash W^{u,\varepsilon}(\bar{u})$. If $u \notin \overline{B(\bar{u}, \varepsilon)}$ then, since (5.3) holds it follows that $v^0 = u$ and, by continuity, there exists $t > 0$ such that $S(t)v^t = u$ and $v^t \in \Gamma$. On the other hand, if $u \in \overline{B(\bar{u}, \varepsilon)}$ then $\exists t > 0, v^t \in V : S(t)v^t = u$ with $v^t \in \Gamma$ since otherwise we have $u \in W^{u,\varepsilon}(\bar{u})$. Thus (5.4) holds.

Now we show that

$$W^u(\bar{u}) \supseteq W^{u,\varepsilon}(\bar{u}) \cup \bigcup_{t>0} S(t)\Gamma.$$

If

$$u \in \bigcup_{t>0} S(t)\Gamma$$

then there exists $w \in \Gamma$ such that $S(t)w = u$. Furthermore, since $w \in W^{u,\varepsilon}(\bar{u})$ it follows that $u \in W^u(\bar{u})$ by (5.3) and the result is proved.

Finally we assume that $W^u(\bar{u})$ is contained in a bounded set $B \subset V$. Thus $S(t)\Gamma \subset B$. By applying the compactness implied by Assumptions 2.3 on overlapping intervals $[-T, T]$, $[0, 2T]$, $[T, 3T]$, $[2T, 4T], \ldots$ and noting that $S(mT)\Gamma \in B$ for all integers $m \geq -1$ we deduce that there exists $K = K(B, T) > 0, \eta > \beta$ such that

$$|y|_\eta \leq K \quad \forall y \in \bigcup_{t\geq 0} S(t)\Gamma;$$

this establishes the relative compactness. □

5.3 Local unstable manifolds

Throughout the remainder of Section 5, Assumption 4.5 is assumed to hold. We consider the mapping (4.28) where L and G are given by (4.2), (4.26). Let

$$\bar{Y} = \{p \in Y : \|p\| \leq \varepsilon\}$$

and seek $\Phi \in C(\bar{Y}, Z)$:

$$q_n = \Phi(p_n) \Longleftrightarrow q_{n+1} = \Phi(p_{n+1}) \quad \forall n : \|p_n\|, \|p_{n+1}\| \leq \varepsilon. \tag{5.5}$$

As we shall see, the graph of the function Φ gives the local unstable manifold of $\bar{u}$. We shall look for Φ lying in the space

$$\begin{aligned}\Gamma(\varepsilon, \alpha) = \{\Phi \in C(\bar{Y}, Z) : \|\Phi\|_C = \sup\nolimits_{p\in\bar{Y}} \|\Phi(p)\| \leq \varepsilon, \\ \|\Phi(p_1) - \Phi(p_2)\| \leq \alpha\|p_1 - p_2\| \quad \forall p_1, p_2 \in \bar{Y}\}.\end{aligned}$$

The subscript C in the norm on $C(\bar{Y}, Z)$ is simply to denote the space of continuous functions in which Φ lies. Recalling T^* from Lemma 4.11, we can now prove

Theorem 5.3 *Let Assumption 4.5 hold and assume that $T \in [T^*, 2T^*]$. Then there exists $\varepsilon^* > 0$ such that, for all $\varepsilon \in (0, \varepsilon^*]$ and for all $\alpha \in [1, 2]$ there exists a unique $\Phi = \Phi_{\varepsilon,\alpha} \in \Gamma(\varepsilon, \alpha)$ such that (5.5) holds for (4.28) and, furthermore,*

$$\|q_m - \Phi(p_m)\| \leq \left(\frac{1+a}{2}\right)^m \|q_0 - \Phi(p_0)\| \tag{5.6}$$

for all m such that $v_n \in B(0,\varepsilon)$ for $1 \leq n \leq m$. Finally $\Phi_{\varepsilon,\alpha}$ is independent of ε, α in the sense that, if $\varepsilon_1 \leq \varepsilon_2$ and $\alpha_1 \leq \alpha_2$,

$$\Phi_{\varepsilon_1,\alpha_1}(p) \equiv \Phi_{\varepsilon_2,\alpha_2}(p) \quad \forall p : \|p\| \leq \varepsilon_1.$$

Proof We apply Theorem 12.3 with $r = \gamma = \varepsilon, b = a^{-1} > 1, \mu = (1+a)/2, B_1 = 2K_1k(2\varepsilon)$ and $B_2 = 4K_2k(2\varepsilon)\varepsilon$. As in Section 4, B_1 and B_2 have been doubled to allow for the incorporation of perturbation error at a later stage.

Note that $r \geq (b-1)^{-1}B_2$ by choice of ε sufficiently small, since $k(\varepsilon) \to 0$ as $\varepsilon \to 0$. Points (C1)–(C4) follow similarly. Since $\Gamma(\varepsilon_1, \alpha_1) \subseteq \Gamma(\varepsilon_2, \alpha_2)$ if $\varepsilon_1 \leq \varepsilon_2$ and $\alpha_1 \leq \alpha_2$, the independence of Φ from ε and α follows from the uniqueness implied by the contraction mapping principle. □

Corollary 5.4 (Local Unstable Manifold) *Let Assumption (4.5) hold. Then there exists $\varepsilon_c > 0$ such that, for all $\varepsilon \in (0, \varepsilon_c]$, the origin for (4.2) has local unstable manifold representable as*

$$W^{u,\varepsilon}(0) = \mathcal{M} := \{v \in V : \mathcal{P}v \in \bar{Y}, \mathcal{Q}v = \Phi(\mathcal{P}v)\}$$

where $\Phi \in \Gamma(\varepsilon, 1)$ satisfies (5.5) and (5.6).

Proof For this proof it is necessary simply to verify: (i) that the manifold constructed by means of the graph in Theorem 5.3 is actually invariant for all $t > 0$ and not just $T \in [T^*, 2T^*]$; (ii) that the invariant manifold is the unstable manifold.

For (i) note that Φ is constructed as a fixed point of the mapping T given by

$$\begin{aligned} p &= L\xi + \mathcal{P}G(\xi + \Psi(\xi)), \\ (T\Psi)(p) &= L\Psi(\xi) + \mathcal{Q}G(\xi + \Psi(\xi)). \end{aligned} \tag{5.7}$$

Note that T depends on a parameter t and when this dependence is important we shall denote it by $T^{(t)}$. Using the fact that L and G are constructed through one parameter semigroups e^{-Ct} and $\bar{S}(t)$ it may be shown that $T^{(t)} \bullet T^{(s)} = T^{(s)} \bullet T^{(t)}$. Hence the fixed point constructed in Theorem 5.3 is independent of $t \in [T^*, 2T^*]$ by Theorem 11.3(i). To apply Theorem 11.3(ii) and deduce invariance for all sufficiently small t it is sufficient to prove that

$$T^{(t)}\Phi_{\varepsilon_c,1} \in \Gamma(\varepsilon^*, 2), \tag{5.8}$$

provided that $\varepsilon_c < \varepsilon^*$. Straightforward analysis of (5.7) for t sufficiently small shows that (5.8) holds by continuity.

For (ii), to show that the graph of Φ defines the local unstable manifold, consider the map

$$p_{n+1} = Lp_n + \mathcal{P}G(p_n + \Phi(p_n)).$$

It is straightforward to show that $\Phi(0) = 0$ by performing the contraction argument used in the construction of Φ in Theorem 12.3 in the space Γ appended with the condition $\Psi(0) = 0$; it is necessary to use the fact that $G(0) = 0$. Thus, recalling $B_1 = 2K_1k(2\varepsilon)$ from Theorem 5.3 we have

$$\|p_n\| \leq a\|p_{n+1}\| + 2K_1(1+\alpha)k(2\varepsilon)\|p_n\|.$$

Hence, for ε sufficiently small,

$$\|p_n\| \leq \frac{1+a}{2}\|p_{n+1}\|.$$

Thus, if $p_0 \in \bar{Y}$ then $p_n \in \bar{Y}$ for all $n \leq 0$ so that, by induction, $\|p_n\| \to 0$ as $n \to \infty$. Since $\Phi(0) = 0$ it follows that $v_n \to 0$ as $n \to -\infty$ for all v_0 on the invariant manifold. □

Now we consider the effect of approximation on the invariant manifold. Specifically we study (4.51) and try to find $\Phi \in C(\bar{Y}, Z)$ such that

$$Q_n = \Phi^h(P_n) \Longleftrightarrow Q_{n+1} = \Phi^h(P_{n+1}) \quad \forall n : \|P_n\|, \|P_{n+1}\| \leq \varepsilon. \tag{5.9}$$

Theorem 5.5 *Let Assumption* 4.5 *hold and assume that* $T \in [T^*, 2T^*]$. *Then there exists* $\varepsilon^* > 0$ *such that, for all* $\varepsilon \in (0, \varepsilon^*]$ *and for all* $\alpha \in [1,2]$ *there exists a unique* $\Phi^h = \Phi^h_{\varepsilon,\alpha} \in \Gamma(\varepsilon, \alpha)$ *such that (5.9) holds for (4.51) and, furthermore,*

$$\|Q_m - \Phi^h(P_m)\| \leq \left(\frac{1+a}{2}\right)^m \|Q_0 - \Phi(P_0)\| \tag{5.10}$$

for all m *such that* $V_n \in B(0,\varepsilon)$ *for* $n = 0, \ldots, m$. *Futhermore* $\Phi^h_{\varepsilon,\alpha}$ *is independent of* ε, α *in the sense that, if* $\varepsilon_1 \leq \varepsilon_2, \alpha_1 \leq \alpha_2$,

$$\Phi_{\varepsilon_1,\alpha_1}(p) \equiv \Phi_{\varepsilon_2,\alpha_2}(p) \quad \forall p : \|p\| \leq \varepsilon_1.$$

Finally Φ^h *is close to* Φ *from Theorem* 5.5 *in the sense that there exists* $K > 0$ *such that*

$$\sup_{p \in \bar{Y}} \|\Phi(p) - \Phi^h(p)\|_C \leq Kh. \tag{5.11}$$

Proof Using Lemma 4.16 we deduce that the same method can be used to construct the local unstable manifold as that leading to Corollary 5.4 — note that the proof of Theorem 5.3 was constructed to be robust under enlargement of K_1 and K_2 by a factor of 2 and this is all that is required by comparing Lemmas 4.11, 4.16.

It remains to establish convergence of the graphs. The graph Φ is a fixed point of T given by (5.7). Likewise Φ^h is a fixed point of T^h given by

$$\begin{aligned} P &= L\xi + \mathcal{P}\tilde{G}(\xi + \Psi(\xi)), \\ (T^h\Psi)(P) &= L\Psi(\xi) + \mathcal{Q}\tilde{G}(\xi + \Psi(\xi)), \end{aligned} \tag{5.12}$$

where $\tilde{G}$ satisfies (4.49).

We now establish convergence. By the uniform contraction principle we have that, for $\Gamma = \Gamma(\varepsilon, \alpha)$,

$$\|\Phi - \Phi^h\|_C \leq \sup_{\Psi \in \Gamma} \tfrac{2}{1+a} \|T\Psi - T^h\Psi\|_C.$$

Thus it remains to estimate $T - T^h$. Clearly

$$\|(T^h\Psi)(P) - (T\Psi)(P)\| \leq \|(T^h\Psi)(P) - (T\Psi)(p)\| + \|(T\Psi)(p) - (T\Psi)(P)\|.$$

Since $T\Psi \in \Gamma$ we deduce that

$$\|(T^h\Psi)(P) - (T\Psi)(P)\| \leq \|(T^h\Psi)(P) - (T\Psi)(p)\| + \alpha\|p - P\|.$$

Comparing (5.7) and (5.12) and applying Lemma 4.16 we deduce that

$$\|(T^h\Psi)(P) - (T\Psi)(P)\| \leq Ch$$

and the result follows. □

Corollary 5.6 (Convergence of Local Unstable Manifold) *Let* 4.5 *hold. Then there exists $\varepsilon^* > 0$ such that, for all $\varepsilon \in (0, \varepsilon^*]$ and all $\alpha \in [1, 2]$ the fixed point $\bar{u}^h - \bar{u}$ for $\bar{S}^h(t)$ has local unstable manifold representable as*

$$W^{u,\varepsilon}(\bar{u}^h - \bar{u}) = \mathcal{M}^h := \{v \in V : \mathcal{P}v \in \bar{Y}, \mathcal{Q}v = \Phi^h(\mathcal{P}v)\}$$

where $\Phi^h \in \Gamma(\varepsilon, 1)$ and satisfies (5.9) and (5.10). Furthermore, there exists $C > 0$ such that, for any ε sufficiently small, there exists a positive $\delta' < \delta$ such that

$$dist\{W_{\mathrm{h}}{}^{u,\delta}(\bar{u}^h - \bar{u}), W^{u,\delta'}(0)\} \leq Ch,$$
$$dist\{W^{u,\delta}(0), W_{\mathrm{h}}{}^{u,\delta'}(\bar{u}^h - \bar{u})\} \leq Ch.$$

Proof Existence and convergence of a manifold invariant under the time T map follows from Theorem 5.5. It is necessary to show that the graph constructed for $T \in [T^*, 2T^*]$ is invariant for all $t > 0$. To do this it is sufficient to show invariance for all t sufficiently small. The same method as used in Corollary 5.4 cannot be used since Lemma 4.16 is not valid for $T \notin [T^*, 2T^*]$. We proceed by using a contradiction argument instead. Assume that $q(0) = \Phi(p(0))$ for some $p(0), q(0)$ such that $v(0) = p(0) + q(0)$ and consider the solution of $v(\tau) = S^h(\tau)v(0)$. Assume for contradiction that, for each $\tau \in (0, T)$, there exists $\eta = \eta(\tau) > 0$ such that

$$\|q(\tau) - \Phi^h(p(\tau))\| = \eta, \tag{5.13}$$

where $p(0), p(\tau) \in \bar{Y}$. Since $v(0) \in \mathcal{M}^h$ it follows that $v(-nT) \in \mathcal{M}^h$ for all positive integers n provided $p(-nT) \in \bar{Y}$. Furthermore, if $p_n = p(nT)$ then, by (4.51),

$$\|p_n\| \leq a\|p_{n+1}\| + 2aB_2,$$

where $B_2 = 4K_2k(2\varepsilon)\varepsilon$ as in Theorem 5.3. Hence we deduce that

$$\|p_{-n}\| \leq a^n\|p_0\| + (1-a^n)\frac{2aB_2}{1-a}.$$

Hence, for ε so small that $2aB_2 \leq (1-a)\|p_0\|$, we deduce that $p_{-n} \in \bar{Y}$ for all $n \geq 0$, so that $v(-nT) \in \mathcal{M}^h$ for all $n \geq 0$. Now let $t = -nT + \tau$ and note that, since $q(-nT) = \Phi^h(p(-nT))$, we have

$$\begin{aligned}
\|q(t) - \Phi^h(p(t))\| &\leq \|q(-nT) - \Phi^h(p(-nT))\| + \|q(-nT) - q(t)\| \\
&\qquad + \|\Phi^h(p(-nT)) - \Phi^h(p(t))\| \\
&\leq \|q(-nT) - q(t)\| + \alpha\|p(-nT) - p(t)\| \\
&\leq \max\{1, \alpha\}\|v(-nT) - v(t)\|_V \\
&\leq C\|v(-nT) - v(t)\|.
\end{aligned}$$

But $\|v(t) - v(-nT)\| \leq K(t, \|v(-nT)\|)$. Thus

$$\|q(t) - \Phi^h(p(t))\| \leq K(t, \|v(-nT)\|).$$

Also

$$\begin{aligned}
\|v(-nT)\| &\leq K\|v(-nT)\|_V \\
&\leq K\max\{\|p(-nT)\|, \|q(-nT)\|\} \qquad (5.14)\\
&\leq K\max\{\|p(-nT)\|, \varepsilon\}.
\end{aligned}$$

Since $\|p(-nT)\|$ is bounded uniformly in n in terms of $\|p(0)\|$ and $\tau \in (0, T)$, we obtain

$$\|q(t) - \Phi^h(p(t))\| \leq C(T, \|p_0\|).$$

Now let $t_m = -mT + \tau$; since τ may be arbitrarily small and since $v(-mT) \in B(0, \varepsilon)$ we can ensure $v(t_m) \in B(0, \varepsilon^*)$ provided $\varepsilon < \varepsilon^*$. Thus, by (5.6),

$$\|q(\tau) - \Phi^h(p(\tau))\| \leq \left(\frac{1+a}{2}\right)^n \|q(t) - \Phi^h(p(t))\|$$

since $\tau = t + nT$ and since the manifold is attractive. Thus, by choice of n sufficiently large we obtain a contradiction to (5.13). Hence Φ^h is invariant under $S^h(t)$ for all $t > 0$. The closeness follows from (5.11). □

5.4 Global unstable sets

In this section we examine the continuity of the global unstable sets of (2.9) with respect to numerical perturbation using our knowledge about the effect of perturbation on the local unstable manifold.

For the following theorem note that boundedness of the unstable manifold implies relative compactness by Lemma 5.2 so that the hypotheses could be weakened.

Theorem 5.7 (Lower Semicontinuity of the Unstable Set) *Assume that (4.2) has an equilibrium point $\bar{v}$ and that $\bar{v}^{\mathrm{h}}$ is the equilibrium point of $S^h(t)$ which converges to $\bar{v}$ as $h \to 0$. Further, suppose that Assumptions* 4.5 *holds. Then, if $W^u(\bar{v})$ is contained in a compact set $B \subset V$, it follows that*

$$dist(\overline{W^u(\bar{v})}, \overline{W_{\mathrm{h}}{}^u(\bar{v}^{\mathrm{h}})}) \to 0 \ \ as \ \ h \to 0.$$

Proof It is sufficient to prove that, given any $\varepsilon > 0$, there exists $\Delta > 0$ such that for every $y \in W^u(\bar{v})$ there exists $y^{\mathrm{h}} \in W_{\mathrm{h}}{}^u(\bar{v}^{\mathrm{h}})$ with the property that $\|y - y^{\mathrm{h}}\| \leq 2\varepsilon$ for $h \in (0, \Delta]$.

Recall $\partial B(\bar{v}, r)$ and Γ given by (2.6), (5.1) and (5.2). Now set

$$\mathcal{W} = W^u(\bar{v}) \backslash W^{u,\varepsilon}(\bar{v}). \tag{5.15}$$

Then, for ε sufficiently small,

$$\mathcal{W} = \bigcup_{t>0} S(t)\Gamma,$$

by Lemma 5.2. It follows that $\overline{\mathcal{W}}$ is compact by assumption. Note that $\{B(x;\epsilon) : x \in \mathcal{W}\}$ is an ε-cover for $\overline{\mathcal{W}}$ and hence, since $\overline{\mathcal{W}}$ is compact, we may extract a finite subcover. Denote this subcover by $\{B_i(\varepsilon)\}_{i=1}^{I}$ and note that each $B_i(\varepsilon)$ contains a point $y_i \in \mathcal{W}$, where $B_i(\varepsilon) = B(y_i, \varepsilon)$. By construction there exists $x_i \in \Gamma$ and $T_i > 0$ such that $S(T_i)x_i = y_i$ for each $y_i \in \mathcal{W}$. Now, by Corallary 5.6, it follows that for any $\eta > 0$ there exists ${x^{\mathrm{h}}}_i \in W_{\mathrm{h}}{}^u(\bar{v}^{\mathrm{h}})$ and $\Delta(i) > 0$ such that

$$\|x_i - {x^{\mathrm{h}}}_i\| \leq \eta \quad \forall h \in (0, \Delta(i)]; \tag{5.16}$$

by the invariance of the unstable manifold (see Lemma 5.2) it follows that ${y^{\mathrm{h}}}_i = S^h(T_i){x^{\mathrm{h}}}_i \in W_{\mathrm{h}}{}^u(\bar{v}^{\mathrm{h}})$.

Note that

$$\begin{aligned} \|y_i - {y^{\mathrm{h}}}_i\| &= \|S(T_i)x_i - S^h(T_i){x^{\mathrm{h}}}_i\| \\ &\leq \|S(T_i){x^{\mathrm{h}}}_i - S^h(T_i){x^{\mathrm{h}}}_i\| + \|S(T_i){x^{\mathrm{h}}}_i - S(T)x_i\|. \end{aligned}$$

It now follows from Assumptions 3.2 and (5.16) that, by the continuity of

$S(T_i)$ and appropriate choice of η,

$$\|y_i - y^{\mathrm{h}}{}_i\| \le \frac{\varepsilon}{2} + C(T_i, \bar{v})\eta$$

for $h \in (0, \Delta(i)]$. Thus, by further reduction of $\Delta(i)$ if necessary, we find that

$$\|y_i - y^{\mathrm{h}}{}_i\| \le \varepsilon, \quad \forall h \in (0, \Delta(i)].$$

Since I is finite, we deduce that there exists $\{y^{\mathrm{h}}{}_i\}_{i=1}^{I}$ each lying in $W_{\mathrm{h}}{}^u(\bar{v}^{\mathrm{h}})$ and $\Delta > 0$ such that

$$\max_{1 \le i \le I} \|y_i - y^{\mathrm{h}}{}_i\| \le \varepsilon \quad \forall h \in (0, \Delta].$$

Thus, since y_i is the center of $B_i(\varepsilon)$, we deduce that for every $y \in B_i(\varepsilon)$ and i such that $1 \le i \le I$ there exists $y^{\mathrm{h}}{}_i \in W_{\mathrm{h}}{}^u(\bar{v}^{\mathrm{h}})$ such that

$$\|y - y^{\mathrm{h}}{}_i\| \le 2\varepsilon \quad \forall h \in (0, \Delta].$$

Since the $B_i(\varepsilon)$, $i = 1, \ldots, I$, form a cover of $\overline{\mathcal{W}}$, we deduce that

$$\operatorname{dist}(\overline{\mathcal{W}}, W_{\mathrm{h}}{}^u(\bar{v}^{\mathrm{h}})) \le 2\varepsilon \quad \forall h \in (0, \Delta]. \tag{5.17}$$

Now, by Corollary 5.6 there exists $\delta' > 0$ such that

$$\operatorname{dist}(W^{u,\delta}(\bar{v}), W_{\mathrm{h}}{}^{u,\delta'}(\bar{v}^{\mathrm{h}})) \le 2\varepsilon \quad \forall h \in (0, \Delta], \tag{5.18}$$

possibly by further reduction of Δ. Putting (5.17) and (5.18) together, the first result follows by (5.15). □

5.5 Bibliography

Two important approaches to the construction of unstable manifolds are the Lyapunov–Perron technique (basically the method used for stable sets in Section 4) and the Hadamard graph transform technique employed in this Section. See Babin and Vishik [4], Hale [50], Wells [97] and Wiggins [98] for material on unstable manifolds.

The existence and convergence of unstable manifolds for approximation of ordinary differential equations was first considered by Beyn [8]; the generalization to partial differential equations was studied by Alouges and Debussche [1] and by Larsson and Sanz-Serna [75], the latter using Lyapunov–Perron techniques. The approach given here is based on the Hadamard graph transform and is similar to the technique used to construct perturbations of center unstable manifolds of ordinary differential equations in Beyn and Lorenz [10].

6 Inertial manifolds

6.1 Introduction

The inertial manifold is a finite-dimensional set (in terms of the number of eigenfunctions of A needed to represent it) which has a manifold structure and exponentially attracts all solutions of (2.9). Roughly speaking it exists provided the spectrum of A has sufficiently large gaps compared to the size of the nonlinearity F. In Section 6.2 we make some precise definitions of the inertial manifold and, in Example 6.6, give an explicit construction of the inertial manifold for a non-local reaction-diffusion equation. Theorem 6.4 gives a construction of the inertial manifold for a mapping derived by considering the time T flow of the semigroup and applying a contraction argument very similar to that used in Section 5 to construct the local unstable manifold; Theorem 6.5 shows that this inertial manifold is also invariant for the underlying partial differential equation. In Section 6.3 we incorporate the effect of the approximation, once again using the uniform contraction principle.

6.2 Existence theory

We assume that there exists $R > 0$ such that

$$\begin{aligned} &\exists T = T(\rho, R) : S(t)B(0,\rho) \subseteq B(0,R) \quad \forall t \geq T, \\ &S(t)B(0,R) \subseteq B(0,R) \quad \forall t \geq 0. \end{aligned} \tag{6.1}$$

Note that typical equations under consideration do not satisfy (2.11) uniformly in V but only on bounded sets of V; use of conditions such as (6.1) can be used to achieve (2.11) by smooth modification of $F(\bullet)$ outside $B(0,R)$. Thus throughout Section 6 we will make the following assumption.

Assumption 6.1 *The function F in (2.9) satisfies (6.1) and (2.11).*

Recall that A has eigenvalues λ_j and eigenfunctions φ_j ordered so that (2.1) holds. In this section we let $\mathbb{P}^m$ denote the projection of X onto the first m eigenfunctions of A so that

$$v = \sum_{j=1}^{\infty} v_j \varphi_j \Rightarrow \mathbb{P}^m v = \sum_{j=1}^{m} v_j \varphi_j.$$

We also set $\mathbb{Q}^m = I - \mathbb{P}^m$ and $Y = \mathbb{P}^m X, Z = \mathbb{Q}^m X$.

Definition 6.2 *An inertial manifold for a semigroup $S(t)$ is a positively invariant set $\mathcal{M}$, defined through a Lipschitz graph $\Phi : Y \mapsto Z$ and constant $R > 0$, and satisfying both of the following:*

(i) $\mathcal{M}$ may be expressed in the form

$$\mathcal{M} := \{u \in V : \mathbb{Q}^m v = \Phi(\mathbb{P}^m v)\} \bigcap B(0,R), \tag{6.2}$$

(ii) there exists $\nu > 0$ such that, for each u_0 in V,

$$\exists C = C(u_0) : dist(S(t)u_0, \mathcal{M}) \leq Ce^{-\nu t} \quad \forall t \geq 0. \tag{6.3}$$

The reason why inertial manifolds are of theortical importance is that they show that the large time behavior of certain partial differential equations is governed by a finite-dimensional ordinary differential equation. Specifically all solutions are attracted to $\mathcal{M}$ exponentially and, on $\mathcal{M}$, the solutions are governed by the equation

$$p_t + Ap = \mathbb{P}^m F(p + \Phi(p)).$$

This is equivalent to a system of m ordinary differential equations in the m coefficients of $p(t)$ in an expansion in terms of the φ_i's.

Example 6.3 Consider eqn (2.9) with A given in Example 2.1 and F given by $F(u)(x) := f(|u|^2)u(x)$ for some smooth f satisfying

$$f(x) \leq \lambda \quad \forall x \in \mathbb{R}^+. \tag{6.4}$$

Recall that $|\bullet|$ denotes the norm on $L_2((0,1))$ for Example 2.1. Thus (6.4) is equivalent to the non-local reaction-diffusion equation

$$\begin{aligned} u_t = u_{xx} + f(\textstyle\int_0^1 u^2(s,t)ds)u, \quad (x,t) \in (0,1) \times (0,\infty), \\ u(0,t) = u(1,t) = 0, \quad t > 0, \\ u(x,0) = u_0(x). \end{aligned} \tag{6.5}$$

Under further conditions on the behavior of f at infinity, (2.11) will be satisfied. It will also be useful to consider the problem

$$\begin{aligned} v_t = v_{xx}, \quad (x,t) \in (0,1) \times (0,\infty), \\ v(0,t) = v(1,t) = 0, \quad t > 0, \\ v(x,0) = u_0(x). \end{aligned} \tag{6.6}$$

Recall φ_k defined in Example 2.1. We let $a(t) = |u(\cdot,t)|^2, b(t) = |v(\cdot,t)|^2$. If

$$u_0(x) = \sum_{j=1}^{\infty} a_k \varphi_k, \quad a_k = \langle u_0, \varphi_k \rangle$$

then separation of variables shows that

$$u(x,t) = \sum_{k=1}^{\infty} a_k e^{-k^2\pi^2 t} e^{\int_0^t f(a(s))ds} \varphi_k. \tag{6.7}$$

Here $a(t)$ satisfies the nonlinear integral equation

$$a(t) = \exp\left\{2\int_0^t f(a(s))ds\right\} b(t).$$

Now we let

$$\mathcal{M} = \{v \in V : \langle v, \varphi_k \rangle = 0 \quad \forall k \geq m\}$$

and show that $\mathcal{M}$ is an inertial manifold for appropriately chosen m. First note that $\mathcal{M}$ is invariant for any m such that if $u_0(x) \in \mathcal{M}$ then $a_k = 0$ for all $k \geq m$; by (6.7) it follows that $u(x,t) \in \mathcal{M}$. It remains to show that $\mathcal{M}$ attracts exponentially. Note that from (6.4)

$$\int_0^t f(a(s))ds \leq \lambda t.$$

Hence

$$\langle u(x,t), \varphi_k \rangle = a_k e^{-k^2\pi^2 t} e^{\int_0^t f(a(s))ds} \leq a_k e^{(\lambda - k^2\pi^2)t}.$$

Choosing m such that $\lambda - m^2\pi^2 \leq -\mu < 0$ we deduce that

$$\langle u(x,t), \varphi_k \rangle \leq a_k e^{-\mu t} \quad \forall k \geq m.$$

Since the a_k are uniformly bounded in k for initial data in $V \equiv H_0^1((0,1))$ it follows that $\mathcal{M}$ is exponentially attracting with rate at least $e^{-\mu t}$.

A similar explicit construction of an inertial manifold is given in Bloch and Titi [12]. That paper concerns a nonlinear beam equation where the nonlinearity occurs only through the appearance of the L_2-norm of the unknown. A closely related analysis to that given here allows for construction of an inertial manifold with structure similar to that given here.

Our basic approach is to use (2.15) to formulate a mapping and prove that the mapping has an inertial manifold; we then show that the inertial manifold for the map is also positively invariant under (2.9) and hence an inertial manifold for (2.9). We define

$$\begin{gathered} L(t) = e^{-At}, \quad G(u,t) = \textstyle\int_0^t L(t-s)F(S(s)u)ds. \\ L := L(T), G(\bullet) := G(\bullet, T). \end{gathered} \tag{6.8}$$

Now consider the mapping

$$u_{n+1} = Lu_n + G(u_n), \tag{6.9}$$

where $u_n = S(nT)u_0$. We set $p_n = \mathbb{P}^m u_n, q_n = \mathbb{Q}^m u_n$ and seek an invariant manifold for (6.9) which hence satisfies

$$q_n = \Phi(p_n) \iff q_{n+1} = \Phi(p_{n+1}) \quad \forall n \geq 0 \tag{6.10}$$

and is attractive in the sense that there exists $\mu \in (0,1)$ such that

$$\|q_n - \Phi(p_n)\| \leq \mu^n \|q_0 - \Phi(p_0)\| \quad \forall n \geq 0. \tag{6.11}$$

Let $\Gamma = \Gamma(\epsilon, \delta)$ denote the closed subset of $C(Y, Z)$ satisfying

$$\|\Psi\|_\Gamma := \sup_{p \in Y} \|\Psi(p)\| \leq \epsilon,$$

$$\|\Psi(p_1) - \Psi(p_2)\| \leq \delta \|p_1 - p_2\| \qquad \forall p_1, p_2 \in Y.$$

We may now prove the preliminary existence result for (6.9).

Theorem 6.4 *Let Assumption* 6.1 *hold. Suppose that for any* $K_3, K_4 > 0$ *there exists an integer* $q_0 > 0$ *such that*

$$\lambda_{q+1}^{1-\beta} \geq K_3, \qquad \lambda_{q+1} - \lambda_q \geq K_4 \lambda_{q+1}^\beta$$

for all $q \geq q_0$. *Then for any* $\zeta > 1$ *and* $\epsilon', \delta' > 0$, *there exists* $T > 0$, *integer* $q_0 > 0$ *and* $\mu \in (0, 1)$ *such that, if* $q \geq q_0$, $\epsilon \in [\varepsilon', \zeta\varepsilon']$, $\delta \in [\delta', \zeta\delta']$ *then there exists* $\Phi \in \Gamma(\varepsilon, \delta)$ *such that (6.10), (6.11) holds for (6.9); thus there exists* $C = C(u_0) > 0$ *such that, if* $\mathcal{M}$ *is given by (6.2) with* R *given by (6.1), then*

$$dist(u_n, \mathcal{M}) \leq C\mu^n. \tag{6.12}$$

Proof We show that the mapping (6.9) satisfies $(G1)$–$(G3)$ and $(C1)$–$(C4)$ of Theorem 12.3; the existence (6.10) and attractivity (6.11) then follow from Theorem 12.3 with $r = \infty$, $\gamma = \epsilon \in [\varepsilon', K\varepsilon']$, $\alpha = \delta \in [\delta', K\delta']$.

We define $\lambda = \lambda_q, \Lambda = \lambda_{q+1}$. Then L satisfies $(G1), (G2)$ with

$$b = e^{-\lambda T}, \qquad a = e^{-\Lambda T}, \qquad c = e^{-\lambda_1 T}.$$

We set $\Lambda T = \sigma \in (0, \infty)$. From the definition of $G(u)$ we have, using Lemma 10.6 and Assumption 6.1 that there exists $C_1 > 0$ such that

$$\begin{aligned} \|G(u)\| &\leq \int_0^T |A^\beta L(T-s) F(S(s)u)| ds \\ &\leq \int_0^T \frac{CK}{(T-s)^\beta} ds \\ &\leq \frac{T^{1-\beta}}{1-\beta} C_1 K. \end{aligned} \tag{6.13}$$

Similarly, using (2.16) (Lipschitz continuity of the semigroup $S(t)\bullet$, which follows from Assumption 6.1) and the construction of F, there exists $C_2 = C_2(T) > 0$ such that

$$\|G(u) - G(v)\| \leq \frac{T^{1-\beta}}{1-\beta} C_2 K \|u - v\|.$$

Thus, $G(u)$ satisfies $(G3)$ with

$$B_1 = B_2 = 2T^{1-\beta} K_5, \tag{6.14}$$

where $K_5 = K \max\{C_1, C_2\}/(1-\beta)$. We have doubled B_1, B_2 to allow for incorporation of perturbation error in Section 6.3.

We verify $(C1)$–$(C4)$. Let $\epsilon, \delta > 0$, $\sigma \in (0, \infty)$, $\mu \in (e^{-\sigma}, 1)$ be given. Define K_3, K_4 and T by

$$T = \frac{\sigma}{\Lambda}$$

$$K_3 = 2 \max\left\{\frac{e^{\sigma} K_5 \sigma^{1-\beta}(1+\delta)}{\mu - e^{-\sigma}}, \frac{(1+\sigma)K_5}{\epsilon\sigma^{\beta}}\right\}, \tag{6.15}$$

$$K_4 = \frac{K_5(1+\delta)^2 e^{\sigma}}{\delta\sigma^{\beta}}. \tag{6.16}$$

Choose q such that

$$\Lambda^{1-\beta} \geq K_3, \quad \Lambda - \lambda \geq K_4 \Lambda^{\beta}. \tag{6.17}$$

By (6.17) and (6.15) we deduce that $\Lambda^{1-\beta} \geq 2e^{\sigma} K_5 \sigma^{1+\beta}(1+\delta)/\mu$. Putting $\sigma = \Lambda T$ and $\Lambda^{1-\beta} = (\sigma/T)^{1-\beta}$ we obtain

$$2e^{\Lambda T} K_5 T^{1-\beta}(1+\delta) \leq \mu.$$

Since $\lambda \leq \Lambda$, we have

$$2e^{\lambda T} K_5 T^{1-\beta}(1+\delta) \leq \mu;$$

this is simply $b^{-1}B_1(1+\delta) \leq \mu$, as required to establish $(C1)$.

By (6.16) and (6.15) we deduce that $\Lambda^{1-\beta} \geq 2(1+\sigma)K_5/(\sigma^{\beta}\epsilon)$ so that, putting $1 + \sigma = 1 + \Lambda T$, we have

$$2(1+\Lambda T)K_5 T \frac{\Lambda^{\beta}}{\sigma^{\beta}} \leq \epsilon \Lambda T.$$

Adding ϵ to both sides and dividing by $1 + \Lambda T$, we obtain

$$\frac{\epsilon}{1+\Lambda T} + 2K_5 T^{1-\beta} \leq \epsilon.$$

Since $e^{-x} \leq 1/(1+x)$ for all $x > 0$, we have $a\epsilon + B_2 \leq \epsilon$ and $(C2)$ is established.

To prove $(C3)$ we note that, by (6.17),

$$\Lambda - \lambda \geq K_4 \Lambda^{\beta} \geq 2K_5(1+\delta)^2 e^{\sigma} \frac{\Lambda^{\beta}}{\delta\sigma^{\beta}}.$$

Since $e^x - 1 \geq x$ for positive x, we may bound $(\Lambda - \lambda)T$ by $e^{(\Lambda-\lambda)T} - 1$. Multiplying the previous inequality by T and using $\sigma = \Lambda T$, we find

$$\delta e^{-\Lambda T} + 2K_5(1+\delta)^2 T^{1-\beta} \leq \delta e^{-\lambda T}.$$

Since $(1+\delta)^2 = (1+\delta)+\delta(1+\delta)$, this last inequality becomes $\delta a+(1+\delta)B_1 \leq \delta b - \delta(1+\delta)B_1$, which is $(C3)$.

$(C4)$ is obtained by noting that, from (6.15),

$$\Lambda^{1-\beta} \geq K_3 \geq 2K_5\sigma^{1-\beta}\frac{(1+\delta)}{\mu - e^{-\sigma}}.$$

Rearranging this gives $2K_5T^{1-\beta}(1+\delta) + e^{-\Lambda T} \leq \mu$. Using the definitions of a, B_1, this last expression becomes $a + 2(1+\delta)B_1 \leq \mu$, which is $(C4)$.

Note that q_0 depends on K_3 and K_4, and hence on ϵ and δ. Taking the supremum of q over all $\epsilon \in [\epsilon', \zeta\epsilon']$, $\delta \in [\delta', \zeta\delta']$ yields q_0 such that the result holds.

It remains to establish (6.12). Recall that $u_n = p_n + q_n$. Note that

$$\text{dist}(u_n, \mathcal{M}) \leq \|u_n - \tilde{u}_n\|$$

where $\tilde{u}_n = p_n + \Phi(p_n)$. Thus, by (12.4) of Theorem 12.3,

$$\text{dist}(u_n, \mathcal{M}) \leq \|q_n - \Phi(p_n)\| \leq \mu^n \|q_0 - \Phi(p_0)\|$$

and the result follows since (6.1) of Assumption 6.1 ensures that u_n enters $B(0, R)$ after a finite number of iterations. □

Theorem 6.4 shows that the time T flow of the semigroup for (2.9) has an attractive invariant manifold. We show now that this manifold is in fact an inertial manifold for eqn (2.9).

Theorem 6.5 (Inertial Manifolds) *Under the assumptions of Theorem* 6.4, *eqn (2.9) has an inertial manifold $\mathcal{M}$ satisfying (6.2) and (6.3) for $\Phi \in \Gamma(\varepsilon', \delta')$ and $\mu > 0$ given in Theorem* 6.4 *and R given by (6.1).*

Proof It remains to show that the manifold $\mathcal{M}$ constructed as a consequence of Theorem 6.4 is in fact invariant for the underlying partial differential equation, rather than just the time T flow of the semigroup. To do this set $\Omega = S(\tau)\mathcal{M}$ for some $\tau \in (0, T)$. We show that, for all τ sufficiently small, $\Omega \equiv \mathcal{M}$ and hence that $\mathcal{M}$ is invariant for $S(t)$. The required result then follows.

Notice that $S(T)\Omega = S(T)S(\tau)\mathcal{M} = S(\tau)S(T)\mathcal{M} = S(\tau)\mathcal{M} = \Omega$ so that Ω is invariant under the time T discrete map. In addition, we can show that Ω is the graph of a global function: recall the definitions of $L(t)$ and $G(u,t)$ given in (6.8); by applying the method of proof of Lemma 12.4 it follows that, for every $p \in Y$ there exists a unique ξ so that

$$p = L(\tau)\xi + \mathbb{P}^m G(\xi + \Phi(\xi), \tau),$$

for any τ sufficiently small. Thus Ω can be expressed as a graph $q = \Psi(p)$ where $\Psi : Y \mapsto Z$ is given by

$$\Psi(p) := L(\tau)\Phi(\xi) + \mathbb{Q}^m G(\xi + \Phi(\xi), \tau).$$

Recall from Theorem 6.4 that $\Phi \in \Gamma(\delta', \epsilon')$. Similarly to the calculation of (6.13) we have

$$\begin{aligned}\|\Psi\| &\leq e^{-\Lambda\tau}\epsilon' + \frac{\tau^{1-\beta}}{1-\beta}C_1K \\ &\leq e^{\Lambda(T-\tau)}a\epsilon' + B\left(\frac{\tau}{T}\right)^{1-\beta}.\end{aligned}$$

Notice that for $\tau = 0, \tau = T$, $\|\Psi\| \leq \epsilon'$. In general, for $0 \leq \tau \leq T$ there is some $\eta \geq 0$ such that

$$\|\Psi\| \leq \epsilon' + \eta.$$

In a similar manner, one obtains

$$\|\Psi(p_1) - \Psi(p_2)\| \leq (\delta' + \eta)\|p_1 - p_2\|$$

for $0 \leq \tau \leq T$. Note that, by continuity, η can be made arbitrarily small by requiring τ to be small. Thus we see that for any $\eta > 0$ there exists a $\tau^*(\eta) > 0$ such that

$$\Psi \in \Gamma(\epsilon' + \eta, \delta' + \eta)$$

for all $\tau \in (0, \tau^*(\eta))$. Thus we choose ζ so large that $(C1)$–$(C4)$, verified in the Theorem 6.4, hold for all $\delta \in [\delta', \delta' + \eta], \epsilon \in [\epsilon', \epsilon' + \eta]$. For such δ, ϵ we have $\Psi \in \Gamma(\delta, \epsilon)$ and we may again apply Lemma 12.4 to obtain that for every $p \in Y$ there exists a $p_0 \in Y$ such that

$$p = Lp_0 + \mathbb{P}^m G(p_0 + \Psi(p_0)),$$

$$q = L\Psi(p_0) + \mathbb{Q}^m G(p_0 + \Psi(p_0)).$$

However, since $S(T)\Omega = \Omega$, $u = p + q \in \Omega$ and $\Omega = \text{Graph}(\Psi)$, we must have $q = \Psi(p)$. Thus Ψ is a fixed point of the map T constructed in (12.8), and by the uniqueness of the fixed point under the conditions of Theorem 6.4, it follows that $\Psi \equiv \Phi$. One may now repeat the argument on the intervals $t \in (k\tau, (k+1)\tau)$ for integer $k \geq 1$. It follows that $\mathcal{M} = \text{Graph}(\Phi)$ is an invariant manifold for (2.9) for all time.

To see that this manifold exponentially attracts all solutions as in (6.3) let $t > 0$ and u_0 be given. Set $t_n(s) = nT + s, s \in [0, T]$. Then from (6.12)

$$\text{dist}(u(t_n), \mathcal{M}) = \text{dist}(S(nT)u(s), \mathcal{M}) \leq \mu^n C(|u(s)|_\gamma).$$

Since $u(s)$ depends continuously on $u(0)$ for all $s \in (0, T)$, the result follows by choosing ν appropriately. □

6.3 Perturbation theory

In the previous section an inertial manifold was constructed for (2.9) under (2.11) on F. Recall that in practice assumptions such as (2.11) only hold within a bounded set $B(0, R)$ satisfying (6.1); the function F can then be modified outside $B(0, R)$ to ensure that (2.11) holds. We make similar assumptions about $S^h(t)$: we assume that there exists $R > 0$ such that

$$\begin{aligned} \exists T = T(\rho, R) : S^h(t)B(0,\rho) \subseteq B(0,R) \quad \forall t \geq T, \\ S^h(t)B(0,R) \subseteq B(0,R) \quad \forall t \geq 0. \end{aligned} \tag{6.18}$$

Thus throughout this section we make

Assumption 6.6 *There exists $R > 0$ such that (6.18) holds for $S^h(t)$ of Assumption* 3.2.

We now define a new map

$$\tilde{S}^h(t)u := \theta(\|u\|^2)S^h(t)u + (1 - \theta(\|u\|^2))S(t)u, \tag{6.19}$$

where $\theta \in C^\infty(\mathbb{R}^+, \mathbb{R}^+)$ satisfies

$$\theta(x) = 1 \, \forall x : x^2 \leq R, \quad \theta(x) = 0 \, \forall x : x^2 \geq 2R$$

and $\theta'(x)$ is uniformly bounded. The derivative of $\tilde{S}^h(t)$ with respect to $u \in V$ and evaluated at $v \in V$ is denoted by $d\tilde{S}^h(v,t)$. Note that the new map does not satisfy the semigroup property on V but, by virtue of (6.18), it does on $B(0,R)$ since it is coincident with $S^h(t)$ on that set. The reason for considering $\tilde{S}^h(t)$ is that it has been constructed to be C^1 close to $S(t)$ *uniformly* on V :

Lemma 6.7 *Define*

$$\tilde{E}(u;t) := S(t)u - \tilde{S}^h(t)u,$$

$$d\tilde{E}(u;t) := dS(u;t) - d\tilde{S}^h(u;t).$$

Then for all $t \in \mathbf{S}$, $t > 0$, there exist constants $C_i(t) < \infty, i = 1, 2,$ and a function $\kappa : \mathbb{R}^+ \mapsto \mathbb{R}^+$ such that the maps $S(\bullet)\bullet \in C^1(\mathbb{R}^+ \times V, V)$ and $\tilde{S}^h(\bullet)\bullet \in C^1(\mathbb{R}^+ \times V, V)$ satisfy

$$\|\tilde{E}(u;t)\| \leq C_1 h \quad \forall u \in V,$$

$$\|d\tilde{E}(u;t)\| \leq C_2 \kappa(h) \quad \forall u \in V,$$

where $\kappa(h) \to 0$ as $h \to 0_+$.

We now let $U_n = \tilde{S}^h(nT)U_0$ so that

$$U_{n+1} = LU_n + \tilde{G}(U_n), \tag{6.20}$$

where

$$\tilde{G}(u) = G(u) + \tilde{E}(u;T). \tag{6.21}$$

We set $P_n = \mathbb{P}^m U_n, Q_n = \mathbb{Q}^m U_n$ and seek an invariant manifold for (6.20) which hence satisfies

$$Q_n = \Phi^h(P_n) \Longleftrightarrow Q_{n+1} = \Phi^h(P_{n+1}) \quad \forall n \geq 0 \tag{6.22}$$

and is attractive in the sense that

$$\|Q_n - \Phi(P_n)\| \leq \mu^n \|Q_0 - \Phi(P_0)\| \quad \forall n \geq 0. \tag{6.23}$$

Theorem 6.8 *Let Assumptions* 6.1 *and* 6.6 *hold. Suppose that for any* $K_3, K_4 > 0$ *there exists an integer* $q_0 > 0$ *such that*

$$\lambda_{q+1}^{1-\beta} \geq K_3, \qquad \lambda_{q+1} - \lambda_q \geq K_4 \lambda_{q+1}^{\beta}$$

for all $q \geq q_0$. *Then for any* $K > 1$ *and* $\epsilon', \delta' > 0$, *there exists* $h_c, T > 0$, *integer* $q_0 > 0$ *and* $\mu \in (0,1)$ *such that, if* $q \geq q_0$, $\epsilon \in [\varepsilon', K\varepsilon']$, $\delta \in [\delta', K\delta']$ *and* $h \in (0, h_c)$ *then there exists* $\Phi^h \in \Gamma(\varepsilon, \delta)$ *such that (6.22), (6.23) hold for (6.20) so that there exists* $C = C(U_0) > 0$ *such that*

$$\mathit{dist}(U_n, \mathcal{M}) \leq C\mu^n.$$

Furthermore, there exists $K > 0$ *such that*

$$\sup_{p \in Y} \|\Phi(p) - \Phi^h(p)\| \leq Kh. \tag{6.24}$$

Proof Using (6.20) and Lemma 6.7 it can be shown that, for h sufficiently small, $\tilde{G}$ satisfies the same estimates (6.14) as G. (Note that those bounds were doubled to allow incorporation of perturbation error at this stage). Thus existence and attractivity follow as in Theorem 6.4.

The convergence result follows from the uniform contraction principle, Theorem 11.2: note that Φ and Φ^h are fixed points of T and T^h respectively, where

$$\begin{aligned} p &= L\xi + \mathbb{P}^m G(\xi + \Psi(\xi)) \\ (T\Psi)(p) &= L\Psi(\xi) + \mathbb{Q}^m G(\xi + \Psi(\xi)), \\ P &= L\xi + \mathbb{P}^m \tilde{G}(\xi + \Psi(\xi)) \\ (T^h\Psi)(P) &= L\Psi(\xi) + \mathbb{Q}^m \tilde{G}(\xi + \Psi(\xi)). \end{aligned} \tag{6.25}$$

Both Φ and Φ^h lie in $\Gamma = \Gamma(\varepsilon', \delta')$ and are constructed with contraction constant μ. Thus

$$\|\Phi - \Phi^h\|_\Gamma \leq \frac{1}{1-\mu} \sup_{\Psi \in \Gamma} \|(T\Psi) - (T^h\Psi)\|_\Gamma$$

where

$$\|\Psi\|_\Gamma := \sup_{p \in Y} \|\Psi(p)\|.$$

Now

$$\|T\Psi - T^h\Psi\|_\Gamma = \sup_{p\in Y} \|(T\Psi)(p) - (T^h\Psi)(p)\|.$$

Hence, by (6.25) and Lemma 6.7 we have

$$\begin{aligned}
&\|(T\Psi)(p) - (T^h\Psi)(p)\| \\
&\quad\leq \|(T\Psi)(p) - (T^h\Psi)(P)\| + \|(T^h\Psi)(P) - (T^h\Psi)(p)\| \\
&\quad\leq \|\mathbb{Q}^m G(\xi + \Psi(\xi)) - \mathbb{Q}^m \tilde{G}(\xi + \Psi(\xi))\| + \delta\|P - p\| \\
&\quad\leq (1+\delta)\|G(\xi + \Psi(\xi)) - \tilde{G}(\xi + \Psi(\xi))\| \qquad (6.26) \\
&\quad\leq (1+\delta)\|\tilde{E}(\xi + \Psi(\xi), T)\| \\
&\quad\leq C(T)(1+\delta)h.
\end{aligned}$$

The convergence result follows. □

An inertial manifold for (6.20) is a set, defined through a Lipschitz graph Φ and constant $R > 0$, of the form

$$\mathcal{M}^h := \{v \in V : \mathbb{Q}^m v = \Phi^h(\mathbb{P}^m v)\} \bigcap B(0, R). \qquad (6.27)$$

The set is assumed to be positively invariant under (6.20) and to exponentially attract all solutions of (2.9) at a uniform rate; that is, there exists $\nu > 0$ such that, for all u_0 in V,

$$\exists C = C(u_0) : \operatorname{dist}(S^h(t)u_0, \mathcal{M}^h) \leq Ce^{-\nu t} \quad \forall t \geq 0. \qquad (6.28)$$

Theorem 6.9 (Continuity of Inertial Manifolds) *Under the assumptions of Theorems* 6.4 *and* 6.8 *there exists $h_c > 0$ such that, for $h \in (0, h_c)$, the semigroup $S^h(t)$ has an inertial manifold $\mathcal{M}^h$ satisfying (6.27) and (6.28) for $\Phi^h \in \Gamma(\varepsilon', \delta')$ and $\mu > 0$ given in Theorem* 6.8 *and R given in (6.27). Furthermore there exists $K > 0$ such that*

$$d_{\mathrm{H}}(\mathcal{M}^h, \mathcal{M}) \leq Kh.$$

Proof The existence part of the proof may be proved from (6.8) in the same way that Theorem 6.5 follows from Theorem 6.4, using the fact that $\tilde{S}^h(t)$ coincides with the semigroup $S^h(t)$ in the positively invariant set $B(0, R)$. The convergence result follows from (6.24): if $x \in \mathcal{M}^h$ then $x = p + \Phi^h(p)$ for some p; set $z = p + \Phi(p)$. Then

$$\operatorname{dist}(\mathcal{M}^h, \mathcal{M}) \leq \sup_{x\in\mathcal{M}^h} \|x - z\| \leq \sup_{p\in Y} \|\Phi(p) - \Phi^h(p)\| \leq Kh$$

as required. Similarly the result follows for $\text{dist}(\mathcal{M}, \mathcal{M}^h)$. □

6.4 Bibliography

Inertial manifolds were first introduced by Foias *et al.* [38]. However, the basic idea that certain partial differential equations behave finite-dimensionally for large time can be traced back considerably earlier — see Foias and Prodi [36] and Constantin *et al.* [20], for example. A variety of analytical results concerning inertial manifolds may be found in Constantin *et al.* [21], Foias *et al.* [39], Mallet-Paret and Sell [78] and Chow *et al.* [17]. The original paper of Foias *et al.* [38] employs an existence technique known as the Lyapunov–Perron approach and, using this theory, the effect of perturbation due to a spectral approximation is studied. Demengel and Ghidaglia [22] use similar analytical techniques to study a particular time discretization. Jones and Stuart [65] use a Hadamard graph transform technique (related to that used by Mallet-Paret and Sell [78] in their existence theory) to construct a perturbation theory general enough to allow consideration of a variety of space and time approximations. It is this theory which is outlined here.

It is also worth mentioning that the idea of inertial manifolds has been used in the development of numerical methods, sometimes referred to as nonlinear Galerkin methods, whereby some attempt is made to approximate the inertial manifold by a graph Φ_{approx} and compute with a spectral method of the form

$$\frac{du^{\mathrm{N}}}{dt} + Au^{\mathrm{N}} = \mathbb{P}F(u^{\mathrm{N}} + \Phi_{approx}(u^{\mathrm{N}})), t > 0, \quad u(0) = \mathbb{P}u_0^{\mathrm{N}}. \tag{6.29}$$

Such numerical methods are studied in, for example, Devulder *et al.* [24], Foias *et al.* [37], Foias *et al.* [39], Jolly *et al.* [60], Jones *et al.* [63], Russel *et al.* [83][84], Temam [93] and Titi [94]. Note that standard spectral methods correspond to taking $\Phi_{approx} \equiv 0$.

7 Attractors

7.1 Introduction

In Section 4 we considered the effect of perturbation on a very simple invariant set: the equilibrium point. We also discussed the effect of perturbation on trajectories in the neighborhood of the equilibrium point. Similar analyses can also be carried out for periodic solutions as mentioned in the bibliography of that section. In this section it is our purpose to study the very complicated invariant sets that arise in systems often loosely termed *chaotic.* The notion of an attractor formalizes the concept of a general object which captures the (possibly chaotic) long time dynamics of a system.

Section 7.2 contains the basic definitions of attractors and some of their properties. In particular, Theorem 7.3 gives a useful method for constructing global attractors and Theorem 7.6 is a very useful characterization of the global attractor as the union of all solutions of (2.9) which are defined and bounded for all positive and negative time. Further results of particular importance in Section 7.2 are the characterization of gradient systems and their attractors: see Theorem 7.7 and Corollary 7.8.

The definitions of upper- and lower-semicontinuity from Section 2.3 are required to understand fully the remainder of Section 7. In Section 7.3 we study the upper-semicontinuity of attractors — see Theorem 7.9. This shows that, after a sufficiently long time, every computed point is close to a point on the true attractor. It does not show, however, that every point on the true attractor has a nearby counterpart in the approximate attractor. In other words parts of the attractor may disappear under perturbation. However, if the true and perturbed attractors are uniformly exponentially attracting then the whole attractor perturbs smoothly under the approximation and the attractor is both upper- and lower-semicontinuous — see Theorem 7.10 in Section 7.4. In Section 7.5 we look at another situation where both upper- and lower-semicontinuity can be proved; this arises when the attractor is the union of the closure of unstable manifolds of equilibria. The simplest situation where this arises is for gradient systems, but other possibilities are also included.

7.2 Background theory

Definition 7.1 *A set A attracts a set B under $S(t)$ if, for any $\varepsilon > 0$, there exists $t^* = t^*(\varepsilon, B, A)$ such that $S(t)B \subset \mathcal{N}(A, \epsilon)$ $\forall t > t^*$. A compact invariant set A is said to be an attractor if A attracts an open neighborhood of itself. A global attractor is an attractor which attracts every bounded set in V.*

Example 7.2 Consider the equation

$$u_t + Au = \lambda u$$

with the operator A given in Example 2.1 with $\Omega = (0, 1)$. If $\lambda \in (\pi^2, 2\pi^2)$ then 0 attracts the set

$$B = \{v \in V : \int_0^1 v(x)\sin(\pi x)dx = 0\}.$$

However 0 is not an attractor since any open neighborhood contains points $u_0 = a\sin(\pi x)$ for which $\|u(t)\| \propto \exp\{(\lambda - \pi^2)t\}$. If $\lambda \in (0, \pi^2)$ then 0 attracts an open neighborhood of itself and is hence an attractor. It is in fact a global attractor.

Attractors are often constructed by applying the following theorem.

Theorem 7.3 *Assume that there exists $\tau \geq 0$ and $B \subset V$, a bounded open set, such that $S(t)\bar{B} \subset B$ $\forall t \geq \tau$ and $\bigcup_{t \geq \tau} S(t)B$ is relatively compact.*

Then $\omega(B)$ is an attractor which attracts B. Furthermore

$$\mathcal{A} := \omega(B) = \bigcap_{t \geq 0} S(t)B.$$

Proof Since $S(t)\bar{B} \subset B$ for $t \geq \tau$ it follows that

$$\omega(B) = \bigcap_{s \geq \tau} \overline{\bigcup_{t \geq s} S(t)B} \subset \bigcap_{s \geq \tau} \overline{\bigcup_{t \geq s} B} = \bar{B}. \tag{7.1}$$

Thus $\omega(B)$ is bounded. Note that, in fact, $\omega(B)$ is compact and invariant by Theorem 2.11.

We now show that $\mathcal{A} := \omega(B)$ attracts B. Assume that it does not. Then there exist $\epsilon > 0$ and sequences $x_k \in B$ and $t_k \to \infty$ such that $S(t_k)x_k \not\subset \mathcal{N}(\mathcal{A}, \epsilon)$. But $S(t_k)x_k$ is a sequence contained in a compact set and hence has a convergent subsequence $S(t_{k_i})x_{k_i} \to y \in \overline{B}$. By Definition 2.10 $y \in \omega(B) = \mathcal{A}$ and this is a contradiction.

Note that $\omega(B) \subset \bar{B}$ by (7.1). We show that, in fact, $\omega(B) \subset B$. Assume for the purposes of contradiction that $\exists y \in \omega(B) \bigcap \partial B$ (where $\partial B = \bar{B} \backslash B$). Since $\omega(B)$ is invariant it follows that, for any $t > 0$ $\exists x \in \omega(B) : S(t)x = y$. But, since $\omega(B) \subset \bar{B}$ we have $x \in \bar{B}$ and hence, by assumption, $y = S(t)x \in B$ for $t \geq \tau$. This is a contradiction and thus no such y exists. Thus $\omega(B) \subset B$.

Now, since $\omega(B) \subset B$ is closed it follows that, for ε sufficiently small, $\mathcal{N}(\omega(B), \varepsilon) \subset B$. Since $\omega(B)$ attracts B it follows that $\omega(B)$ attracts an open neighborhood of itself and is hence an attractor.

Finally, since $\omega(B) \subset B$ and $\omega(B)$ is invariant we have $\omega(B) \subset S(t)B$ for all $t \geq 0$. Hence

$$\omega(B) \subset \bigcap_{t \geq 0} S(t)B.$$

Furthermore, since

$$S(s)B \subset \overline{\bigcup_{t \geq s} S(t)B}$$

it follows that

$$\bigcap_{s \geq 0} S(s)B \subset \bigcap_{s \geq 0} \overline{\bigcup_{t \geq s} S(t)B} = \omega(B).$$

The final result follows. □

Recall the Definition 2.13 of dissipativity. It follows that a dissipative dynamical system which has some smoothing properties giving compactness will have a global attractor, by Theorem 7.3. Thus we have

Corollary 7.4 *Assume that $S(t)$ generates a dissipative dynamical system with absorbing set B. Then $\omega(B)$ is an attractor which attracts B.*

Furthermore

$$\mathcal{A} := \omega(B) = \bigcap_{t \geq 0} S(t)B.$$

Another corollary concerns the abstract sectorial evolution equation of Section 2.5.

Corollary 7.5 *Consider the dynamical system generated by eqn (2.9) under (2.11). There exists $R > 0$ such that $\omega(B(0,R)) = \bigcap_{t \geq 0} S(t)B(0,R)$ is a global attractor for (2.9).*

Proof From (2.9), (2.11), the variation of constants formula (2.15) and Lemma 10.6 we have

$$\|u(t)\| \leq e^{-\delta t}\|u(0)\| + \int_0^t \frac{Ce^{-\delta(t-s)}}{(t-s)^{\beta}} ds.$$

Hence

$$\|u(t)\| \leq e^{-\delta t}\|u(0)\| + K \tag{7.2}$$

where

$$K := \int_0^\infty \frac{Ce^{-\delta\tau}}{\tau^\beta} d\tau.$$

Choosing $R = K + \varepsilon$ and setting $B = B(0,R)$ we deduce from (7.2) that $S(t)\overline{B} \subset B$ for $t \geq t_0$, where t_0 is chosen so that

$$e^{-\delta t_0}(K + \varepsilon) < \varepsilon.$$

Furthermore, by applying the compactness estimate of Assumption (2.3), which bounds $|u(t)|_\eta$, on the time intervals $[0, 2t_0], [t_0, 3t_0], [2t_0, 4t_0], \ldots$, we deduce that $\bigcup_{t \geq t_0} S(t)B$ is relatively compact. Thus $\mathcal{A}$ attracts B by Theorem 7.3. Furthermore, (7.2) shows that for any bounded set E there exists $T = T(E)$ such that $S(t)E \subset B$ for all $t \geq T$. Hence $\mathcal{A}$ is a global attractor. □

The following characterization of the global attractor is very useful:

Theorem 7.6 *Consider a dynamical system $S(t)$ with global attractor $\mathcal{A}$. The set $\mathcal{A}$ is equivalent to the union of all complete bounded orbits of $S(t)$.*

Proof Let $x \in \mathcal{A}$. Since $\mathcal{A}$ is invariant it follows that $S(t)x \in \mathcal{A}$ and $\exists y \in \mathcal{A} : S(t)y = x$, for every $t > 0$. Thus a complete orbit through x exists and is bounded since $\mathcal{A}$ is compact. This shows that $\mathcal{A}$ is contained in the union of all bounded complete orbits.

Now let x be a point on a complete bounded orbit H. Note that H is invariant and hence, for any $t > 0$ $\exists y^t \in H : S(t)y^t = x$. We prove that $H \subseteq \mathcal{A}$. Assume that it is not, for contradiction; then, for any ε sufficiently small, $\exists x \in H : x \notin \mathcal{N}(\mathcal{A}, \varepsilon)$. But, since $\mathcal{A}$ is a global attractor and H is bounded, $\exists t^* > 0$:

$$x = S(t)y^t \in S(t)H \subseteq \mathcal{N}(\mathcal{A}, \varepsilon) \quad \forall t \geq t^*;$$

thus $x \in \mathcal{N}(\mathcal{A}, \varepsilon)$, a contradiction. This completes the proof. □

The most general results concerning the effect of perturbation on attractors are quite weak and structure must be placed on the attractor to obtain stronger results. In this context a particular class of systems of interest to us are gradient systems. Recall the definition of $\mathcal{E}$ given in (4.1) and the Definition 2.15. Recall also the Definition 5.1 of the unstable set. The following theorem and corollary elucidate the behavior of trajectories and the structure of the attractor for gradient systems.

Theorem 7.7 *If $S(t)$ defines a gradient system then, for any $u_0 \in V$, $\omega(u_0) \subseteq \mathcal{E}$ and for any negative orbit $\{\varphi(t), t \leq 0\}$ through u_0 for which $\bigcup_{t \leq t_1} \varphi(t)$ is relatively compact, $\alpha(u_0) \subseteq \mathcal{E}$. If, in addition, the set $\mathcal{E}$ comprises only isolated points then, for any $u_0 \in V$, there exists $x \in \mathcal{E}$ such that $\omega(u_0) = x$ and, for any negative orbit $\{\varphi(t), t \leq 0\}$ through u_0 for which $\bigcup_{t \leq t_1} \varphi(t)$ is relatively compact, there is $y \in \mathcal{E}$ such that $\alpha(u_0) = y$.*

Proof Consider the case of ω limit sets; the argument for α limit sets is similar. As in the proof of Theorem 7.5 we deduce that, for any $\tau > 0$, $\bigcup_{t \geq \tau} S(t)u_0$ is relatively compact. Hence $\omega(u_0)$ is non-empty, compact, invariant and connected by Theorem 2.11. Now let x and y be two points in $\omega(u_0)$ so that there are sequences $t_i, \tau_i \to \infty$ with $t_i < \tau_i < t_{i+1}$ so that $S(t_i)u_0 \to x$ and $S(\tau_i)u_0 \to y$. Since

$$\mathcal{V}(S(t_{i+1})u_0) \leq \mathcal{V}(S(\tau_i)u_0) \leq \mathcal{V}(S(t_i)u_0)$$

and there exists $c > 0$ such that $\mathcal{V}(S(t_i)u_0) \to c$ by continuity of $\mathcal{V}$, we deduce that $\mathcal{V}(S(\tau_i)u_0) \to c$ also. Thus $\mathcal{V}(y) = c$. Since $\omega(u_0)$ is invariant, for each $t \in \mathbb{R}$ we may choose a y such that $y = S(t)x$ — here we have extended $S(\bullet)$ to negative arguments for brevity — and deduce that $\mathcal{V}(S(t)x) = \mathcal{V}(x)$ for all $t \in \mathbb{R}$. By (iv) of Definition 2.15 we have $x \in \mathcal{E}$. By connectedness of the limit set we deduce that $\omega(u_0)$ must be a single point if $\mathcal{E}$ comprises only isolated points. □

Corollary 7.8 *If $S(t)$ defines a gradient system then it has a global attractor given by*

$$\mathcal{A} = W^u(\mathcal{E}) := \{u_0 \in V : \textit{ a negative orbit } \{\varphi(t), t \leq 0\} \textit{ exists through } u_0 \\ \textit{and } dist(\varphi(t), \mathcal{E}) \to 0 \textit{ as } t \to -\infty\}.$$

If, in addition, $\mathcal{E}$ comprises isolated points, then

$$\mathcal{A} = \bigcup_{v \in \mathcal{E}} W^u(v).$$

Proof By Corollary 7.6 the set $\mathcal{A}$ comprises all bounded complete orbits. Since $\mathcal{A}$ is compact by definition, each such complete orbit is compact. Applying Theorem 7.7 the result follows. □

7.3 Upper-semicontinuity of attractors

In this section we prove a basic result concerning the upper semicontinuity of attractors. We assume that $S(t)$ and $S^h(t)$ have global attractors. That the perturbation has a global attractor can often be proved directly for many particular approximation schemes. We discuss this briefly in Section 9. However, note that if $S(t)$ has a global attractor, the general Assumptions 3.2 made here would generally only imply the existence of a local attractor; a theory can be developed to cater for this case also.

Theorem 7.9 (Upper-Semicontinuity of Attractors) *Consider the approximation of the semigroup $S(t)$ by $S^h(t)$. Assume that $S(t)$ has a global attractor $\mathcal{A}_0$ and assume that there exist $h_0 > 0$ and a bounded $B \subset V$ such that $S^h(t)$ has a global attractor $\mathcal{A}^h$ for each $h \in (0, h_0]$ and*

$$\bigcup_{h\in[0,h_0]} \mathcal{A}^h \subseteq B.$$

Then

$$dist(\mathcal{A}_h, \mathcal{A}) \to 0 \quad as \quad h \to 0.$$

Proof First note that under Assumptions 3.2 we may assume, without loss of generality, that there exists $t_0 > 0$ and $g(t)$, bounded, continuous and monotonic increasing for $t \geq t_0$, such that

$$\|S^h(t)u - S(t)u\| \leq C(B)g(t)h \quad \forall u \in B, t \geq t_0. \tag{7.3}$$

Furthermore, since $\mathcal{A}_0$ attracts B, it follows that there exists a bounded, continuous function $f(t)$, defined for $t > 0$ and decreasing monotonically to zero, such that

$$\text{dist}(S(t)B, \mathcal{A}_0) \leq C(B)f(t) \quad \forall t \geq 0. \tag{7.4}$$

Note also that, since $\mathcal{A}^h$ is invariant under $S^h(t)$ and since $v \in S(t)\mathcal{A}^h$ implies that there exists $w \in \mathcal{A}^h$ such that $v = S(t)w$, we have from (7.3)

$$\begin{aligned} \operatorname{dist}(S^h(t)\mathcal{A}^h, S(t)\mathcal{A}^h) &= \sup_{u \in S^h(t)\mathcal{A}^h} \left\{ \inf_{v \in S(t)\mathcal{A}^h} \|u - v\| \right\} \\ &= \sup_{u \in \mathcal{A}^h} \left\{ \inf_{v \in S(t)\mathcal{A}^h} \|S^h(t)u - v\| \right\} \\ &= \sup_{u \in \mathcal{A}^h} \left\{ \inf_{w \in \mathcal{A}^h} \|S^h(t)u - S(t)w\| \right\} \\ &\leq \sup_{u \in \mathcal{A}^h} \|S^h(t)u - S(t)u\| \\ &\leq C(B)g(t)h. \end{aligned} \tag{7.5}$$

Also we have, since $\mathcal{A}^h$ is invariant under $S^h(t)$, since (7.4) holds and since $\mathcal{A}^h \subseteq B$,

$$\begin{aligned} \operatorname{dist}(\mathcal{A}^h, \mathcal{A}) &\leq \operatorname{dist}(\mathcal{A}^h, S(t)\mathcal{A}^h) + \operatorname{dist}(S(t)\mathcal{A}^h, \mathcal{A}) \\ &\leq \operatorname{dist}(S^h(t)\mathcal{A}^h, S(t)\mathcal{A}^h) + C(B)f(t) \\ &\leq C(B)g(t)h + C(B)f(t). \end{aligned} \tag{7.6}$$

Thus

$$\operatorname{dist}(\mathcal{A}^h, \mathcal{A}) \leq C(B)[hg(t) + f(t)] \quad \forall t > 0. \tag{7.7}$$

By the properties of $f(\bullet)$ and $g(\bullet)$ we may choose $h_c > 0$ and $t^* = t^*(h)$ such that

$$hg(t^*) = f(t^*) \quad \forall h \in (0, h_c]. \tag{7.8}$$

Figure 2 illustrates this situation. Furthermore, by the monotonicity of f and g it follows that $t^*(h) \to \infty$ as $h \to 0$. Thus, by (7.7),

$$\operatorname{dist}(\mathcal{A}^h, \mathcal{A}) \leq 2C(B)f(t^*(h)). \tag{7.9}$$

By the properties of $f(\bullet)$ and $t^*(\bullet)$ the required result follows. □

7.4 Continuity for exponentially attracting attractors

Note that Theorem 7.9 does not necessarily give a rate of convergence for the quantity $\operatorname{dist}(\mathcal{A}_h, \mathcal{A})$ which is a power of h. This is since nothing is assumed about the *rate of attraction* of the attractor. If the rate is assumed exponential then a stronger result can be proved and we obtain the error bound given in Theorem 7.10 below. Note that the bound is less than the rate of convergence of individual trajectories, unstable manifolds or inertial

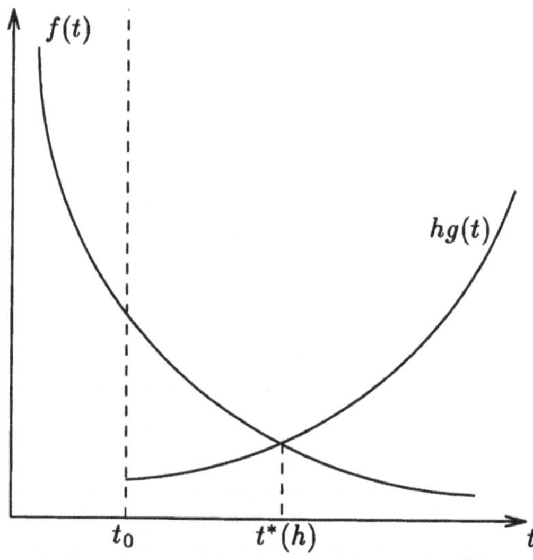

FIG. 2.

manifolds and reflects the competition between the exponential attraction to $\mathcal{A}$ (which determines η) and the exponential divergence of trajectories on $\mathcal{A}$ (which determines α).

Theorem 7.10 (Continuity of Exponentially Attracting Attractors) *Consider the approximation of the semigroup $S(t)$ by $S^h(t)$ satisfying (3.2). Let $S(t)$ have a global attractor $\mathcal{A}_0$ and assume that there exist $h_0 > 0$ and a bounded $B \subset V$ such that $S^h(t)$ has a global attractor $\mathcal{A}^h$ for each $h \in (0, h_0]$ and*

$$\bigcup_{h\in[0,h_0]} \mathcal{A}^h \subseteq B.$$

Assume also that there exists $\alpha, \eta, t_0 \in \mathbb{R}^+$ such that the approximation error satisfies (7.3) with $g(t) = e^{\alpha t}, t \geq t_0$ and that the attractors $\mathcal{A}^h$ are uniformly exponentially attracting so that (7.4) holds with $f(t) = e^{-\eta t}$. Then there exists $K > 0$ such that

$$d_{\mathrm{H}}(\mathcal{A}_h, \mathcal{A}) \leq Kh^{\beta} \quad \forall h \in (0, h_c],$$

where $\beta = \eta/(\alpha + \eta)$.

Proof Consider $\mathrm{dist}(\mathcal{A}^h, \mathcal{A})$ first. By (7.8) we have

$$he^{\alpha t^*} = e^{-\eta t^*}$$

so that $e^{-t^*} = h^{1/(\alpha+\eta)}$. Hence $f(t^*) = h^\beta$ and the bound follows from (7.9). The bound on $\text{dist}(\mathcal{A}, \mathcal{A}^h)$ follows by reversing the roles of $\mathcal{A}$ and $\mathcal{A}^h$ in the proof of Theorem 7.9. This can be done since the rate of convergence is uniform in $h \in (0, h_c]$. □

7.5 Lower-semicontinuity of attractors

The fact that lower-semicontinuity is hard to prove in general is not an artefact of the analysis. Simple examples exist which indicate that lower-semicontinuity is not true in general. Roughly the difficulty is that parts of the attractor which are not exponentially attracting may disappear under perturbation. Unfortunately the uniformly-in-h exponentially attracting attractors of Theorem 7.10 do not arise that often in applications and, even when they do, establishing that they have the right properties can be very hard. Instead we proceed in this section to prove lower-semicontinuity by making assumptions on the nature of the flow on the attractor $\mathcal{A}$. One important case where this is possible is when the dynamical system $S(t)$ is in gradient form and the set $\mathcal{E}$ of equilibria given by (4.1) is a bounded set containing only hyperbolic equilibria. A natural generalization of this is to make the following assumption:

Assumption 7.11 *The dynamical system (2.9) has a global attractor $\mathcal{A}$ where*

$$\mathcal{A} = \bigcup_{x \in \mathcal{E}} \overline{W^u(x)}$$

and $\mathcal{E}$ comprises a finite number of hyperbolic equilibrium points of (2.9).

Note that this assumption is a consequence of the system being in gradient form and having hyperbolic equilibria — see Corollary 7.8; however, Assumption 7.11 is weaker. For example, Assumption 7.11 admits equations with a single unstable equilibrium point and a unique limit cycle attracting all initial data except that starting at the equilibrium point.

Note also that, since the attractor is compact and contains all equilibria and since the hyperbolic fixed points are isolated by Theorem 4.3, the number of fixed points is automatically finite if they are hyperbolic. Under Assumption 7.11 we may prove lower-semicontinuity of the attractor as well as upper-semicontinuity:

Corollary 7.12 (Lower-Semicontinuity of Attractors) *Assume that Assumption 7.11 holds and consider the approximation of the semigroup $S(t)$ by $S^h(t)$. Denote the global attractor of $S(t)$ by $\mathcal{A}_0$ and assume that there exist $h_0 > 0$ and a bounded $B \subset V$ such that $S^h(t)$ has a global attractor $\mathcal{A}^h$ for each $h \in (0, h_0]$ and*

$$\bigcup_{h \in [0, h_0]} \mathcal{A}^h \subseteq B.$$

Then

$$d_{\mathrm{H}}(\mathcal{A}_h, \mathcal{A}) \to 0 \quad \text{as} \quad h \to 0.$$

Proof It follows from Theorem 7.9 that there exists an approximate attractor $\mathcal{A}_h$ satisfying

$$\mathrm{dist}(\mathcal{A}_h, \mathcal{A}) \to 0 \quad \text{as} \quad h \to 0.$$

Thus it remains to establish the lower-semicontinuity result that

$$\mathrm{dist}(\mathcal{A}, \mathcal{A}_h) \to 0 \quad \text{as} \quad h \to 0. \tag{7.10}$$

Let $\bar{u} \in \mathcal{E}$. By Theorem 5.7 we deduce that there exists a fixed point $\bar{u}^h$ of S^h such that

$$\mathrm{dist}(\overline{W^u(\bar{u})}, \overline{W^u_h(\bar{u}^h)}) \to 0 \quad \text{as} \quad h \to 0. \tag{7.11}$$

Now we prove that

$$\overline{W_{\mathrm{h}}{}^u(\bar{u}^h)} \subseteq \mathcal{A}_h. \tag{7.12}$$

Let $x \in W_{\mathrm{h}}{}^u(\bar{u}^h)$; then, by definition, there exists $x_i \to \bar{u}^h$ and $t_i \to \infty$ such that $x = S^h(t_i)x_i$ and hence it follows that $x \in \mathcal{A}^h$ since $\mathcal{A}^h$ is a global attractor and $\{x_i\}_{i=1}^{\infty}$ are contained in a bounded set. Thus (7.11), (7.12) prove that

$$\mathrm{dist}(\overline{W^u(\bar{u}^h)}, \mathcal{A}_h) \to 0 \quad \text{as} \quad h \to 0.$$

Since Assumption 7.11 holds it is clear that (7.10) follows and the proof is complete. □

7.6 Bibliography

For background theory on attractors and related material see Babin and Vishik [4], Bhatia and Szego [11], Hale [50] and Temam [92]. In particular the construction of global attractors given in Theorem 7.3 is closely related to the presentation in Temam [92]. Gradient systems are discussed extensively in Hale [50]; aside from their physical significance in problems modeled by dynamic energy minimization (see, for example, Elliott [28]), gradient systems are of fundamental importance in the theory of dynamical systems because of the simple characterization of the attractor given in Corollary 7.8 and the robustness to perturbations which follows from this.

The proof of continuity for exponentially attracting attractors given in Theorem 7.10 is taken from similar results in Babin and Vishik [4] and in Hale *et al.* [51]. The proof of upper-semicontinuity of attractors given in Theorem 7.9 is motivated by the proof in Babin and Vishik [4] concerning exponentially attracting attractors; slightly different proofs of upper semicontinuity are available in, for example, Hale *et al.* [51] and Temam [92]. These two general works generated a number of specific applications such as those studied by Dettori [23] and Shen [86]. Closely related results are proved for uniformly asymptotically stable sets (which are positively invariant sets containing the attractor) in Kloeden and Lorenz [67] [68]. Relationships between the results on attractors and the results on uniformly asymptotically stable sets are given in Hill and Süli [57]. For partial differential equations proof of upper-semicontinuity often requires error bounds for non-smooth initial data; see Elliott and Larsson [30], Larsson [74] and Yin-Yan [99]. This can be overcome in certain cases where the underlying equation, and its approximation, have a strong property known as Gevrey regularity [35] — see Lord and Stuart [77]. For analysis of upper-semicontinuity in the context of time-discrete, multistep methods see Hill and Suli [56].

Explicit and simple examples showing why lower-semicontinuity is not true in general may be found in Humphries *et al.* [59] and in Kapitanskii and Kostin [66]. Note, however, that if the stable and unstable manifolds of the equilibrium points for a gradient system intersect transversally then the attractor is exponentially attracting and Theorem 7.10 may be applied to establish upper- and lower-semicontinuity. The first general proofs of lower-semicontinuity for gradient systems, without requiring the transversal intersection property, appear in Hale and Raugel [52]. Related results may also be found in Babin and Vishik [4] and in Kostin [70]. These approaches assume that the attractor is in gradient form with hyperbolic equilibria. A subsequent generalization may be found in Humphries [58] where the attractor for an ordinary differential equation is assumed to be the union of unstable manifolds of hyperbolic equilibria; this includes the assumption of Hale *et al.* [51] but is more general than it. Results closely related to those of Humphries [58] may be found in Kapitanskii and Kostin [66] and in Humphries *et al.* [59] where partial differential equations are studied. Extensions of this approach to non-hyperbolic equilibrium points may be found in Kostin [71] and in Elliott and Kostin [29].

For estimates of the dimension of the attractor of time-discretized quasilinear partial differential equations, and comparison to dimension estimates for the underlying partial differential equation, see Eden *et al.* [26].

8 Error analysis for gradient systems

8.1 Introduction

So far we have derived three basic types of error bound for trajectories. The first appears in Assumptions 3.2 and detailed derivations are given for spectral methods in Theorem 3.6 (and its C^1 counterpart Theorem 3.7). The Assumptions 3.2 concern a uniform error bound on compact time intervals disjoint from the origin and bounded sets in V (see the remark following the proof of Theorem 3.6). The second appears in Corollary 4.10 and concerns an error bound uniform in $t \geq 2\tau$ for solutions asymptotic to a stable equilibrium point; this bound is not uniform in a bounded set of initial data essentially because initial data from a bounded set can take arbitrarily long to reach a neighborhood of the equilibrium — see the remark following the proof. The third appears in Theorem 4.18 and Theorem 4.19 and concerns an error bound for trajectories comprising a local phase portrait near an equilibrium point. It is uniform in time and uniform across a sufficiently small neighborhood of an equilibrium point. In this section we put these results together in various ways to derive error bounds for gradient systems and other classes of problems with similar properties.

What we would ideally like is to find an error bound which is uniform in large time and uniform across bounded sets of initial data. In general this does not appear to be possible. If, however, we restrict attention to gradient (and other closely related) systems and weaken the notion of approximation to allow piecewise continuous solutions with a finite number of discontinuities, then the goal is attainable. This we show in the following section — see Theorem 8.8.

8.2 The result

We start by making an assumption about solutions of (2.9) which is then shown to be satisfied by a certain class of gradient systems. Recall the set $\mathcal{E}$ defined in (4.1).

Assumption 8.1 *The dynamical system generated by (2.9) satisfies the following properties*

(i) the set $\mathcal{E}$ comprises a finite number of hyperbolic equilibria $\{z_i\}_{i=1}^N$ and there is a constant K_1 such that, for all sufficiently small $\rho > 0$, there is a bounded open set $\mathcal{U} \supset \mathcal{E}$ such that

$$\mathcal{U} = \bigcup_{i=1}^{N} Q_i, \quad Q_i \cap Q_j = \emptyset \text{ for } i \neq j, \quad Q_i \subseteq \mathcal{N}(z_i, K_1\rho); \tag{8.1}$$

(ii) for each bounded set $B \subset V$ there is a time T such that, for any $u_0 \in B$ there is $\tau \in [0,T]$ for which $S(\tau)u_0 \in \mathcal{U}$;

(iii) there is a constant $K_2 > 0$ such that, for each $u_0 \in V$ and each $\rho > 0$ sufficiently small, there is a subset of $\mathcal{E}$, relabelled $\{z_i\}_{i=1}^M$, times $\{t_0^+, \{t_i^\pm\}_{i=1}^M\}$ satisfying $t_0^+ = 0 \leq t_1^-$, $t_i^- < t_i^+ < t_{i+1}^-$, $i = 1, \ldots, N-1$, $t_{N-1}^- < t_N^+ = \infty$ and $\varepsilon > 0$ such that

(a) $S(t)u_0 \in \mathcal{N}(z_i, K_2\varepsilon)$ for all $t \in (t_i^-, t_i^+)$, $i = 1, \ldots, M$;
(b) $\mathcal{N}(z_i, K_2\varepsilon) \bigcap \mathcal{N}(z_j, K_2\varepsilon) = \emptyset$ for $i \neq j$;
(c) $S(t)u_0 \notin \mathcal{U}$ $\forall t \in [t_i^+, t_{i+1}^-]$, $i = 0, \ldots, M-1$;
(d) $S(t)u_0 \to z_N$ as $t \to \infty$.

Roughly, this states that the solutions of the dynamical system pass through a finite number of small neighborhoods of equilibria before finally entering and remaining in one such neighborhood for all t sufficiently large.

Definition 8.2 *The semigroup $S(t)$ generated by (2.9) is said to define a standard gradient system if it is a gradient system and*

(i) there exists $G \in C(V, \mathbb{R})$ such that, for all functions $w \in C^1(\Lambda, V)$ for some open interval $\Lambda \subset \mathbb{R}$,

$$\frac{d}{d\lambda}\{G(w(\lambda))\} = \langle \frac{dw}{d\lambda}, F(w) \rangle \quad \forall \lambda \in \Lambda; \tag{8.2}$$

(ii) $\mathcal{V}(\varphi) := \frac{1}{2}|\varphi|^2_{\frac{1}{2}} - G(\varphi)$;

(iii) $\mathcal{V}(\theta) - \mathcal{V}(\varphi) \leq \langle A\theta - F(\theta), \theta - \varphi \rangle + C|\theta - \varphi|^2 \;\; \forall \theta, \varphi \in X^1 = D(A)$.

Important remark By taking the inner product of (2.9) with u_t it follows that a standard gradient system satisfies

$$\frac{d}{dt}\{\mathcal{V}(u(t))\} = -|u_t|^2 \tag{8.3}$$

and we will use this equation explicitly in the following. Roughly speaking, the constant C in (iii) is a bound from above on the quadratic form constructed from the second derivative of $\mathcal{V}(\bullet)$ and then (iii) follows from Taylor expansion.

Example 8.3 Consider the reaction-diffusion equation (2.12) of Example 2.5, under the assumptions (2.13). It is shown to be a gradient system in Example 2.16. Further study of (2.12), (2.13) shows that it is in fact a standard gradient system with

$$G(\varphi) = \int_0^1 h(\varphi(x))dx,$$

where h is given by (2.26). All that is not immediately obvious and remains to be checked is (iii). To establish (iii) note that

$$\begin{aligned} \mathcal{V}(\theta) - \mathcal{V}(\varphi) &= \langle A\theta - F(\theta), \theta - \varphi \rangle - \tfrac{1}{2}\langle A(\theta - \varphi), \theta - \varphi \rangle \\ &+ \langle h(\varphi) - h(\theta), 1 \rangle + \langle F(\theta), \theta - \varphi \rangle. \end{aligned}$$

Now

$$\langle h(\varphi) - h(\theta), 1 \rangle + \langle F(\theta), \theta - \varphi \rangle =$$

$$\int_0^1 \{h(\varphi(x)) - h(\theta(x)) + f(\theta(x))(\theta(x) - \varphi(x))\}\, dx.$$

But

$$|h(\varphi) - h(\theta) - f(\theta)(\varphi - \theta)| \leq \frac{C}{2}|\theta - \varphi|^2 \quad \forall \theta, \varphi \in \mathbb{R}$$

by Taylor expansion, (2.26) and (2.13). Hence, since A is positive definite, the result follows. □

We now show that standard gradient systems satisfy Assumptions 8.1 before moving on to prove appropriate error estimates under those assumptions.

Theorem 8.4 *Assume that the semigroup generated by (2.9) defines a standard gradient system with a finite number of hyperbolic equilibria and that there exists $K_1 > 0$ such that*

$$\mathcal{U} := \{\eta \in X^1 = D(A) : |A\eta - F(\eta)| < \rho\}$$

satisfies (8.1) with $\delta = K_1\rho$ for all ρ sufficiently small. Then there exists $K_2 > 0$ such that Assumptions 8.1 *are satisfied with $\varepsilon = K_2\rho$ for all ρ sufficiently small.*

Proof The theorem makes Assumptions 8.1(i) a hypothesis. We turn to (ii). Let $u_0 \in B\backslash\mathcal{U}$, with B bounded. Let

$$\tau = \sup\left\{t \in \mathbb{R}^+ | S(s)u_0 \in B\backslash\mathcal{U}\, \forall s \in [0,t]\right\},$$

and note that $\tau > 0$ by continuity. By (8.3) we have

$$\mathcal{V}(u(s_2)) - \mathcal{V}(u(s_1)) = -\int_{s_1}^{s_2} |u_t(s)|^2 ds. \tag{8.4}$$

Hence, in particular,

$$\mathcal{V}(u(\tau)) - \mathcal{V}(u_0) = -\int_0^{\tau} |u_t(s)|^2 ds \leq -\tau\rho^2.$$

But $\mathcal{V}(u_0)$ is bounded by a constant $K(u_0)$ and $\mathcal{V}(u(\tau)) \geq 0$ since $\mathcal{V} \in C(V, \mathbb{R}^+)$ by definition. Hence we deduce that

$$\tau \leq K(u_0)/\rho^2.$$

Point (ii) follows with

$$T = \frac{1}{\rho^2} \sup_{u_0 \in B} K(u_0).$$

In the following the K_i denote constants independent of E which arise in the course of the analysis. We now establish point (iii) of Assumption 8.1. We assume that there exist times t_i and $u_i := u(t_i), i = 1,2$ such that $u(t) \notin \mathcal{U}$ for all $t \in [t_1, t_2]$ and $u(t_i) \in \partial Q_{i_k}$, the boundary of Q_{i_k}, for some integers i_1, i_2 between 1 and N. We will show first that there exists $K_3 > 0$ such that

$$u(t) \in B(z_l, K_3\rho) \quad \forall t \in [t_1, t_2] \quad \text{if} \quad i_1 = i_2 = l. \tag{8.5}$$

Secondly we will show that there exists $K_4 > 0$ such that

$$\mathcal{V}(z_m) - \mathcal{V}(z_l) \leq -K_4\rho \quad \text{if} \quad i_1 = l \neq i_2 = m. \tag{8.6}$$

Let us consider the case (8.5). By (8.4) and Definition 8.2(iii) it follows that, for $t \in [t_1, t_2]$,

$$\begin{aligned}
\textstyle\int_{t_1}^{t} |u_t(s)|^2 ds &\leq \int_{t_1}^{t_2} |u_t(s)|^2 ds \\
&\leq \mathcal{V}(u_1) - \mathcal{V}(u_2) \\
&\leq |Au_1 - F(u_2)|\,|u_1 - u_2| + C|u_1 - u_2|^2 \\
&\leq K_1[1 + CK_1]\rho^2 \\
&\leq K_5\rho^2.
\end{aligned}$$

Now we deduce that

$$(t - t_1)\rho^2 \leq \int_{t_1}^{t} |u_t(s)|^2 ds \leq K_5\rho^2$$

and hence that $t - t_1 \leq K_5$.

From the variation of constants formula (2.15) we have

$$u(t) = e^{-At}u(t_1) + \int_{t_1}^{t} e^{-A(t-s)}F(u(s))ds,$$

$$z_l = e^{-At}z_l + \int_{t_1}^{t} e^{-A(t-s)}F(z_l)ds.$$

Using Assumption 2.3, which shows that $F \in C(X^\eta, X)$ for some $\eta < 1$, and the Lemma 10.6, we obtain

$$\|u(t) - z_l\| \leq \|u(t_1) - z_l\| + \int_{t_1}^{t} \frac{K_6\|u(s) - z_l\|}{(t-s)^\eta} ds.$$

Using the facts that $\eta < 1$, $t - t_1 \leq K_5$ and $\|u(t_1) - z_l\| \leq K_1\rho$ the result (8.5) follows by application of the Gronwall Lemma 10.11.

Now consider (8.6). We have from (8.4) that

$$\mathcal{V}(u_2) - \mathcal{V}(u_1) \leq -\int_{t_1}^{t_2} |u_t|^2 ds \leq -\rho \int_{t_1}^{t_2} |u_t| ds \leq -\rho|u_2 - u_1|.$$

But

$$|u_2 - u_1| \geq |z_m - z_l| - |z_m - u_2| - |z_l - u_1|.$$

Since the equilibria are isolated we have that there exists $\zeta > 0$ such that

$$\min_{i \neq j} |z_l - z_m| > \zeta$$

and so, for sufficiently small ρ, since

$$|z_m - u_2|, |z_l - u_1| \leq K_1 \rho, \tag{8.7}$$

it follows that $|u_i - u_j| > \zeta/2$. Hence

$$\mathcal{V}(u_2) - \mathcal{V}(u_1) \leq -\frac{\rho\zeta}{2}$$

and, finally,

$$\mathcal{V}(z_m) - \mathcal{V}(z_l) \leq \mathcal{V}(u_2) - \mathcal{V}(u_1) + \mathcal{V}(z_m) - \mathcal{V}(u_2) + \mathcal{V}(z_l) - \mathcal{V}(u_1).$$

But $\mathcal{V}(z_m) - \mathcal{V}(u_2)$ and $\mathcal{V}(z_l) - \mathcal{V}(u_1)$ are both $\mathcal{O}(\rho^2)$ by Definition 8.2 (ii) and (8.7), so the required result follows.

To complete the proof of (iii) note that (c) follows by Theorem 7.7 since the system is in gradient form. Clearly $S(t)u_0$ passes through a finite number $M \leq N$ of the Q_i and we order these so that $\mathcal{V}(z_i) \leq \mathcal{V}(z_j), 1 \leq j \leq i \leq M$ without loss of generality. Define

$$t_i^- = \inf\{t : u(t) \in Q_i\}, \quad t_i^+ = \sup\{t : u(t) \in Q_i\}.$$

By (8.5) we have $u(t) \in \mathcal{N}(z_i, K_3\rho)$ for all $t \in [t_i^-, t_i^+]$. By (8.6) we deduce that if $I_i = [t_i^-, t_i^+]$ then $I_i \cap I_j = \emptyset$ for $i \neq j$ and the result follows. □

Example 8.5 We know that the reaction-diffusion equation (2.12) of Example 2.5 subjected to (2.13) yields a standard gradient system. Generically all the equilibria are isolated (see Babin and Vishik [4]) and application of the implicit function theorem in this case shows that the equilibria satisfy the remaining hypothesis of Theorem 8.4.

Definition 8.6 *The function $\tilde{u}(t)$ is said to be a piecewise continuous solution generated by a dynamical system $S(t)$, if there exist an integer N, non-negative numbers $\{T_i\}_{i=0}^N$ and elements $\{U_i\}_{i=0}^{N-1}$ of V such that $0 = T_0 < T_1 < T_2 < \cdots < T_N = \infty$ and for $i = 1, \cdots N$*

$$\tilde{u}(t) = S(t - T_{i-1})U_{i-1}, \; T_{i-1} \leq t < T_i.$$

Definition 8.7 *A piecewise continuous solution of (2.9) is said to be a combined stabilised trajectory of S if there exists $\rho > 0$ and $\{\bar{u}_j\}_{j=0}^{N-1} \in \mathcal{E}$ such that $B(\bar{u}_i, \rho) \bigcap B(\bar{u}_k, \rho) = \emptyset$ for $i \neq k$, with $U_j \in B(\bar{u}_j; \rho)$ for $j = 0, \cdots N-1$ and $\mathcal{V}(\bar{u}_j) < \mathcal{V}(\bar{u}_{j-1})$ for $j = 1, \cdots N-1$.*

Theorem 8.8 *Let Assumptions 8.1 be satisfied and let $E \subset V$ be bounded. Then, for any $u_0 \in E$, there exists a constant $C = C(E, \tau)$ and a combined* *stabilised trajectory $\tilde{u}^h(t)$ of $S^h(t)$, such that for any $u_0 \in E$*

$$\sup_{t \geq \tau} \|S(t)u_0 - \tilde{u}^h(t)\| \leq Ch.$$

Proof For simplicity assume that $u_0 \in \mathcal{U}$; the case $u_0 \notin \mathcal{U}$ can be handled similarly. Let

$$I_i = (t_i^-, t_i^+), \quad i = 1, \ldots, M$$

and remove all such intervals with $|I_i| \leq T_i^*$ where $T_i^* = T^*(z_i)$ from Theorem 4.13 and Corollary 4.14. Relabel the remaining $\{I_i\}_{i=1}^{M_0}$ where $M_0 \leq M$. Define

$$I_i^* = (t_i^- + T_i^*, t_i^+), i = 1, \ldots, M_0, \quad J_i = [t_i^+, t_{i+1}^- + T_{i+1}^*], i = 0, \ldots, M_0 - 1.$$

Note that

$$|J_i| \leq T_0 + \sum_{j=1}^{N} (T + T_j^*)$$

since, between intervals I_i^* and I_{i+1}^* the solution can pass through at most N other equilibria and can spend at most T_j^* time units in the neighborhood of each z_j. Whilst outside $\mathcal{U}$, Assumption 8.1(ii) shows that the solution can spend at most T time units. Thus we have shown that $|J_i|$ is bounded above in terms of E, but independently of the specific choice of $u_0 \in E$. Set

$$T_i = t_i^- + T_i^*, \quad U_i = S^h(T_i^*)u_i^h(0)$$

where $u_i^h(T) = u^h(T)$ given in Theorem 4.18 with $\bar{u} = z_i$ for $i = 1, \ldots, M_0 - 1$ and $u_{M_0}^h(T) = u^h(T)$ given in Theorem 4.19 with $\bar{u} = z_{M_0}$.

On the interval I_i^* we apply Theorems 4.18, 4.19 to get the required error bounds and on J_i we apply Assumptions 3.2 together with (2.16) to get the required error bound. The constants in Theorems 4.18, 4.19 are independent of E and depend only upon the equilibria $\{z_i\}_{i=1}^N$. The constants in Assumption 3.2 and (2.16) depend only upon E through the time interval J_i which is bounded above in terms of E, independently of the specific choice of u_0 in E. □

Important remark Note that the total number of discontinuities is bounded above by the total number of equilibria. Note also that if the solution $S(t)u_0 \to z_i$ as $t \to \infty$ and does not pass through any Q_j, $j \neq i$, then the number of discontinuities is at most one; it will in fact be zero if z_i is stable for then the discontinuity on the boundary of Q_i disappears as Theorem 4.19 becomes equivalent to Theorem 4.8 and the uniformly valid approximate solution has the same initial condition as the underlying solution.

8.3 Bibliography

The idea that trajectories of gradient systems can be uniformly approximated in time by a piecewise continuous trajectory with a finite number of discontinuities is contained in the book of Babin and Vishik [4]. Their approach is to make all but the first segment of the piecewise approximation lie on the unstable manifold of an equilibrium point. The price to pay for this is that the order of approximation is $\mathcal{O}(h^\lambda)$ for some $\lambda \in (0,1)$; however, since the unstable manifolds are finite-dimensional this is a finite-dimensional approximation of an infinite-dimensional dynamical system. In this article we have proved a weaker result, in the sense that the pieces of approximating solution are not finite-dimensional in nature, but a stronger rate of approximation, namely $\mathcal{O}(h)$. The approach taken here to prove the results such as Theorem 8.4 originates in Stuart and Humphries [90] where the effect of error control on gradient dynamical systems in finite dimensions is considered. Similar techniques were then used in Elliott and Stuart [32] to study the viscous Cahn–Hilliard equation (3.10).

9 Practical numerical stability

The main purpose of the results in Sections 3–8 in the context of computation is to enable interpretation of data gleaned from long-time simulations. In particular we have shown that certain invariant objects persist under numerical perturbations and we have obtained a variety of error estimates. Amongst other results these enable us to: (i) state with confidence the sense in which computations near an equilibrium point make sense; (ii) state with confidence the sense in which data gleaned from numerical simulations on (possibly chaotic) attractors should be interpreted; (iii) state with confidence the sense in which error bounds for trajectories of gradient systems can be viewed as being uniformly valid in time and across a bounded set of initial data.

However, in the context of numerical approximation, the results described in Sections 3–8 do not distinguish between the relative merits of different approximation methods other than in their rate of convergence. It is of some importance to gain an understanding of which numerical methods work well in practice and the concept of *practical numerical stability* is relevant here. For our purposes we shall take this to mean the construction of schemes which preserve some important features of the underlying semigroup under mild or no restrictions on the mesh parameters.

The first illustration is to consider the equation (2.9) under the assumption that there exist $\alpha, \beta \geq 0$ such that

$$\frac{1}{2}|u|^2_{\frac{1}{2}} - \langle F(u), u\rangle \geq \beta |u|^2 - \alpha.$$

It then follows from (2.9) that

$$\frac{1}{2}\frac{d}{dt}|u|^2 \leq \alpha - \beta|u|^2$$

and hence that the ball $B = \{u \in X : |u|^2 \leq R\}$ is positively invariant for any $R > \alpha/\beta$. Furthermore any bounded set of initial data is mapped inside B in a finite time so that B is absorbing and the system is dissipative in the sense of Definition 2.13. Such results are crucial stepping-stones to establishing the global bounds on the nonlinearity that we *assume* in Section 2.5 and Section 6, but which must typically be proven a priori for many equations arising in applications. One of the earliest works to look at the preservation of dissipativity in the numerical approximation of equations like (2.9) is Foias *et al.* [34], where finite difference and spectral approximations of the Kuramoto–Sivashinsky equation were studied. Elliott and Stuart [32] address similar issues for a finite difference approximation of a reaction-diffusion equation and temporal discretization by a variety of one-step methods. In the paper of Armero and Simo [3], preservation of dissipativity is studied for the Navier–Stokes equations under finite element time approximation and a variety of one-step temporal approximations. The reviews of Stuart and Humphries [89] and of Humphries *et al.* [59] contain surveys of the literature concerning preservation of dissipativity.

A second illustration of the concept of practical numerical stability is the preservation of the gradient structure of Definition 2.15. An early example containing explicit reference to the importance of retaining the Lyapunov functional under approximation is contained in Elliott [28] where finite element spatial approximation, together with some one-step time approximations of the Cahn–Hilliard equation, are considered. Further studies are contained in Elliott and Stuart [31] where finite difference, one-step and multi-step methods are analysed for a reaction-diffusion equation. See also Stuart and Humphries [89] for a review of this subject.

Bibliography

1. Alouges, F. and Debussche, A. (1991). On the qualitative behavior of the orbits of a parabolic partial differential equation and its discretization in the neighborhood of a hyperbolic fixed point. *Num. Funt. Anal. and Opt.*, **12**, 253–269.
2. Alouges, F. and Debussche, A. (1993). On the discretization of a partial differential equation in the neighborhood of a periodic orbit. *Num. Math.*, **65**, 143–175.
3. Armero, F. and Simo, J. Unconditional stability and long-term behavior of transient algorithms for the incompressible Navier–Stokes and Euler equations. To appear in *Comp. Meth. Appl. Mech. and Eng.*
4. Babin, A. and Vishik, M.I. (1992). *Attractors of evolution equations.*

Studies in Mathematics and its Applications, North-Holland, Amsterdam.

5. Bai, F., Spence, A. and Stuart, A.M. (1993). Numerical computation of heteroclinic connections in systems of gradient partial differential equations. *SIAM J. Appl. Math.*, **53**, 743–769.
6. Bates, P.W. and Lu, K. A Hartman–Grobman theorem for Cahn–Hilliard and Phase-Field equations. To appear *J. Dyn. Diff. Eq.*
7. Beyn, W.-J. (1987). On invariant closed curves for one-step methods. *Numer. Math.*, **51**, 103–122.
8. Beyn, W.-J. (1987). On the numerical approximation of phase portraits near stationary points. *SIAM J. Num. Anal.*, **24**, 1095–1113.
9. Beyn, W.-J. (1992). Numerical methods for dynamical systems. *Numerical Analyis; Proceedings of the SERC Summer School, Lancaster, 1990*, edited by W.A. Light. Clarendon Press, Oxford.
10. Beyn, W.-J. and Lorenz, J. (1987). Center manifolds of dynamical systems under discretization. *Num. Func. Anal. and Opt.*, **9**, 381–414.
11. Bhatia, N.P. and Szego, G.P. (1970). *Stability Theory of Dynamical Systems.* Springer-Verlag, New York.
12. Bloch, A.M. and Titi, E.S. (1990). On the dynamics of rotating elastic beams. *New Trends in Systems Theory*, edited by G. Conte, A. Perdon and B.F. Wyman. Birkhauser, Berlin, 1990.
13. Braun, M. and Hershenov, J. (1977). Periodic solution of finite difference equations. *Quart. Appl. Math.*, **35**, 139–147.
14. Broomhead, D. and Iserles, A. (1992). *Proceedings of the IMA Conference on the dynamics of numerics and the numerics of dynamics, 1990.* Cambridge University Press, Cambridge.
15. Caginalp, G. (1986). An analysis of a phase field model of a free boundary. *Arch. Rat. Mech.*, **92**, 205–245.
16. Chow, S.-N. and Hale, J.K. (1982). *Methods of Bifurcation Theory.* Springer.
17. Chow, S.-N., Lu, K. and Sell, G.R. (1992). Smoothness of inertial manifolds. *J. Math. Anal. Appl.*, **169**, 283–321.
18. Chueshow, I.D. (1993). Global attractors for nonlinear problems of mathematical physics. *Russian Math. Surv.*, **48**, 133–161.
19. Constantin, P. and Foias, C. (1988). *Navier–Stokes Equations.* Chicago University Press.
20. Constantin, P., Foias, C. and Temam, R. (1985). Attractors representing turbulent flows. *Mem. Amer. Math. Soc.*, **314**.
21. Constantin, P., Foias, C., Nicolaenko, B. and Temam, R. (1989). *Integral manifolds and Inertial Manifolds for Dissipative Partial Differential Equations.* Appl. Math. Sciences, Springer Verlag, New York.

22. Demengel, F. and Ghidaglia, J.M. (1989). Time-discretization and inertial manifolds. *Math. Mod. and Num. Anal.*, **23**, 395–404.
23. Dettori, L. (1990). Spectral approximations of attractors of a class of semilinear parabolic equations. *CALCOLO*, **27**, 139–168.
24. Devulder, C., Marion, M. and Titi, E.S. (1993). On the rate of convergence of the nonlinear Galerkin methods. *Math. Comp.*, **60**, 495–514.
25. Doan, H.T. (1985). Invariant curves for numerical methods. *Quart. Appl. Math.*, **3**, 385–393.
26. Eden, A., Michaux, B. and Rakotoson, J.M. (1990). Semi-discretized nonlinear evolution equations as discrete dynamical systems and error analysis. *Ind. J. Math.*, **39**, 737–784.
27. Eirola, T. (1988). Invariant curves for one-step methods. *BIT*, **28**, 113–122.
28. Elliott, C.M. (1989). The Cahn–Hilliard model for the kinetics of phase separation. *Mathematical models for phase change problems*, edited by J.F. Rodrigues. Birkhauser, Berlin.
29. Elliott, C.M. and Kostin, I. (1994). Lower semicontinuity of a non-hyperbolic attractor for the viscous Cahn–Hilliard equation. Submitted to *Nonlinearity.*
30. Elliott, C.M. and Larsson, S. (1992). Error estimates with smooth and nonsmooth data for a finite element method for the Cahn–Hilliard equation. *Math. Comp.*, **58**, 603–630.
31. Elliott, C.M. and Stuart, A.M. (1993). Global dynamics of discrete semilinear parabolic equations. *SIAM J. of Num. Anal.*, **30**, 1622–1663.
32. Elliott, C.M. and Stuart, A.M. (1994). The viscous Cahn–Hilliard equation. Part II: analysis. Submitted.
33. Eriksson, K., Estep, D., Hansbo, P. and Johnson, C. (1995). Adaptive Finite Element Methods. To appear in *Acta Numerica*, Cambridge University Press, Cambridge.
34. Foias, C., Jolly, M.S., Kevrekidis, I.G. and Titi, E.S. (1991). Dissipativity of numerical schemes. *Nonlinearity*, **4**, 591–613.
35. Ferrari, A.B. and Titi, E.S. (1994). Gevrey regularity of solutions of a class of analytic nonlinear parabolic equations. Submitted to *Communications in PDEs.*
36. Foias, C. and Prodi, G. (1967). Sur le comportement global des solutions non stationnaires des equations de Navier–Stokes en dimension 2. *Rend. Sem. Mat. Univ. Padova*, **39**.
37. Foias, C., Manley, O.P. and Temam, R. (1988). Modelization of the interaction of small and large eddies in two dimensional turbulent flows. *Math. Mod. and Num. Anal.* M^2AN, **22**, 93–114.

38. Foias, C., Sell, G. and Temam, R. (1988). Inertial manifolds for nonlinear evolutionary equations. *J. Diff. Eq.*, **73**, 309–353.

39. Foias, C., Sell, G. and Titi, E.S. (1989). Exponential tracking and approximation of inertial manifolds for dissipative nonlinear equations. *J. Dynamics and Diff. Eq.*, **1**, 199–243.

40. Friedman, A. (1964). *Partial Differential Equations of Parabolic Type.* Prentice Hall.

41. Garay, B.M. (1993). Discretization and some qualitative properties of ordinary differential equations about equilibria. *Acta Math. Univ. Comenianae*, **62**, 249–275.

42. Garay, B.M. (1994). Discretization and Morse-Smale dynamical systems on planar discs. *Acta Math. Univ. Comenianae*, **63**, 25–38.

43. Garay, B.M. (1994). Discretization and normal hyperbolicity. *Z. Angew. Math. Mech.*, **74**, T662–T663.

44. Aulbach, B., and Garay, B.M. (1994). Discretization of semilinear differential equations with an exponential dichotomy. *Computers Math. Applic.*, **28**, 23–35.

45. Garay, B.M. (1994). On structural stability of ordinary differential equations with respect to numerical methods. Submitted to *Numer. Math.*

46. Garay, B.M. (1994). The discretized flow on domains of attraction: a structural stability result. Submitted to *Fund. Math.*

47. Garay, B.M. (1994). On C^j-closeness between the solution flow and its numerical approximations. Submitted to *J. Difference Eq. Appl.*

48. Garay, B.M. (1994). On various closeness concepts in numerical ODEs. Submitted to *Computers Math.*

49. Hale, J.K. (1969) *Orindary Differential Equations.* Wiley, Chichester.

50. Hale, J.K. (1988). *Asymptotic Behavior of Dissipative Systems.* AMS Mathematical Surveys and Monographs 25, Rhode Island.

51. Hale J.K., Lin, X.-B. and Raugel, G. (1988). Upper Semicontinuity of Attractors for Approximations of Semigroups and Partial Differential Equations. *Math. Comp.*, **50**, 89–123.

52. Hale, J.K. and Raugel, G. (1989). Lower Semicontinuity of Attractors of Gradient Systems and Applications. *Annali di Mat. Pura. Applic.*, **CLIV**, 281–326.

53. Hartman, P. (1969). *Ordinary Differential Equations.* Wiley, Chichester.

54. Henry, D. (1981). *Geometric Theory of Semilinear Parabolic Equations.* Lecture Notes in Mathematics, Springer-Verlag, New York.

55. Heywood, J.G. and Rannacher, R. (1986). Finite element approximations of the Navier–Stokes problem. Part II: Stability of solutions and

error estimates uniform in time. *SIAM J. Num. Anal.*, **23**, 750–777.

56. Hill, A.T. and Süli, E. (1993). Upper semicontinuity of attractors for linear multistep methods approximating sectorial evolution equations. Submitted to *Math. Comp.*
57. Hill, A.T. and Süli, E. (1994). Set convergence for discretizations of the attractor. Submitted to *Numer. Math.*
58. Humphries, A.R. (1994). Approximation of attractors and invariant sets by Runge–Kutta methods. In preparation.
59. Humphries, A.R., Jones, D.A. and Stuart, A.M. (1994). Approximation of dissipative partial differential equations over long time intervals. *Numerical Analysis, Dundee, 1993*, edited by D.F. Griffiths and G.A. Watson. Longman, New York.
60. Jolly, M.S., Kevrekidis, I.G. and Titi, E.S. (1990). Approximate inertial manifolds for the Kuramoto–Sivashinsky equation: analysis and computations. *Physica D*, **D44**, 38–60.
61. Johnson, C., Larsson, S., Thomee, V. and Wahlbin, L.B. (1987). Error estimates for spatially discrete approximations of semilinear parabolic equations with non-smooth initial data. *Math. Comp.*, **49**, 331–357.
62. Jones, D.A. (1994). On the Behavior of Attractors under Finite Difference Approximation. Submitted to *Nonlinearity.*
63. Jones, D.A., Margolin, L.G. and Titi, E.S. (1994). On the effectiveness of the approximate inertial manifold — a computational study. To appear in *J. Thero. Comp Fluid Mech.*
64. Jones. D.A. and Titi, E.S. (1994). C^1 Approximations of Inertial Manifolds for Dissipative Nonlinear Equations. Submitted to *J. Diff. Eq.*
65. Jones, D.A. and Stuart, A.M. (1995). Attractive Invariant Manifolds Under Approximation. To appear in *J. Diff. Eq.*
66. Kaptanskii, L.V. and Kostin, I.N. (1991). Attractors of nonlinear evolution equations and their approximations. *Leningrad Math. J.*, **1**, 97–117.
67. Kloeden, P. and Lorenz, J. (1986). Stable attracting sets in dynamical systems and their one-step discretizations. *SIAM J. Num. Anal.*, **23**, 986–995.
68. Kloeden, P. and Lorenz, J. (1989). Liapunov stability and attractors under discretization. *Differential Equations, proceedings of the equadiff conference*, edited by C.M. Dafermos, G. Ladas and G. Papanicolaou. Marcel-Dekker, New York.
69. Kloeden, P. and Palmer, K.J. (1993). *Chaotic Numerics*, American Mathematical Society, Contemporary Mathematics no. 172, Providence.
70. Kostin, I.N. (1992). A regular approach to a problem on the attractors

of singularly perturbed equations. *J. Sov. Math.*, **62**, 2664–2688.

71. Kostin, I.N. (1994). Lower semicontinuity of a non-hyperbolic attractor. Submitted to *J. London Math. Soc.*
72. Ladyzhenskaya, O. (1991). *Attractors for Semigroups and Evolution Equations.* Cambridge University Press, Cambridge, 1991.
73. Larsson, S. (1989). The long-time behavior of finite element approximations of solutions to semilinear parabolic problems. *SIAM J. Num. Anal.*, **26**, 348–365.
74. Larsson, S. (1992) *Non-smooth data error estimates with applications to the study of long-time behavior of finite element solutions of semilinear parabolic problems.* Pre-print, Chalmers University, Sweden.
75. Larsson, S. and Sanz-Serna, J.M. (1994). The behavior of finite element solutions of semilinear parabolic problems near stationary points. *SIAM J. Num. Anal.*, **31**, 1000–1018.
76. Lions, J.L. (1969). *Quelques Methodes de Resolution des Problemes aux Limites Non Lineaires.* Dunod, Paris.
77. Lord, G.J. and Stuart, A.M. (1994) Discrete Gevrey Regularity and Attractors for a Finite Difference Approximation of the Ginzburg–Landau equation. Submitted to *Num. Func. Anal. Opt.*
78. Mallet-Paret, J. and Sell, G.R. (1988). Inertial manifolds for reaction-diffusion equations in higher space dimensions. *J. Amer. Math. Soc.*, **1**, 805–864.
79. Miklavčič, M. (1985). Stability for semilinear equations with noninvertible linear operator. *Pac. J. Math.*, **118**, 199–214.
80. Orszag, S.A. (1970). Transform method for calculation of vector coupled sums: Application to the spectral form of the vorticity equation. *J. Atmos. Sci.*, **27**, 890–895.
81. Pazy, A. (1983) *Semigroups of Linear Operators and Applications to Partial Differential Equations.* Springer-Verlag, New York.
82. Pugh, C. and Shub, M. (1988). C^r stability of periodic solutions and solution schemes. *Appl. Math. Lett.*, **1**, 281–285.
83. Russell, R.D., Sloan, D.M. and Trummer, M.R. (1992). On the structure of Jacobians for spectral methods for nonlinear PDEs. *SIAM J. Sci. Stat. Comp.*, **13**, 541–549.
84. Russell, R.D., Sloan, D.M. and Trummer, M.R. (1993). Some numerical aspects of computing inertial manifolds. *SIAM J. Sci. Stat. Comp.*, **14**, 19–43.
85. Sanz-Serna, J.M. and Stuart, A.M. (1992). A note on uniform in time error estimates for approximations to reaction-diffusion equations. *IMA J. Num. Anal.*, **12**, 457–462.
86. Shen, J. (1989). Convergence of approximate attractors for a fully

discrete system for reaction-diffusion equations. *Numer. Funct. Anal. and Opt.*, **10**, 1213–1234.

87. Stetter, H. (1973). *Analysis of Discretization Methods for Ordinary Differential Equations.* Springer-Verlag, New York.
88. Stuart, A.M. (1994). Numerical Analysis of Dynamical Systems. *Acta Numerica 1994*, Cambridge University Press, Cambridge.
89. Stuart, A.M. and Humphries, A.R. (1994). Model problems in numerical stability theory for initial value problems. *SIAM Review*, **36**, 226–257.
90. Stuart, A.M. and Humphries, A.R. (1995). Analysis of local error control for dynamical systems. To appear in *SIAM J. Num. Anal.*
91. Temam, R. (1979). *Navier–Stokes equations.* North-Holland, Amsterdam.
92. Temam, R. (1988). *Infinite Dimensional Dynamical Systems in Mechanics and Physics.* Springer, New York.
93. Temam, R. (1989). Attractors for the Navier–Stokes equations: localization and approximation. *J. Fac. of Sci., The Univ. of Tokyo, IA*, **36**, 629-647.
94. Titi, E.S. (1990). On approximate inertial manifolds to the Navier–Stokes equations. *J. Math. Anal. Appl.*, **149**, 540-557.
95. Titi, E.S. (1991). Un critère pour l'approximation des solutions périodiques des équations de Navier–Stokes. *C.R. Acad. Sci. Paris*, **312**, 41–43.
96. Thomée, V. (1984) *Galerkin Finite Element Methods for Parabolic Problems.* Springer-Verlag, New York.
97. Wells, J.C. (1976). Invariant manifolds of nonlinear operators. *Pac. J. Math.*, **62**, 285–293.
98. Wiggins, S. (1994). *Normally Hyperbolic Invariant Manifolds in Dynamical Systems.* Springer-Verlag, New York.
99. Yin-Yan (1993). Attractors and error estimates for discretizations of incompressible Navier–Stokes equations. Submitted to *SIAM J. Num. Anal.*

10 Appendix A — Sectorial evolution equations

In this appendix we outline the basic theory of sectorial evolution equations in a separable Hilbert space X with norm $|\bullet|$. The results are all taken from Henry [54] and Pazy [81] with the exception of Lemma 10.12. Note however that the precise definition of "solution" used here is that given in Miklavčič [79] since that used in Henry [54] does not necessarily yield uniqueness. The results in Henry are not changed in any essential way by this change of definition Hale [50]. We let A denote a linear, densely defined operator in a Hilbert space X with compact inverse and eigenvalue and eigenfunction pairs $\{\lambda_i, \varphi_i\}$ ordered so that

$$\text{Re}\{\lambda_i\} \le \text{Re}\{\lambda_{i+1}\}.$$

Recall that the *resolvent set* of A is the set of λ in the complex plane for which $(A - \lambda I)^{-1}$ is a bounded linear operator in X. The norm and inner product on X are denoted by $|\bullet|, \langle\bullet,\bullet\rangle$ respectively.

Definitions 10.1, 10.2 and Lemma 10.3 are Definitions 1.3.1, 1.3.3 and and Theorem 1.3.4 of Henry respectively.

Definition 10.1 *A linear, closed, densely defined operator A in the Hilbert space X is said to be sectorial if, for some $\varphi \in (0, \pi/2), M \ge 1$ and $a \in \mathbb{R}$, its resolvent set $\mathcal{R}$ satisfies*

$$\mathcal{R} \supseteq \mathcal{S} := \{\lambda | \varphi \le |arg(\lambda - a)| \le \pi, \lambda \ne a\}$$

and, furthermore

$$|(\lambda I - A)^{-1}| \le M/|\lambda - a| \quad \forall \lambda \in \mathcal{S}.$$

Definition 10.2 *An analytic semigroup on a Hilbert space X is a family of continuous linear operators on X, $\{T(t)\}_{t\ge 0}$, satisfying*

(i) $T(0) = I, T(t)T(s) = T(t+s), \quad \forall t, s \ge 0;$
(ii) $T(t)x \to x$ as $t \to 0^+$ for each $x \in X$;
(iii) $t \mapsto T(t)x$ is real analytic on $0 < t < \infty$ for each $x \in X$.

The infinitesimal generator L of $T(t)$ is defined by

$$Lx = \lim_{t\to 0^+} \left\{\frac{T(t)x - x}{t}\right\}$$

with domain $D(L)$ consisting of all $x \in X$ for which this limit exists.

Lemma 10.3 *If A is a sectorial operator, then $-A$ is the infinitesimal generator of an analytic semigroup $T(t)$.*

Formally we may think of $T(t) = e^{-At}$. Indeed if A is self-adjoint with respect to the inner product $\langle\bullet,\bullet\rangle$, so that any $v \in X$ can be represented as

$$v = \sum_{j=1}^{\infty} v_j \varphi_j, \quad v_j = \langle v, \varphi_j \rangle, \tag{10.1}$$

then it is appropriate to consider $T(t)$ as being given by

$$T(t)v = e^{-At}v = \sum_{j=1}^{\infty} e^{-\lambda_j t} v_j \varphi_j. \tag{10.2}$$

In the case where A is not sectorial an analogous definition can be made. Henceforth we will use e^{-At} to denote $T(t)$. In the non self-adjoint case the precise definition of $T(t)$ is through a contour integral, evaluation of which coincides with the approach outlined here in the self-adjoint case.

With these definitions we may define fractional powers of the operator A; the following definition (from Henry, Definition 1.4.1) can be shown to make sense by use of the properties of sectorial operators; $\Gamma(\bullet)$ denotes the Γ-function.

Definition 10.4 *If A is a sectorial operator with* $\mathrm{Re}\{\lambda_1\} > 0$ *then, for $\alpha > 0$, define the fractional powers of A by*

$$A^{-\alpha} = \frac{1}{\Gamma(\alpha)} \int_0^{\infty} t^{\alpha-1} e^{-At} dt,$$

$A^0 = I$ and $A^{\alpha} = (A^{-\alpha})^{-1}$ with $D(A^{\alpha}) = R(A^{-\alpha})$.

Returning to the case where A is self-adjoint so that any $v \in X$ can be represented as in (10.1) we find that

$$A^{\alpha} v = \sum_{j=1}^{\infty} \lambda^{\alpha} v_j \varphi_j \;\; \forall \alpha \in \mathbb{R}. \tag{10.3}$$

The following two results follow from Henry, Definition 1.4.7 and Theorems 1.4.2, 1.4.3 and 1.4.8.

Lemma 10.5 *If A is a sectorial operator, then the space $X^{\alpha} = D(A_1^{\alpha})$ is a Hilbert space with norm $|\bullet|_{\alpha} = |A_1^{\alpha} \bullet|$, where $A_1 = A + aI$ for any a such that A_1 has positive eigenvalues. If A has compact resolvent and $\alpha > \beta \geq 0$ then the inclusion $X^{\alpha} \subset X^{\beta}$ is compact.*

Lemma 10.6 *If A is a sectorial operator then for any $\alpha \leq 0$ there exists $K = K(\alpha) < \infty$ such that*

$$|A^{\alpha}| \leq K.$$

Furthermore, if $\mathrm{Re}\{\lambda_1\} > \delta > 0$ *then, for any $\alpha > 0$ there exists $C = C(\alpha) < \infty$ such that*

$$|e^{-At}|_{\alpha} \leq C t^{-\alpha} e^{-\delta t} \quad \forall t > 0.$$

We sketch the proof of a related result for the case when A is self-adjoint. From (10.3) and (10.2) we have that

$$A^{\alpha}e^{-At}v = \sum_{j=1}^{\infty} \lambda_j^{\alpha} e^{-\lambda_j t} v_j \varphi_j.$$

Hence

$$|e^{-At}v|_{\alpha}^2 = \sum_{j=1}^{\infty} \lambda_j^{2\alpha} e^{-2\lambda_j t} v_j^2,$$

assuming the normalization $|\varphi_j|^2 = 1 \;\forall j$. If $t = 0$ and $\alpha \leq 0$ we have

$$|v|_{\alpha} \leq \lambda_1^{\alpha}|v|$$

so that A^{α} is bounded on X.

A simple calculation reveals that

$$\max_{y \geq 0} y^{2\alpha} e^{-2yt} = \alpha^{2\alpha} e^{-2\alpha}/t^{2\alpha}.$$

Hence there exists $C = C(\alpha)$ such that

$$|e^{-At}v|_{\alpha} \leq C/t^{\alpha}.$$

It is also of use to note that

$$|e^{-At}| \leq 1 \quad \forall t \geq 0 \tag{10.4}$$

if A is self-adjoint.

Now we consider solving the equation

$$\frac{du}{dt} + Au = f(t, u), \; t > 0, \quad u(0) = u_0 \tag{10.5}$$

Here, for U an open subset of $\mathbb{R}^+ \times X^{\beta}$, $f : U \mapsto X$ satisfies the following: for every $(t, x) \in U$ there is a neighborhood $V \subset U$ and constants $L \geq 0$, $0 < \theta \leq 1$, such that

$$|f(t_1, x_1) - f(t_2, x_2)| \leq L(|t_1 - t_2|^{\theta} + |x_1 - x_2|_{\beta}). \tag{10.6}$$

For simplicity we will denote

$$V \equiv X^{\beta} \quad \text{and} \quad |\bullet|_{\beta} \equiv \|\bullet\|.$$

Formally we see then that the equation

$$\frac{du}{dt} + Au = 0, \quad u(0) = u_0 \tag{10.7}$$

has solution $u(t) = T(t)u_0 = e^{-At}u_0$. This can be made precise by using the definition of "solution" given in Definition 10.8. With this in mind we will frequently use the variation of constants formula obtained formally from (10.5) by using e^{-At} as an integrating factor to yield the integral equation

$$u(t) = e^{-At}u(0) + \int_0^t e^{-A(t-s)} f(s, u(s))ds. \tag{10.8}$$

In this context we define (see Pazy, Definition 4.2.3):

Definition 10.7 *A mild local solution of eqn (10.5) is a function u in $C([0,T), X)$ satisfying (10.8). A mild solution of (10.5) is a mild local solution for each $T > 0$.*

Having constructed such solutions it is important to understand when they are classical solutions of the equation. With this in mind we define (see Miklavčič [79] and Hale [50])

Definition 10.8 *A local solution of (10.5) is a function $u : [0,T) \mapsto V$ such that $u(t) \in C([0,T), X)$, $u_t(t)$ exists in X for $t > 0$, $u(t) \in D(A), t > 0$, $f(\bullet, u(\bullet)) \in C([0,T), X)$ and (10.5) is satisfied on $t \in [0,T)$. A solution of (10.5) is a local solution of (10.5) for each $T > 0$.*

The following theorem is given in Pazy, Theorems 6.3.1 and 6.3.3.

Theorem 10.9 *Assume that A is sectorial, that $-A$ generates a semi-group $T(t)$ satisfying*

$$|T(t)| \leq M$$

and that $f(t,x)$ satisfies (10.6) for some $\beta \in [0,1)$. Then for any $u_0 \in V$ there exists $\tau = \tau(u_0)$ and, for each $t^* < \tau$, a constant $C = C(\|u_0\|, t^*)$ such that (10.5) has a solution on $[0,\tau)$ and

$$|\frac{du}{dt}(t)|_\alpha \leq Ct^{\beta-\alpha-1}, \quad \forall \alpha \in [\beta - 1, 1),\ t \in (0, t^*).$$

Furthermore, if there exists continuous, non-decreasing, real-valued $k(t)$ such that

$$|f(t,x)| \leq k(t)(1 + \|x\|) \quad \forall t \geq 0, x \in V, \tag{10.9}$$

then $\tau = \infty$.

For convenience we denote the solution operator for the nonlinear problem (10.5) by $S(u_0, t)$ so that $u(t) = S(u_0, t)$. This indicates the dependence of the solution on the time t and initial data u_0. The following theorem concerns the regularity of the operator $S(\bullet, \bullet)$. Let $dS(\bullet, t)$ denote the Frećhet derivative of $S(x,t)$ with respect to x and let $df(t, \bullet)$ denote the Frećhet derivative of $f(t,x)$ with respect to x. The next result is contained in Henry, Theorem 3.4.4 and Corollary 3.4.5.

Theorem 10.10 *Suppose that A is a sectorial operator in a Hilbert space X and that (10.6) holds. Suppose further that $f : U \mapsto X$ is C^r with* *derivatives continuous on U uniformly in t for (t, x) in a neighborhood of each point in U. Then the map $(x, t) \mapsto S(x, t)$ is C^r for each $t > 0$ on its interval of existence. Furthermore, $v(t) = dS(x, t)\xi$ is a mild solution of the equation*

$$v_t + Av = df(t, u)v, \quad v(0) = \xi.$$

We frequently use the variation of constants formula (10.8), together with Lemma 10.6, to analyse the solution of (10.5) and its approximations. In this context the next lemma, from section 1.2.2 of Henry (see also Elliott and Larsson [30]) is fundamental.

Lemma 10.11 *Assume that $B, C \geq 0$, $\alpha, \beta \in [0, 1)$ and $T \in (0, \infty)$. Then there exists $M = M(B, \alpha, \beta, T) < \infty$ such that for any integrable function $u : [0, T] \mapsto \mathbb{R}$ satisfying*

$$0 \leq u(t) \leq Ct^{-\alpha} + B \int_0^t (t-s)^{-\beta} u(s) ds$$

for t a.e. in $[0, T)$, we have

$$0 \leq u(t) \leq CMt^{-\alpha}, \quad t \text{ a.e. in } [0, T].$$

The following specific case of the Gronwall lemma will also be of importance to us; a related result is proved in Henry, Theorem 7.1.1.

Lemma 10.12 *Assume that $B, C, \gamma > 0$ and $\nu \in (0, 1]$. Then there is a constant $K = K(\nu) > 0$ such that for any bounded function $u : [0, \infty) \mapsto \mathbb{R}$ satisfying*

$$0 \leq u(t) \leq Ce^{-\gamma t} + B \int_0^t \frac{e^{-\gamma(t-s)}}{(t-s)^{1-\nu}} u(s) ds,$$

it follows that

$$u(t) \leq 2C \exp\{(KB^{1/\nu} - \gamma)t\}.$$

Proof By setting

$$q(t) = \exp\{(\gamma - KB^{1/\nu}t\}u(t)/C$$

we see that it is sufficient to prove that

$$\|q\|_\infty := \sup_{t \geq 0} q(t) \leq 2.$$

Now if $\delta KB^{1/\nu} = 1$ then

$$q(t) \leq e^{-t/\delta} + B \int_0^t (t-s)^{-1+\nu} \exp\{-(t-s)/\delta\} q(s) ds.$$

Hence

$$\begin{aligned} q(t) &\leq 1 + B\|q\|_\infty \int_0^t \frac{e^{-u/\delta}}{u^{1-\nu}} du \\ &\leq 1 + B\|q\|_\infty \int_0^\infty \frac{e^{-u/\delta}}{u^{1-\nu}} du \\ &\leq 1 + B\|q\|_\infty \int_0^\infty \frac{e^{-v}}{v^{1-\nu}} dv \\ &= 1 + \frac{\|q\|_\infty}{K^\nu} \int_0^\infty \frac{e^{-v}}{v^{1-\nu}} dv \\ &= 1 + \frac{\|q\|_\infty I}{K^\nu}, \end{aligned}$$

where

$$I = \int_0^\infty \frac{e^{-v}}{v^{1-\nu}} dv.$$

Since $\nu > 0$ we may choose K such that $2I = K^\nu$ to obtain

$$q(t) \leq 1 + \|q\|_\infty / 2.$$

Since this is true for all $t \geq 0$ we have the required result. □

The following three results are useful when the independent variable u is translated to $v = u - \bar{u}$, hence introducing a new linear operator C. This occurs, for example, when studying properties of (10.5) in the neighborhood of an equilibrium point. Let $\sigma(A)$ denote the spectrum of an operator A. The next result follows from Corollary 1.4.5 and Theorem 1.4.8 of Henry.

Lemma 10.13 *If A is sectorial with* $\mathrm{Re}\{\sigma(A)\} > 0$, *and if C is a linear operator with $(C - A)A^{-\alpha}$ bounded for some $\alpha \in [0, 1)$, then C is sectorial. Furthermore $D(C_1^\beta) = D(A^\beta)$ if c is chosen so that $C_1 = C + cI$ has* $\mathrm{Re}\{\sigma(C_1)\} > 0$; *the norms $|A^\beta \bullet|$ and $|C_1^\beta \bullet|$ are equivalent.*

The following theorem is Theorem 1.5.2 in Henry:

Theorem 10.14 *Let C be a closed linear operator in X and let $\sigma_1(C)$ denote a bounded spectral set of C and $\sigma_2(C)$ its complement in $\sigma(C) \cup \infty$. Then $X = Y \oplus Z$ where Y, Z are the projections of X associated with the two spectral sets σ_1 and σ_2. Furthermore, Y and Z are invariant under C.*

Theorem 10.15 *Let A and C satisfy the conditions of Lemma* 10.13 *and Theorem* 10.14. *Let σ_1 satisfy $-\delta <$* $\mathrm{Re}\{\sigma_1(C)\}$ *$< -\gamma < 0$ and* $\mathrm{Re}\{\sigma_2(C)\} > \gamma > 0$. *Then there exists $K_1 > 0$ such that for all $\alpha \in [0, 1]$*

and $\alpha - \beta \in [0,1]$

$$|A^{\alpha}e^{-Ct}y| \leq K_1 e^{\delta t}|y|, \quad \forall y \in Y, t > 0;$$
$$|A^{\alpha}e^{Ct}y| \leq K_1 e^{-\gamma t}|y|, \quad \forall y \in Y, t > 0;$$
$$|A^{\alpha}e^{-Ct}z| \leq K_1 t^{-(\alpha-\beta)}e^{-\gamma t}|A^{\beta}z|, \quad \forall z \in Z, t > 0.$$

Finally, there is a constant $K_2 = K_2(T) > 0$ *such that*

$$|A^{\alpha}e^{-Ct}v| \leq K_2 t^{-(\alpha-\beta)}|A^{\beta}v|, \quad \forall v \in V,\ t \in (0,T].$$

Proof The first two results follow from the fact that C restricted to Y is a bounded linear operator and all norms are equivalent on Y; see Henry Theorems 1.5.2 and 1.5.3.

For the third result, let a be such that $C_1 = C + aI$ has spectrum with positive real part. By Henry, Theorem 1.4.4 and 1.5.3, we have for $\alpha \in [0,1]$,

$$|C_1^{\alpha}e^{-Ct}z| \leq \frac{c}{t^{\alpha}}e^{-\gamma t}|z| \quad \forall z \in Z.$$

Hence, if $0 \leq \alpha - \beta \leq 1$, then

$$|C_1^{\alpha-\beta}e^{-Ct}C_1^{\beta}z| \leq \frac{c}{t^{\alpha-\beta}}e^{-\gamma t}|C_1^{\beta}z| \quad \forall z \in Z.$$

Thus

$$|C_1^{\alpha}e^{-Ct}z| \leq \frac{c}{t^{\alpha-\beta}}e^{-\gamma t}|C_1^{\beta}z| \quad \forall z \in Z.$$

By the norm equivalence of A and C_1 given in Henry, Theorem 1.4.8, the third point follows. The final point follows by combining the first and third points and noting that C_1 is a bounded operator on Y. □

11 Appendix B — Contraction principles and Taylor expansions

In this appendix we recall the contraction mapping theorem and two important corollaries. The first result is standard and its proof can be found in numerous texts on analysis.

Theorem 11.1 (Contraction Mapping Theorem) *Suppose that* $F : B \mapsto B$ *where* B *is a closed subset of a Banach space* X *with norm* $\| \bullet \|$. *Suppose also that* $F(\bullet)$ *is a contraction on* B *with constant* $\mu < 1$, *so that*

$$\|F(v) - F(w)\| \leq \mu\|v - w\|, \quad \forall v, w \in B.$$

Then there is exactly one point $u \in B$ *such that* $u = F(u)$.

The next two simple corollaries of the contraction mapping theorem are extremely useful.

Theorem 11.2 (Uniform Contraction Principle) *Let* B *be a closed subset of a Banach space* X *with norm* $\|\bullet\|$. *Consider a contraction mapping*

$F : B \mapsto B$ with contraction constant $\mu < 1$ and fixed point u and a one parameter family of contraction mappings $F_\lambda : B \mapsto B$ with contraction constant μ and fixed point u_λ for $\lambda \in (0, \lambda_c)$. If, for any $\lambda \in (0, \lambda_c)$ there exists $\varepsilon(\lambda) > 0$, such that

$$\|F_\lambda(v) - F(v)\| \leq \epsilon(\lambda) \quad \forall v \in B$$

then

$$\|u_\lambda - u\| \leq \frac{\epsilon(\lambda)}{1-\mu}.$$

Proof We have

$$F(u) = u, \quad F_\lambda(u_\lambda) = u_\lambda.$$

Thus, using contractivity and the error bound, we have

$$\begin{aligned} \|u - u_\lambda\| &= \|F(u) - F_\lambda(u_\lambda)\| \\ &\leq \|F(u) - F(u_\lambda)\| + \|F(u_\lambda) - F_\lambda(u_\lambda)\| \\ &\leq \mu\|u - u_\lambda\| + \varepsilon(\lambda). \end{aligned} \tag{11.1}$$

This gives the desired result. □

Theorem 11.3 (Commuting Contraction Principle) *Let B be a closed subset of a Banach space X with norm $\| \bullet \|$. Consider a one parameter family of contraction mappings $F_\lambda : B \mapsto B$ with contraction constant $\mu < 1$ and fixed point u_λ for $\lambda \in (\lambda_1, \lambda_2)$. Then, if*

$$F_\lambda \bullet F_\eta = F_\eta \bullet F_\lambda \quad \forall \eta, \lambda \in (0, \lambda_2)$$

it follows that

(i) u_λ is independent of $\lambda \in (\lambda_1, \lambda_2)$ and we denote it by $\bar{u}$;
(ii) if $F_\lambda \bar{u} \in B$ for $\lambda \in (0, \lambda_0)$ then $\bar{u}$ is a fixed point of F_λ for all $\lambda \in (0, \lambda_0)$.

Proof Let

$$F_\eta(u_\eta) = u_\eta, \quad F_\lambda(u_\lambda) = u_\lambda \quad \eta, \lambda \in (\lambda_1, \lambda_2).$$

Then

$$\begin{aligned} (F_\lambda \bullet F_\eta)(u_\eta) &= F_\lambda(u_\eta) \\ \Rightarrow \quad (F_\eta \bullet F_\lambda)(u_\eta) &= F_\lambda(u_\eta) \\ \Rightarrow \quad F_\lambda(u_\eta) &= u_\eta. \end{aligned} \tag{11.2}$$

Thus $u_\eta = u_\lambda$. This proves (i). We denote the fixed point by $\bar{u}$.

For (ii) let $\eta \in (\lambda_1, \lambda_2)$ and $\lambda \in (0, \lambda_0)$. Then

$$\begin{aligned} F_\eta(\bar{u}) &= \bar{u} \\ \Rightarrow \quad (F_\lambda \bullet F_\eta)(\bar{u}) &= F_\lambda(\bar{u}) \\ \Rightarrow \quad (F_\eta \bullet F_\lambda)(\bar{u}) &= F_\lambda(\bar{u}). \end{aligned} \tag{11.3}$$

Since $F_\lambda(\bar{u}) \in B$ it follows by uniqueness of the fixed point in B that $F_\lambda(\bar{u}) = \bar{u}$ and the result follows. □

Once differentiability of nonlinear operators between Hilbert spaces has been established it is natural to consider Taylor expansions; for a broad introduction to the calculus of nonlinear operators in a Hilbert space see Chow and Hale [16]. The following result may be found in that text. We let Df^j denote the jth Frechet derivative of f, a multilinear operator.

Theorem 11.4 (Taylor Expansions) *Suppose that X, Y are Hilbert spaces and that $U \subset X$ is an open set. Then, if $f \in C^k(U, Y)$,*

$$f(x+h) = f(x) + Df(x)h + \ldots + \frac{1}{(k-1)!} Df^{k-1}(x) h^{n-1}$$
$$+ \frac{1}{(k-1)!} \int_0^1 (1-s)^{n-1} Df^n(x+sh) h^n ds.$$

12 Appendix C — Attractive invariant manifolds

The material in this Appendix is generalized from Jones and Stuart [65]. The generalization is to consider maps which are written as graphs $\Phi \in C(Y, Z)$ over a bounded ball in Y as well as the case where the graph is over the whole of Y. The former case is useful for the study of unstable manifolds and the latter for inertial manifolds.

We consider the general question of the existence of attractive invariant manifolds for the map:

$$W_{m+1} = M(W_m), \quad M(W) = LW + N(W). \tag{12.1}$$

(Actually, as will be apparent from the theorem, the manifold may only be locally invariant in the case where the graph of Φ is defined over a compact ball in the Y coordinate.) We assume that $L : V \mapsto V$ is linear and that $N : V \mapsto V$ is nonlinear. The space V is decomposed into two subspaces Y and Z which are assumed invariant under L. Thus

$$V = Y \oplus Z, \quad LY = Y, \quad LZ = Z. \tag{12.2}$$

We let $\mathcal{P}$ and $\mathcal{Q}$ denote the projections of V onto Y and Z respectively and define $p_m = \mathcal{P}W_m, q_m = \mathcal{Q}W_m$; thus $W_m = p_m + q_m$. Throughout the construction of the attractive invariant manifold we make the following assumptions.

Assumption 12.1 *There exist positive constants a, b, c, B_1, B_2 such that*

$$\|Lz\| \leq a\|z\| \;\; \forall z \in Z; \tag{G1}$$

$$\forall p \in Y, \; \exists! w \in Y \; s.t. \; Lw = p \;\; and \;\; b\|y\| \leq \|Ly\| \leq c\|y\| \;\; \forall y \in Y; \tag{G2}$$

$$\|\mathcal{R}(N(u) - N(v))\| \leq B_1\|u - v\|, \quad \|\mathcal{R}N(u)\| \leq B_2, \tag{G3}$$

for all u, v such that $\|\mathcal{P}u\|, \|\mathcal{P}v\| \leq r$, $\|\mathcal{Q}u\|, \|\mathcal{Q}v\| \leq \gamma$ and where $\mathcal{R}$ equals either $I, \mathcal{P}$ or $\mathcal{Q}$. Furthermore, there exist constants $\alpha \in (0, \infty)$ and $\mu \in (0, 1)$ such that

$$b^{-1}B_1(1 + \alpha) \leq \mu; \tag{C1}$$

$$a\gamma + B_2 \leq \gamma; \tag{C2}$$

$$\theta := a\alpha + B_1(1 + \alpha) \leq \alpha\varphi, \tag{C3}$$

where $\varphi := b - B_1(1 + \alpha) > 0$ by (C1);

$$a + B_1(1 + \alpha) \leq \mu. \tag{C4}$$

Under these assumptions we will seek an invariant manifold for (12.1) which is representable as the graph of a function Φ, acting on a subspace of Y, and satisfying

$$q_m = \Phi(p_m) \Longleftrightarrow q_{m+1} = \Phi(p_{m+1}) \quad \forall m : \|p_m\|, \|p_{m+1}\| \leq r \tag{12.3}$$

and is attractive in the sense that

$$\|q_m - \Phi(p_m)\| \leq \mu^m\|q_0 - \Phi(p_0)\| \;\; \forall m : \|p_n\| \leq r, \|q_n\| \leq \gamma, \; \forall n = 0, \ldots, m. \tag{12.4}$$

Let

$$\bar{Y} = \{p \in Y : \|p\| \leq r\}.$$

The appropriate space in which we seek Φ is now defined:

Definition 12.2 *Let $\Gamma = \Gamma(\gamma, \alpha)$ denote the closed subset of $C(\bar{Y}, Z)$ satisfying*

$$\|\Psi\|_\Gamma := \sup_{p \in \bar{Y}} \|\Psi(p)\| \leq \gamma,$$

$$\|\Psi(p_1) - \Psi(p_2)\| \leq \alpha\|p_1 - p_2\| \;\; \forall p_1, p_2 \in \bar{Y}.$$

Theorem 12.3 (Existence of Attractive Invariant Manifolds) *Suppose that Assumptions 12.1 hold for the mapping (12.1) and that $r \geq (b - 1)^{-1}B_2$ if $b > 1$ and $r = \infty$ if $b \leq 1$. Then there exists a unique $\Phi \in \Gamma(\alpha, \gamma)$ such that (12.3) and (12.4) hold.*

Throughout the remainder of this section we assume that the conditions of this theorem are satisfied without stating this explicitly in every result.

The proof of the theorem will be given in a series of lemmas. We may write (12.1) as

$$p_{m+1} = Lp_m + \mathcal{P}N(p_m + q_m), \tag{12.5}$$

$$q_{m+1} = Lq_m + \mathcal{Q}N(p_m + q_m). \tag{12.6}$$

The graph Φ giving the invariant manifold is a fixed point of the operator $T : C(\bar{Y}, Z) \mapsto C(\bar{Y}, Z)$ defined by

$$p = L\xi + \mathcal{P}N(\xi + \Phi(\xi)) \tag{12.7}$$

$$(T\Phi)(p) = L\Phi(\xi) + \mathcal{Q}N(\xi + \Phi(\xi)). \tag{12.8}$$

We employ the contraction mapping theorem to prove the existence of a fixed point of T. We first show that the map T is well defined.

Lemma 12.4 *For any $\Phi \in \Gamma$ and $p \in \bar{Y}$ there exists a unique $\xi \in \bar{Y}$ satisfying (12.7).*

Proof. We consider the case $b > 1$ and $r \geq (b-1)^{-1}B_2$. The case $b \leq 1$ is similar. Note that by $(G2)$ L^{-1} exists on Y. Thus we may consider the iteration

$$\xi^{k+1} = L^{-1}[p - \mathcal{P}N(\xi^k + \Phi(\xi^k))]. \tag{12.9}$$

If $p \in \bar{Y}$, then this map takes $\xi^k \in \bar{Y}$ into $\xi^{k+1} \in \bar{Y}$ since

$$\|\xi^{k+1}\| \leq b^{-1}r + b^{-1}B_2 \leq b^{-1}r + b^{-1}(b-1)r = r.$$

For any two sequences $\{\xi^k\}, \{\eta^k\}$ generated by (12.9) we have, by (G2), since $\Phi \in \Gamma$,

$$\begin{aligned} \|\xi^{k+1} - \eta^{k+1}\| &\leq b^{-1}B\|\xi^k + \Phi(\xi^k) - \eta^k - \Phi(\eta^k)\| \\ &\leq b^{-1}B(1+\alpha)\|\xi^k - \eta^k\|. \end{aligned}$$

By Condition (C1) the mapping is a contraction and the existence of ξ given any $p \in Y$ follows. □

Thus, by Lemma 12.4, $T\Phi : \bar{Y} \to Z$ is well defined. We now show that T maps $\Gamma(\alpha, \gamma)$ into itself.

Lemma 12.5 *The mapping T defined by (12.8) satisfies $T : \Gamma \mapsto \Gamma$.*

Proof. From (12.8) and (G1), (G3) we have for all $p \in \bar{Y}$ and $\Phi \in \Gamma$

$$\begin{aligned} \|(T\Phi)(p)\| &\leq a\|\Phi(\xi)\| + B_2 \\ &\leq a\gamma + B_2. \end{aligned}$$

Thus we have, by (C2), $\|T\Phi\|_\Gamma \leq \gamma$.

Let $p_1, p_2 \in \bar{Y}$. From Lemma 12.4 there exists $\{\xi_i\}_{i=1}^2$ such that (12.8) is satisfied with $p = \{p_i\}_{i=1}^2$. Subtracting the two equations, we obtain

$$\begin{aligned}\|(T\Phi)(p_1) - (T\Phi)(p_2)\| &\leq a\|\Phi(\xi_1) - \Phi(\xi_2)\| \\ &\qquad + B_1\|(\xi_1 - \xi_2) + \Phi(\xi_1) - \Phi(\xi_2)\| \\ &\leq [a\alpha + B_1(1+\alpha)]\|\xi_1 - \xi_2\| \\ &= \theta\|\xi_1 - \xi_2\|,\end{aligned}$$

where we have used $(G1), (G3)$, $(C3)$ and the properties of $\Phi \in \Gamma$. From $(G2)$, $(G3)$ and (12.7) we have

$$\begin{aligned}b\|\xi_1 - \xi_2\| &\leq \|L(\xi_1 - \xi_2)\| \\ &\leq \|p_1 - p_2\| + B_1(1+\alpha)\|\xi_1 - \xi_2\|.\end{aligned}$$

Using $(C1)$ we deduce that $\varphi > 0$, and hence

$$\|\xi_1 - \xi_2\| \leq \frac{1}{\varphi}\|p_1 - p_2\|.$$

Thus $(C3)$ implies

$$\|(T\Phi)(p_1) - (T\Phi)(p_2)\| \leq \frac{\theta}{\varphi}|p_1 - p_2\| \leq \alpha\|p_1 - p_2\|.$$

Hence $T : \Gamma \mapsto \Gamma$. □

Now we may show that the map T is a contraction on the space Γ.

Lemma 12.6 *For any* $\Phi_1, \Phi_2 \in \Gamma$ *we have*

$$\|T\Phi_1 - T\Phi_2\|_\Gamma \leq \mu\|\Phi_1 - \Phi_2\|_\Gamma.$$

Proof. By Lemma 2.3, for any $p \in \bar{Y}$ and $\{\Phi_i\}_{i=1}^2 \in \Gamma$ we can find $\{\xi_i\}_{i=1}^2$ such that for $i = 1, 2$

$$p = \mathcal{P}M(\xi_i + \Phi_i(\xi_i)) \tag{12.10}$$

$$(T\Phi_i)(p) = \mathcal{Q}M(\xi_i + \Phi_i(\xi_i)).$$

Using $(G1), (G3)$ we have

$$\|(T\Phi_1)(p) - (T\Phi_2)(p)\| \leq (a + B_1)\|\Phi_1(\xi_1) - \Phi_2(\xi_2)\| + B_1\|\xi_1 - \xi_2\|;$$

adding and subtracting $\Phi_2(\xi_1)$, using the triangle inequality and $(C3)$, we majorize the last inequality by

$$\|(T\Phi_1)(p) - (T\Phi_2)(p)\| \leq (a+B_1)\|\Phi_1(\xi_1) - \Phi_2(\xi_1)\| + \theta\|\xi_1 - \xi_2\|. \tag{12.11}$$

Now, using $(G2)$ and (12.10), we have similarly that

$$\begin{aligned} b\|\xi_1 - \xi_2\| &\leq \|L(\xi_1 - \xi_2)\| \leq B_1\|\xi_1 - \xi_2 + \Phi_1(\xi_1) - \Phi_2(\xi_2)\| \\ &\leq B_1(1+\alpha)\|\xi_1 - \xi_2\| + B_1\|\Phi_1(\xi_1) - \Phi_2(\xi_1)\|. \end{aligned}$$

Thus since $\varphi > 0$, by $(C1)$

$$\|\xi_1 - \xi_2\| \leq \frac{B_1}{\varphi}\|\Phi_1(\xi_1) - \Phi_2(\xi_1)\|.$$

Returning to (12.11) and using $(C3), (C4)$, we find

$$\|(T\Phi_1)(p) - (T\Phi_2)(p)\| \leq \mu\|\Phi_1(\xi_1) - \Phi_2(\xi_1)\|. \tag{12.12}$$

The result follows after taking the supremum over $\xi_1 \in \bar{Y}$ and then $p \in \bar{Y}$. □

Proof of Theorem 12.3 The existence of the manifold Φ follows from Lemmas 12.4–12.6. To establish the exponential attraction of solutions to this manifold let $W_m = p_m + q_m$ be an arbitrary trajectory of (12.5), (12.6) with $\|p_m\| \leq r$, $\|q_m\| \leq \gamma$. Set

$$\begin{aligned} p &= Lp_m + \mathcal{P}N(p_m + \Phi(p_m)), \\ \Phi(p) &= L\Phi(p_m) + \mathcal{Q}N(p_m + \Phi(p_m)). \end{aligned}$$

Then using $(G1), (G2)$ we have

$$\begin{aligned} \|q_{m+1} - \Phi(p_{m+1})\| &\leq \|q_{m+1} - \Phi(p)\| + \|\Phi(p) - \Phi(p_{m+1})\| \\ &\leq (a + B_1)\|q_m - \Phi(p_m)\| + \alpha\|p - p_{m+1}\|. \end{aligned}$$

However, by (12.5), we have

$$\|p - p_{m+1}\| \leq B_1\|q_m - \Phi(p_m)\|.$$

Thus by $(C4)$ we obtain

$$\|q_{m+1} - \Phi(p_{m+1})\| \leq \mu\|q_m - \Phi(p_m)\|$$

and the result follows. □

Delay Differential Equations: Theory and Numerics

M. Zennaro

Dipartimento di Scienze Matematiche,
Università di Trieste,
34100 Trieste,
Italy

1 Introduction

Many real life phenomena in physics, engineering, biology, medicine and economics can be modeled by initial value problems (IVPs) for ordinary differential equations (ODEs) of the type

$$\begin{cases} y'(t) = f(t, y(t)), & t \geq t_0, \\ y(t_0) = y_0, \end{cases} \tag{1.1}$$

where the function $y(t)$ represents some physical quantity which evolves in time.

However, in order to make the model more consistent with the real phenomenon, it sometimes is necessary to modify the right hand side of (1.1) to include the dependence of the derivative y' also on y computed at some past value $t - \tau$. According to the complexity of the phenomenon, the *delay* τ, which always is nonnegative, may be just a constant (*constant delay* case), or a function of t (*variable delay* case), or even a function of t and y itself (*state dependent delay* case). In any case eqn (1.1) assumes the form

$$\begin{cases} y'(t) = f(t, y(t), y(t - \tau)), & t \geq t_0, \\ y(t) = \varphi(t), \quad t \leq t_0, \end{cases} \tag{1.2}$$

which is called a *delay differential equation* (DDE).

A first difference between eqns (1.1) and (1.2) is that the solution of the latter is determined by an initial function $\varphi(t)$ rather than by a simple initial value y_0, as happens for the former. As a consequence, even if the functions $f(t, y, x)$, $\tau(t, y)$ and $\varphi(t)$ in (1.2) are C^∞-continuous, in general the solution y is not smoothly linked to the initial function $\varphi(t)$ at the point t_0, where only C^0-continuity can be assured. Such discontinuity is spread forward along the integration interval. More precisely, a set of *discontinuity points* is generated, whose location is determined by the delayed argument $t - \tau$.

The theory of existence and uniqueness of solutions to (1.2) does not present substantial additional difficulties with respect to the ordinary case (1.1) as long as the delay τ is uniformly strictly positive and does not depend on the solution y itself. Indeed, these are the cases to which we are confining ourselves in these notes for a treatment of (1.2) from a numerical point of view.

In our presentation the stress is on general concepts and results rather than on details and proofs. In fact, our aim is to provide a gentle introduction to the theory and numerics of DDEs at such a level that it can be understood by people who are new to the subject.

We also select a number of particular topics, which are not necessarily more interesting or important than others. The criterion we adopted in the selection was mainly to consider material which is as easy as possible, but at the same time representative of more general problems and situations. Nevertheless, we also give some hints and stimulate the reader by referring him to the literature for a deeper knowledge. Moreover, in some cases, our presentation may implicitly suggest some problems which remain open.

We assume that the reader is familiar with the theory of ODEs and with numerical methods for their solution. A good set of books to refer to about these subjects is: Lambert [49], Butcher [21], Hairer, Nørsett and Wanner [31], Hairer and Wanner [32] and Dekker and Verwer [24].

In Section 2 we fix our attention on a particular class of DDEs, making a theoretical study of their solutions, especially in terms of their asymptotic behavior.

In Section 3 we outline a general strategy for the numerical solution of the DDEs considered in Section 2. Moreover, we select some particular methods and analyze them in terms of accuracy.

In Section 4 we analyze the stability properties of the proposed numerical methods. We pay particular attention to this material, since we think it is particularly interesting and stimulating. In fact, in many cases, the stability analysis leads to rather complicated mathematical problems.

Finally, in Section 5 we briefly consider the numerical solution of other types of DDEs, not included in the previous class.

2 Theoretical analysis of a class of delay differential equations

The general theory of DDEs is widely developed and we refer the reader to the classical books by Hale [33], Driver [25] and El'sgol'ts and Norkin [26]. We also quote the recent book by Kolmanovskii and Myshkis [48].

In these notes we fix our attention on systems of DDEs of the type (1.2), where $f : [t_0, t_f] \times \mathbb{R}^m \times \mathbb{R}^m \longrightarrow \mathbb{R}^m$ is uniformly Lipschitz continuous in both the last two variables and the delay τ is either constant or variable, but not state dependent. Moreover, if it is variable, the following properties are assumed to be satisfied

- there exists a constant $\tau_0 > 0$ such that $\tau(t) \geq \tau_0$ for all $t \in [t_0, t_f]$;
- the delayed argument $t - \tau(t)$ is a strictly increasing function for all $t \in [t_0, t_f]$.

2.1 Location of discontinuities and existence of solutions

The discontinuity points mentioned in Section 1 are generated inductively by the recursion

$$\xi_k - \tau(\xi_k) = \xi_{k-1}, \quad k \geq 1, \tag{2.1}$$

where $\xi_0 = t_0$. Because of the hypotheses made, an increasing sequence $\{\xi_k\}_{k\geq 0}$ is determined which can actually be computed a priori by using (2.1). In this way, a sequence of intervals $[\xi_{k-1}, \xi_k]$ is also defined. Moreover, each pair of consecutive discontinuity points satisfies $\xi_k - \xi_{k-1} \geq \tau_0$.

The above hypotheses yield existence and uniqueness of the solution quite easily. In fact, it can be proved just by using induction on the intervals $[\xi_{k-1}, \xi_k]$ and the well known existence and uniqueness theorem for ODEs (1.1) under the hypothesis of uniform Lipschitz continuity of the right hand side.

The discontinuity points ξ_k are sometimes called *primary discontinuities*. If the functions $f(t, y, x)$, $\tau(t)$ and $\varphi(t)$ in (1.2) have some discontinuities with respect to t in some of their derivatives, then such discontinuities are also propagated by the delayed argument $t - \tau(t)$ following the rule (2.1) and are called *secondary discontinuities*. However, in order to simplify the discussion, we assume that all the functions in (1.2) are C^∞-continuous. Therefore, in the interior of each interval $[\xi_{k-1}, \xi_k]$ the solution y is C^∞-continuous as well, and no secondary discontinuities are present. Moreover, it can easily be seen that, at each discontinuity point ξ_k, the solution y is at least k times continuously differentiable, being only C^0-continuous at ξ_0. In other words, there is a smoothing of the solution y at the discontinuity points ξ_k as the index k increases.

2.2 Some important particular equations

In particular, it is interesting to consider the class of *autonomous* DDEs, that is equations of the form

$$\begin{cases} y'(t) = f(y(t), y(t-\tau)), & t \geq t_0, \\ y(t) = \varphi(t), & t \leq t_0, \end{cases} \tag{2.2}$$

with constant delay τ. These are equations whose corresponding functional operator is independent of the variable t.

Moreover, we distinguish the class of scalar linear DDEs, with variable coefficients

$$\begin{cases} y'(t) = \lambda(t)y(t) + \mu(t)y(t-\tau), & t \geq t_0, \\ y(t) = \varphi(t), & t \leq t_0, \end{cases} \tag{2.3}$$

and with constant coefficients

$$\begin{cases} y'(t) = \lambda y(t) + \mu y(t-\tau), & t \geq t_0, \\ y(t) = \varphi(t), & t \leq t_0, \end{cases} \tag{2.4}$$

respectively. Observe that, if the delay τ is constant, then equation (2.4) is a linear autonomous equation.

Example 2.1 A very famous nonlinear autonomous DDE is the following *logistic equation*, which often is assumed in biology as a model for the growth of a population

$$\begin{cases} y'(t) = y(t)[a - y(t-1)], & t \geq 0, \\ y(t) = \varphi(t), & -1 \leq t \leq 0, \end{cases} \tag{2.5}$$

where $a > 0$ is a parameter. Clearly, the discontinuity points of the DDE (2.5) are given by the set of positive integers.

Another important particular class is given by the so called *pure delay* equations. They have the form

$$\begin{cases} y'(t) = f(t, y(t-\tau)), & t \geq t_0, \\ y(t) = \varphi(t), & t \leq t_0. \end{cases} \tag{2.6}$$

Within this class, as a particular case of the linear scalar equations (2.3) and (2.4) we have

$$\begin{cases} y'(t) = \mu(t)y(t-\tau), & t \geq t_0, \\ y(t) = \varphi(t), & t \leq t_0, \end{cases} \tag{2.7}$$

and

$$\begin{cases} y'(t) = \mu y(t-\tau), & t \geq t_0, \\ y(t) = \varphi(t), & t \leq t_0, \end{cases} \tag{2.8}$$

respectively.

Example 2.2 A very simple but curious pure delay equation is:

$$\begin{cases} y'(t) = y(t-1), & t \geq 0, \\ y(t) = 1, & -1 \leq t \leq 0. \end{cases} \tag{2.9}$$

As for the logistic equation (2.5), the discontinuity points of (2.9) are given by the set of positive integers. In each interval $[k-1, k]$ the solution is a polynomial of degree k.

2.3 Stability

For applications, it may sometimes be interesting to consider some delay equations whose solutions show a stable asymptotic behavior (see Example 2.4). Therefore, we now analyze the conditions which ensure such stable behavior. To this end we assume, in this subsection only, that the functions in (1.2) are complex valued.

We begin with the results of Torelli [66], where the general nonlinear complex case (1.2) is treated. However, our formulation is slightly different.

Consider another system, defined by the same function $f(t, y, x)$, but with another initial condition:

$$\begin{cases} z'(t) = f(t, z(t), z(t-\tau)), & t \geq t_0, \\ z(t) = \psi(t), & t \leq t_0. \end{cases} \tag{2.10}$$

Theorem 2.3 *Given an inner product $(\cdot,\cdot)$ in $\mathbf{C}^m$ and the corresponding norm $\|\cdot\|$, let $Y(t)$ and $X(t)$ be continuous functions such that*

$$Y(t) \geq \sup_{x, y_1 \neq y_2} \frac{\mathrm{Re}((f(t, y_1, x) - f(t, y_2, x), y_1 - y_2))}{\|y_1 - y_2\|^2} \tag{2.11}$$

and

$$X(t) \geq \sup_{y, x_1 \neq x_2} \frac{\|(f(t, y, x_1) - f(t, y, x_2)\|}{\|x_1 - x_2\|}. \tag{2.12}$$

If

$$Y(t) + X(t) \leq 0, \quad t \geq t_0, \tag{2.13}$$

then it holds that

$$\|y(t) - z(t)\| \leq \max_{x \leq t_0} \|\varphi(x) - \psi(x)\|, \quad t \geq t_0. \tag{2.14}$$

Example 2.4 (see Hale [34])

$$y'(t) = -ay(t) + \frac{by(t-\tau)}{1 + [y(t-\tau)]^n},$$

where $a > 0$ and b are real parameters and n is an even positive integer. This equation is a model for respiratory diseases, where $y(t)$ represents the concentration of carbon dioxide at time t. For certain values of the parameters and of the delay, the solution is oscillatory, and sometimes it oscillates even chaotically. This is the case in patients with leukemia.

On the other hand, it can be seen that, for $n = 2, 4$, it always holds that $|\frac{\partial f}{\partial x}| \leq |b|$. Therefore, using (2.12) and (2.13), for $n = 2, 4$ we have a stable behavior whenever $a \geq |b|$.

One can prove Theorem 2.3 in an indirect way by using the Razumikhin-type theorems presented in the book by Hale [33]. However, it is interesting to recall very briefly the outline of the direct proof given in [66], since the same idea is applied when analyzing the stability of the numerical methods. There are two main steps: the first is given by the following Proposition 2.5, and the second is just a simple use of induction on the intervals $[\xi_{k-1}, \xi_k]$.

Proposition 2.5 *Consider the following two systems of ODEs with forc-*

ing terms

$$\begin{cases} y'(t) = f(t, y(t), u(t)), & t \geq t_0, \\ y(t_0) = y_0, \end{cases} \tag{2.15}$$

$$\begin{cases} z'(t) = f(t, z(t), v(t)), & t \geq t_0, \\ z(t_0) = z_0, \end{cases} \tag{2.16}$$

where $u(t)$ and $v(t)$ are continuous functions. Moreover, consider $Y(t)$ and $X(t)$ defined by (2.11) and (2.12), respectively. If

$$Y(t) \leq 0, \quad t \geq t_0, \tag{2.17}$$

and if there exists a bounded function $r(t)$ such that

$$X(t) = r(t)Y(t), \quad t \geq t_0, \tag{2.18}$$

then the solutions $y(t)$ and $z(t)$ satisfy

$$\|y(t) - z(t)\| \leq \max\{\|y_0 - z_0\|, \sup_{t_0 \leq x \leq t} (|r(x)| \cdot \|u(x) - v(x)\|)\}, \quad t \geq t_0.$$

We remark that in [66] the hypothesis (2.17) is replaced by the stronger hypothesis that $Y(t) < 0$, $t \geq t_0$. As a consequence, Theorem 2.3 is not proved completely. On the other hand, it can easily be seen that condition (2.13) is entirely equivalent to conditions (2.17) and (2.18) with $|r(t)| \leq 1$.

Also for the particular scalar linear equations (2.3) and (2.4), we consider the case where the coefficients and the initial function are complex. In such a context, condition (2.13) becomes

$$\text{Re}(\lambda(t)) + |\mu(t)| \leq 0, \quad t \geq t_0, \tag{2.19}$$

for eqn (2.3) and

$$\text{Re}(\lambda) + |\mu| \leq 0, \quad t \geq t_0, \tag{2.20}$$

for eqn (2.4), whereas, for both equations, (2.14) reduces to

$$|y(t)| \leq \max_{x \leq t_0} |\varphi(x)|, \quad t \geq t_0. \tag{2.21}$$

It is useful to point out that, for eqn (2.4), the stronger condition

$$\text{Re}(\lambda) + |\mu| < 0, \quad t \geq t_0, \tag{2.22}$$

implies the *asymptotic stability* of the solution $y(t)$, that is $lim_{t \to +\infty} y(t) = 0$ for all initial functions $\varphi(t)$. Since eqn (2.4) is linear and has constant coefficients and constant delay, its stability analysis may be done by studying the roots of its *characteristic equation*

$$\zeta - \lambda - \mu \exp(-\tau\zeta) = 0. \tag{2.23}$$

Now, it is easy to see that condition (2.22) implies that all the characteristic roots ζ of (2.23) are such that $\mathrm{Re}(\zeta) < 0$, which is a necessary and sufficient condition for the asymptotic stability of (2.4).

Remark 2.6 *Condition (2.13) is sufficient for the validity of (2.14), independently of the particular constant delay τ. The same observation applies to condition (2.19) (or (2.20)) for the validity of (2.21), and to condition (2.22) for the asymptotic stability of the solutions of equation (2.4). Indeed, for a given constant delay τ, all of these conditions may be relaxed accordingly to the particular value of τ.*

Nevertheless, if we want (2.14) and (2.21) to hold for all constant delays τ, then conditions (2.13) and (2.19) (or (2.20)), respectively, are also necessary. Moreover, condition (2.20) is necessary for the asymptotic stability of the solutions of eqn (2.4) for all constant delays τ.

In view of the stability analysis of numerical methods made in Section 4, it is also useful to reformulate conveniently Proposition 2.5 in the particular cases of the linear equations (2.3) and (2.4).

Proposition 2.7 *Consider the following scalar ODE with forcing term*

$$\begin{cases} y'(t) = \lambda(t)y(t) + \mathrm{Re}(\lambda(t))g(t), & t \geq t_0, \\ y(t_0) = y_0, \end{cases} \tag{2.24}$$

where $g(t)$ is a continuous function. If $\mathrm{Re}(\lambda(t)) \leq 0$, $t \geq t_0$, then

$$|y(t)| \leq \max\{|y_0|, \max_{t_0 \leq x \leq t} |g(x)|\}, \quad t \geq t_0. \tag{2.25}$$

Proposition 2.8 *Consider the following scalar ODE with forcing term*

$$\begin{cases} y'(t) = \lambda y(t) + \mathrm{Re}(\lambda)g(t), & t \geq t_0, \\ y(t_0) = y_0, \end{cases} \tag{2.26}$$

where $g(t)$ is a continuous function. If $\mathrm{Re}(\lambda) \leq 0$, then (2.25) holds.

It is clear that between the general nonlinear system (1.2) and the scalar linear autonomous case there are a lot of intermediate classes of equations which could be worth considering. Particular attention has been given to the case of a linear autonomous system of the form

$$\begin{cases} y'(t) = Ly(t) + My(t-\tau), & t \geq t_0, \\ y(t) = \varphi(t), & t \leq t_0, \end{cases} \tag{2.27}$$

where L and M are constant $m \times m$-matrices.

Assume that the delay τ is constant. It can be seen that, if the matrices L and M are simultaneously diagonalizable, then a suitable change of variable involving the common eigenvectors leads to a decoupled system of m independent scalar equations of the form (2.4). Therefore the asymptotic stability of the system (2.27) is, in this case, completely described by the eigenvalues of the two matrices.

More complicated is the case when the eigenspaces of L and M are different. It has been proved (see in 't Hout [41]) that a sufficient condition for the asymptotic stability of (2.27) is:

$$\mathrm{Re}(\lambda) < 0 \; \forall \text{ eigenvalues } \lambda \text{ of } L \quad \text{and} \quad \sup_{\mathrm{Re}(\xi)=0} \rho[(\xi I - L)^{-1} M] < 1, \quad (2.28)$$

where $\rho[\cdot]$ denotes the spectral radius of a matrix.

Finally, we consider the stability of the pure delay equations (2.6), (2.7) and (2.8). It is clear that it is not possible to get any information by particularization of the results we presented for (1.2), (2.3) and (2.4). In fact, the sufficient conditions (2.13), (2.19) and (2.20) would imply $X(t) = 0$ in (2.12), $\mu(t) = 0$ in (2.7) and $\mu = 0$ in (2.8), respectively. Hence, (2.6) would reduce to the trivial ODE $y'(t) = f(t)$, whereas both (2.7) and (2.8) would reduce to the trivial ODE $y'(t) = 0$. This is due to the fact that the sufficient conditions mentioned are independent of the particular delay τ (see Remark 2.6). We conclude that, for pure delay equations, we can get conditions for stability only if we take the particular delay τ into account.

By using the method of the characteristic equation, it can be proved that all the solutions of the equation (2.8) are asymptotically stable if

$$\mathrm{Re}(\mu) < 0 \quad \text{and} \quad \pi/2 < \arg(\mu) < 3\pi/2,$$

and

$$0 < \tau|\mu| < \min\{3\pi/2 - \arg(\mu), \arg(\mu) - \pi/2\}.$$

As for equation (2.7) with real variable coefficient, Torelli and Vermiglio [68] proved that the solutions satisfy

$$|y(t)| \leq 2 \max_{x \leq t_0} |\varphi(x)|, \quad t \geq t_0,$$

whenever

$$-1/\tau \leq \mu(t) < 0, \quad t \geq t_0.$$

3 How to solve delay differential equations numerically

In this section we present some techniques for the numerical solution of DDEs. Other surveys of this area are available: see Cryer [23], Bellen [8] and Meinardus and Nürnberger [52]. We refer the reader also to the book by Hairer, Nørsett and Wanner [31].

We begin with one of the first methods, also known as the *method of steps*, developed by Bellman (see [12] and [13]). Then, after giving an outline of a general approach, we propose some particular choices within this general strategy.

In our opinion, Bellman's method is not the most natural approach to a DDE, but it is certainly ingenious. It applies to DDEs subject to the hypotheses made at the beginning of Section 2. However, in order to simplify the discussion, we confine ourselves to the case of constant delay τ. It is clear that the discontinuity points are $\xi_k = t_0 + k\tau$. On the first interval $[t_0, t_0 + \tau]$ the DDE has the form

$$\begin{cases} y'(t) = f(t, y(t), \varphi(t-\tau)), & \xi_0 \le t \le \xi_1, \\ y(t_0) = \varphi(t_0), \end{cases}$$

and it is solved by means of a standard numerical method for ODEs. If we assume by induction to have solved numerically the DDE up to to the discontinuity point ξ_{k-1}, then we can rewrite the DDE up to the next discontinuity point ξ_k as the following system of ODEs

$$\begin{cases} y_i'(t) = f(t-(k-i)\tau, y_i(t), y_{i-1}(t)), & \xi_{k-1} \le t \le \xi_k, \\ y_i(\xi_{i-1}) = y(\xi_{i-1}), & i = 1, ..., k, \end{cases} \tag{3.1}$$

where we have set $y_0(t) = \varphi(t - k\tau)$ and $y_i(t) = y(t-(k-i)\tau)$, $i = 1, ..., k$. Since we have already computed approximations to the initial values $y_i(\xi_{i-1}), i = 1, ..., k$, we can solve the system numerically by means of the chosen method for ODEs, in order to get an approximation also in the current interval $[\xi_{k-1}, \xi_k]$. In this way, we are obliged to solve a system of ever growing dimension and, in principle, to recompute many times the same pieces of solution related to the previous intervals. On the other hand, the reduction to a system of ODEs avoids the typical complications due to the presence of the delayed argument, as we shall see later.

3.1 A general approach

The hypotheses made at the beginning of Section 2 very naturally suggest the following strategy

- choose a *discrete* numerical method for solving ODEs;
- choose an interpolation procedure in order to obtain a uniform approximation $\eta(t)$ of the solution $y(t)$;
- compute one after the other the discontinuity points ξ_k defined by the recurrence relation (2.1) and, in each interval $[\xi_{k-1}, \xi_k]$, use the chosen numerical ODE method to approximate the solution of the (ordinary) differential equation

$$\begin{cases} w'(t) = f(t, w(t), \eta(t-\tau)), & \xi_{k-1} \le t \le \xi_k, \\ w(\xi_{k-1}) = \eta(\xi_{k-1}). \end{cases} \tag{3.2}$$

The first two steps of this procedure may alternatively be replaced by the direct choice of a *continuous* numerical method for ODEs. In this context, the words *discrete* and *continuous* mean that the numerical method gives an approximation of the solution on a discrete set of points or in the whole interval of integration, respectively.

It is clear that a large variety of methods is available, according to the particular choices we can make in the first two steps. Usually, the ODE methods are chosen in the well known classes of *Runge–Kutta* (RK) and *linear multistep* (LM) methods. The interpolation procedures are also chosen mainly by following two alternative strategies: an interpolation of Lagrange or Hermite–Birkhoff type and a so called *one-step interpolation* procedure. As is natural to expect, people using LM methods mostly employ Lagrange or Hermite–Birkhoff interpolation, whereas people using RK methods, which are one-step methods, mostly employ one-step interpolation procedures. Nevertheless, in some papers the use of one-step methods is combined with multistep interpolation procedures.

We shall propose one-step techniques. Here we do not want to support one technique more than the other, but we just mention the fact that, near the discontinuity points, formulae of multistep type might give more problems in terms of accuracy. Usually they have to be modified next to such points, in order that the lack of smoothness in the solution does not cause a break down in the order of accuracy.

As we anticipated in Section 1, for the theory of numerical methods for ODEs we refer the reader to the books by Lambert [49], Butcher [21], Hairer, Nørsett and Wanner [31], Hairer and Wanner [32] and Dekker and Verwer [24].

3.2 Runge–Kutta methods and one-step interpolation

We briefly recall the form of an s-stage RK method for the numerical solution of the ODE (1.1). Given a mesh $\Delta := \{t_0, t_1 \ldots, t_n, \ldots, t_N = t_f\}$, the method is the following:

$$Y_{n+1}^i = y_n + h_{n+1} \sum_{j=1}^{\nu} a_{ij} f(t_{n+1}^j, Y_{n+1}^j), \quad i = 1, ..., \nu, \tag{3.3}$$

$$y_{n+1} = y_n + h_{n+1} \sum_{i=1}^{\nu} b_i f(t_{n+1}^i, Y_{n+1}^i), \tag{3.4}$$

where $t_{n+1}^i := t_n + c_i h_{n+1}, c_i := \sum_{j=1}^{\nu} a_{ij}, i = 1, \ldots, \nu$, $h_{n+1} := t_{n+1} - t_n$. The c_i's are called *abscissae* and, for most of common methods, they belong to $[0,1]$.

The computational complexity of the method is determined mainly by the number ν of *stages* and by the form of the coefficient matrix $A = [a_{ij}]_{i,j=1}^{\nu}$. It is well known that, when the matrix A is lower triangular with zero diagonal elements, the method is called *explicit* and the computational cost is lower, whereas, when the matrix A is full, the method is called *implicit* and the computational cost is higher.

The one-step interpolants are constructed step-by-step by making use of information from the underlying step $[t_n, t_{n+1}]$ only, possibly by including some additional stages. The resulting continuous extension $\eta(t)$ is defined, in each subinterval determined by the mesh Δ, by a one-step continuous quadrature rule of the form

$$\eta(t_n + \theta h_{n+1}) = y_n + h_{n+1} \sum_{i=1}^{s} b_i(\theta) f(t_{n+1}^i, Y_{n+1}^i), \quad 0 \le \theta \le 1, \tag{3.5}$$

where the $b_i(\theta)$'s are polynomials and the number of stages involved is $s \ge \nu$. The additional $s - \nu$ stages, if any, are given by

$$Y_{n+1}^i = y_n + h_{n+1} \sum_{j=1}^{s} a_{ij} f(t_{n+1}^j, Y_{n+1}^j), \quad i = \nu + 1, ..., s, \tag{3.6}$$

with corresponding discrete weights $b_i = 0$, $i = \nu + 1, \ldots, s$. So the original coefficient matrix $A = [a_{ij}]_{i,j=1}^{\nu}$ is embedded into the block lower triangular matrix $\begin{bmatrix} A & 0 \\ [a_{ij}]_{i=\nu+1,j=1}^{s,\nu} & [a_{ij}]_{i,j=\nu+1}^{s} \end{bmatrix}$. In order to assure the continuity of the interpolant, that is

$$\eta(t_n) = y_n \quad \text{and} \quad \eta(t_{n+1}) = y_{n+1}, \tag{3.7}$$

the polynomials $b_i(\theta)$ satisfy $b_i(0) = 0$ and $b_i(1) = b_i$, $i = 1, \ldots, s$.

The set of formulae (3.3)–(3.6)–(3.5) is called a *continuous* RK method, whereas, in contrast, the RK method (3.3)–(3.4) is called *discrete.*

It is worth observing that, in many papers and books, the RK formulae are written in an equivalent different form, the so called *K-notation.* So, for example, the discrete RK method (3.3)–(3.4) takes the form

$$K_{n+1}^i = f(t_{n+1}^i, y_n + h_{n+1} \sum_{j=1}^{\nu} a_{ij} K_{n+1}^j), \quad i = 1, ..., \nu,$$

$$y_{n+1} = y_n + h_{n+1} \sum_{i=1}^{\nu} b_i K_{n+1}^i.$$

Notice that the K-notation is obtained from formulae (3.3)–(3.4) by setting $K_{n+1}^i = f(t_{n+1}^i, Y_{n+1}^i)$, $i = 1, ..., \nu$.

Definition 3.1 *The RK method (3.3)–(3.4) has (nodal or discrete) order p if p is the largest integer such that, for all C^p-continuous right hand side functions $f(t, y)$ in (1.1) and for all mesh points,*

$$\|z_{n+1}(t_{n+1}) - y_{n+1}\| = O(h_{n+1}^{p+1}),$$

where $z_{n+1}(t)$ is the local solution to the local problem

$$\begin{cases} z_{n+1}'(t) = f(t, z_{n+1}(t)), & t \ge t_n, \\ z_{n+1}(t_n) = y_n. \end{cases} \tag{3.8}$$

The one-step interpolant (3.5) has uniform order q if q is the largest integer such that

$$\max_{t_n \leq t \leq t_{n+1}} \|z_{n+1}(t) - \eta(t)\| = O(h_{n+1}^{q+1}).$$

The uniform order q of the interpolant, which clearly is $\leq p$, is also the uniform order of the continuous RK method (3.3)–(3.6)–(3.5).

Remark 3.2 *A necessary condition for the RK method (3.3)–(3.4) actually to perform with order p is that the right hand side function $f(t, y)$ in (3.8) be sufficiently smooth, namely at least C^p-continuous.*

In most applications it is required that the uniform accuracy order q of the continuous extension $\eta(t)$ be equal to the nodal order p of the underlying RK method, or at least to $p-1$. For this reason, we are often obliged to append the $s-\nu$ extra stages. On the other hand, in some applications it is possible to use interpolants of uniform order lower than $p-1$. However, in that case, the interpolant must satisfy certain *asymptotic orthogonality* conditions which define the so called *Natural Continuous Extensions* (NCEs) introduced by Zennaro [79].

Definition 3.3 *The RK method (3.3)–(3.4) of order p has a NCE $\eta(t)$ of degree d if there exist ν polynomials $b_i(\theta), i = 1, \ldots, \nu$, of degree $\leq d$, independent of the function f, such that, for the function defined by (3.5) (with $s = \nu$), (3.7) holds and*

$$\max_{t_n \leq t \leq t_{n+1}} \|z_{n+1}(t) - \eta(t)\| = O(h_{n+1}^{d+1}),$$

$$\| \int_{t_n}^{t_{n+1}} G(t)[z'_{n+1}(t) - \eta'(t)]dt\| = O(h_{n+1}^{p+1})$$

for every sufficiently smooth matrix-valued function G, where $z_{n+1}(t)$ is the local solution to (3.8).

In [79] it is proved that every RK method has at least a NCE of minimal degree $\lfloor \frac{p+1}{2} \rfloor$.

A particular class of continuous RK methods which has been extensively studied is that of the *one-step collocation* methods. They were originally defined as follows. Chosen ν abscissae $c_1, \ldots, c_\nu \in [0, 1]$, along the mesh Δ compute the polynomial $\eta(t)$ of degree $\leq \nu$ satisfying

$$\eta'(t_{n+1}^i) = f(t_{n+1}^i, \eta(t_{n+1}^i)), \quad i = 1, \ldots, \nu, \quad \eta(t_n) = y_n.$$

It is easy to check that such methods can be rewritten as a continuous RK method (3.3)–(3.5) with $s = \nu$ (see, for example, Nørsett and Wanner [57]) and that the interpolant (3.5) is a NCE of the discrete formula (3.3)–(3.4) (see [79]).

In what follows, we illustrate one of the main applications of the interpolants and, in particular, of the NCEs. As we shall see in the next subsection, it includes the application to DDEs.

Consider the following IVP with *driving equation:*

$$\begin{cases} z'(x) = g(x, z(x), y(\alpha(x)), y'(\alpha(x))), & x \geq x_0, \\ z(x_0) = z_0, \end{cases} \tag{3.9}$$

where $y(t)$ is the solution of the ODE (1.1) and $\alpha(x)$ is one-to-one between the interval $[x_0, x_0 + \tilde{H}]$ and $[t_0 = \alpha(x_0), t_0 + H = \alpha(x_0 + \tilde{H})]$. Moreover, denote by $\beta(t)$ the inverse of $\alpha(x)$.

Now assume that we have solved numerically the ODE (1.1) for one step $[t_0, t_0 + h]$, $h \leq H$, by means of a RK method of order p and let $\eta(t)$ be a NCE of degree d. Finally, consider the following IVP perturbed from (3.9)

$$\begin{cases} w'(x) = g(x, w(x), \eta(\alpha(x)), \eta'(\alpha(x))), & x \geq x_0, \\ w(x_0) = z_0, \end{cases} \tag{3.10}$$

in the corresponding interval $[x_0 = \beta(t_0), x_0 + \tilde{h} = \beta(t_0 + h)]$. Then we have the following result, whose proof can be found in [79].

Theorem 3.4 *For the solutions $z(x)$ and $w(x)$ of the IVPs (3.9) and (3.10) the following statements hold*

$$\|z(x_0 + \tilde{h}) - w(x_0 + \tilde{h})\| = O(h^{p+1}),$$

$$\max_{x_0 \leq x \leq x_0 + \tilde{h}} \|z(x) - w(x)\| = O(h^{d+1}),$$

$$\| \int_{x_0}^{x_0 + \tilde{h}} G(x)[z'(x) - w'(x)]dx\| = O(h^{p+1})$$

for every sufficiently smooth matrix-valued function G.

3.3 Runge–Kutta methods for delay differential equations

Given a mesh $\Delta := \{t_0, t_1 \ldots, t_n, \ldots, t_N = t_f\}$, the numerical methods we consider for the numerical solution of the DDE (1.2) are obtained by applying the continuous RK method (3.3)–(3.6)–(3.5) to the ODE (3.2) in each interval $[\xi_{k-1}, \xi_k]$. They have the form

$$Y_{n+1}^i = y_n + h_{n+1} \sum_{j=1}^{s} a_{ij} f(t_{n+1}^j, Y_{n+1}^j, \eta(t_{n+1}^j - \tau)), \quad i = 1, \ldots, s, \tag{3.11}$$

$$\eta(t_n + \theta h_{n+1}) = y_n + h_{n+1} \sum_{i=1}^{s} b_i(\theta) f(t_{n+1}^i, Y_{n+1}^i, \eta(t_{n+1}^i - \tau)), \quad 0 \leq \theta \leq 1, \tag{3.12}$$

where the delay τ has to be evaluated at the relevant points t_{n+1}^j. Of course, for $t \leq t_0$ we define $\eta(t) = \varphi(t)$. Furthermore, we assume for simplicity (but without any practical restriction) that the abscissae c_i lie in $[0, 1]$, $i = 1, \dots, s$.

This method will be called the *RK method for DDEs*. Moreover, we shall say that (3.3)–(3.4) is the *underlying discrete RK method*, whereas (3.5) is the *underlying interpolant*. The pair formed by the underlying discrete RK method and by the underlying interpolant is called the *underlying continuous RK method*. Observe that, for simplicity, (3.11) combines (3.3) for the stages of the discrete RK method with (3.6) for the possible additional stages of the interpolant.

The K-notation for the RK method for DDEs (3.11)–(3.12) is easily obtained by setting $K^i_{n+1} = f(t^i_{n+1}, Y^i_{n+1}, \eta(t^i_{n+1} - \tau(t^i_{n+1})))$, $i = 1, \dots, s$.

Clearly, it would be desirable that the nodal order p of the underlying discrete RK method be inherited by the RK method for DDEs. Therefore, in view of Remark 3.2, it is necessary to include at least the first p discontinuity points ξ_k in the mesh Δ. Moreover, the interpolant (3.12) should have uniform order $\geq p - 1$. In fact, a robust ODE solver is implemented in a variable step size mode. In such a way, it is not possible to predict a priori where the retarded term $\eta(t - \tau)$ will be computed at each step, and a uniform high-order accuracy is consequently required. We can summarize this discussion in the following theorem.

Theorem 3.5 *If the RK method for DDEs (3.11)–(3.12) is applied to (1.2) with a mesh Δ which includes the discontinuity points ξ_k, if the underlying discrete RK method has nodal order p and if the underlying interpolant has uniform order $q \geq p - 1$, then the continuous numerical solution $\eta(t)$ is such that*

$$\max_{t_0 \leq t \leq t_f} \|y(t) - \eta(t)\| = O(h^p),$$

where $h = \max_{1 \leq n \leq N} h_n$.

As is well known, the variable step size strategy is adopted in order to ensure, as far as possible, the proportionality of the global error to a given tolerance and to minimize the computational cost. There are two main ways of doing this. The first consists in the estimation of the local error of the advancing method by means of another higher-order method, usually a method of order $p + 1$. The second (which may be less reliable in terms of error-tolerance proportionality, but in general more efficient in terms of the balance between the attainable accuracy and the computational cost) consists in the estimation of the local error of the advancing method by means of another lower-order method, usually a method of order $p - 1$. The first strategy requires an interpolant of uniform order p, whereas in the second strategy an interpolant of uniform order $p - 1$ is sufficient.

A rough explanation of this fact may be found in the light of Remark 3.2. In fact, the RK method for DDEs (3.11)–(3.12) is obtained by applying the continuous RK method (3.3)–(3.6)–(3.5) to the ODE (3.2). The ODE

(3.2) is not a smooth problem because the interpolant $\eta(t)$ is a piecewise polynomial function, which is only continuous at the mesh points. However, it is easy to prove by induction on the intervals $[\xi_{k-1}, \xi_k]$ that, in the underlying step $[t_n, t_{n+1}]$, the local problem

$$\begin{cases} z'_{n+1}(t) = f(t, z_{n+1}(t), \eta(t-\tau)), & t_n \leq t \leq t_{n+1}, \\ z_{n+1}(t_n) = y_n, \end{cases} \tag{3.13}$$

which is actually solved numerically, is a perturbation of the smooth problem

$$\begin{cases} \tilde{z}'_{n+1}(t) = f(t, \tilde{z}_{n+1}(t), \tilde{\eta}_{n+1}(t-\tau)), & t_n \leq t \leq t_{n+1}, \\ \tilde{z}_{n+1}(t_n) = y_n, \end{cases} \tag{3.14}$$

where $\tilde{\eta}_{n+1}(t-\tau)$ is a suitable *delayed local solution* which is C^∞-continuous and satisfies

$$\max_{t_n \leq t \leq t_{n+1}} \|\tilde{\eta}_{n+1}(t-\tau) - \eta(t-\tau)\| = O(h^{q+1}),$$

where q is the uniform order of accuracy of the interpolant $\eta(t)$. This implies

$$\|\tilde{z}_{n+1}(t_{n+1}) - z_{n+1}(t_{n+1})\| = O(h^{q+2}).$$

Now, given a discrete RK method, let w_{n+1} and $\tilde{w}_{n+1}$ be the numerical solutions of (3.13) and (3.14), respectively. Since the numerical solution depends continuously on the data, we have also

$$\|\tilde{w}_{n+1} - w_{n+1}\| = O(h^{q+2}).$$

Therefore, in view of Remark 3.2 it is easy to prove that, in order to preserve the order p of the advancing method, in the second variable step size strategy it is necessary and sufficient that $q \geq p-1$ whereas, in the first variable step size strategy, $q \geq p$ is required for the successful implementation of the higher-order method used to estimate the local error.

For a deeper analysis of the relationships between local and global error of DDE solvers and for a discussion of the correct order of accuracy to require to the interpolation procedure, we refer the reader to Arndt [2], Buchacker and Filippi [17], Higham [36] (see also Bellen and Zennaro [10]).

When it is known a priori that the solution of the DDE does not change its behavior significantly along the integration interval, a less robust ODE code may be successfully implemented, which does not perform the step size selection at each step. If this is the case, even constant step size might be a reasonable choice.

In particular, a possibility is given by the so called *constrained mesh* strategy, in which all the mesh points of the $(k+1)$st interval $[\xi_k, \xi_{k+1}]$ are the inverse images under the function $t - \tau(t)$ of the corresponding mesh points of the kth interval $[\xi_{k-1}, \xi_k]$. When the delay τ is constant, the mesh Δ can be chosen to be uniform with constant step size as an integer submultiple of the delay τ. When the delay τ is variable, the step size can be chosen to be uniform only in the first interval $[\xi_0, \xi_1]$: in the subsequent intervals the mesh points are automatically determined by the rule above. In this case the computation of the mesh requires the solution of a scalar nonlinear equation like (2.1) per step. The constrained mesh strategy was first introduced for the application of the one-step collocation method at Gaussian points to the numerical solution of DDEs by Bellen [7].

The main advantage of the constrained mesh strategy is that it allows the use of interpolants of uniform order lower than $p-1$, namely of the NCEs. In fact, induction on the intervals $[\xi_{k-1}, \xi_k]$ and standard arguments easily yield the following corollary to Theorem 3.4, which represents the extension of the original results in [7], valid for one-step collocation, to the whole class of RK methods (see also Bellen [8] for further generalizations to more general functional differential equations).

Corollary 3.6 *If the RK method for DDEs (3.11)–(3.12) is applied to (1.2) with a constrained mesh Δ and the underlying interpolant is a NCE of degree d, then the continuous numerical solution $\eta(t)$ is such that*

$$\max_{t_0 \le t \le t_f} \|y(t) - \eta(t)\| = O(h^{d'}), \tag{3.15}$$

where $d' = \min\{d+1, p\}$, and

$$\max_{1 \le n \le N} \|y(t_n) - \eta(t_n)\| = O(h^p). \tag{3.16}$$

Observe that, if $d \ge p-1$, then we (obviously) find again the result stated by Theorem 3.5.

Now assume that the delay τ is constant and that the RK method (3.11)–(3.12) is such that $Y_{n+1}^i = \eta(t_{n+1}^i), i = 1, \ldots, s$. These methods are called *natural* in [81] and are characterized by

$$a_{ij} = b_j(c_i), \quad i, j = 1, \ldots, s. \tag{3.17}$$

It is easy to see that all the one-step collocation methods are natural. More generally, in [81] it is shown that the class of natural RK methods coincides with the class of *one-step projection* methods in some polynomial space.

For such particular methods, it is worth pointing out some connections between Bellman's method of steps and the constrained mesh strategy. It is very easy to see that, if we solve the DDE by means of the natural RK method for DDEs (3.11)–(3.12) with a constrained mesh Δ, we get the same numerical solution given by the Bellman approach which uses the same points of Δ in each interval $[\xi_{k-1}, \xi_k]$. Put in another way, this is an alternative proof of Corollary 3.6 for natural RK methods for DDEs in the case of constant delay. On the other hand, it can be seen that for nonconstant delays the two approaches are not equivalent.

In any case, the two approaches are quite different from the computational point of view. As already mentioned at the beginning of this section, Bellman's method of steps is redundant as it needs to compute the same piece of solution many times. On the other hand, contrary to the other approach, for the integration on the interval $[\xi_{k-1}, \xi_k]$ it is not necessary to store the solution computed in the previous interval $[\xi_{k-2}, \xi_{k-1}]$. Moreover, Bellman's approach never needs interpolation.

To conclude this section, we show what the RK methods for DDEs look like when they are applied to the pure delay equation (2.6)

$$\eta(t_n+\theta h_{n+1}) = y_n+h_{n+1}\sum_{i=1}^{s} b_i(\theta)f(t_{n+1}^i, \eta(t_{n+1}^i-\tau)), \quad 0 \le \theta \le 1. \quad (3.18)$$

Since f is independent of y, the stage values computed by (3.11) are useless, and hence are not computed at all. Indeed, what we need is just a *continuous quadrature rule*. For more details, see Torelli and Vermiglio [68].

3.4 Available methods and software

For equations of the class considered in this section, reasonable codes may be obtained by adapting a standard ODE code, for which there is a large availability. It is only necessary to endow it with an interpolation procedure and, when the delay τ is variable, with a rootfinding routine for solving the discontinuity point equations (2.1). On the other hand, more sophistication is required when we want to treat more general classes of DDEs, like those presented in Section 5.

We conclude this section by giving a list of references where particular methods which have been proposed for the numerical solution of DDEs can be found. Some of them also contain pseudocodes and/or computer programs.

In the history of numerical methods for DDEs, apart from the method of steps developed by Bellman, first investigations were made by Feldstein [27] for the Euler method and soon after by Stetter [62], who suggested the use of Hermite interpolation for the approximation of the retarded argument.

Later on Tavernini [63, 64] developed a one-step technique in order to handle equations in which the delay may vanish. Algorithms based on fourth-order RK methods and two-point Hermite interpolation were presented by Neves [53, 54] and by Oppelstrupp [59].

Later still, for DDEs of the class considered in this section, Oberle and Pesh [58] proposed Runge–Kutta–Fehlberg methods together with multistep Hermite interpolation, whereas Bellen and Zennaro [10] presented some algorithms based on a predictor-corrector version of the one-step collocation method at Gaussian points. Within the constrained mesh strategy introduced by Bellen [7], Vermiglio [70] proposed a one-step subregion method. The numerical solution of state dependent DDEs has been considered by Bock and Schlöder [15], Feldstein and Neves [29], Al-Mutib [1].

More software oriented papers have been written by Weiner and Strehmel [72], Neves and Thompson [56], Paul [60], Willé and Baker [76], Butcher [22], Baker, Butcher and Paul [3] and Hayashi [35].

Finally, we quote the set of DDEs collected by Paul [61] for testing and comparing numerical algorithms.

4 Stability properties of numerical methods

In this section we analyze the stability properties of the numerical methods for DDEs proposed in Section 3.

As we already said in Section 1, we assume that the reader is familiar with the stability analysis of numerical methods for ODEs. To this purpose, we refer him in particular to Dekker and Verwer [24].

Our analysis will be carried out by starting from the simplest test equation (2.4) with constant delay, which leads to the concepts of *P-stability* and *GP-stability* as generalizations of the concept of *A-stability* for ODEs. Then we shall consider the test equation (2.3) with constant delay which leads to the concepts of *PN-stability* and *GPN-stability* as generalizations of the concept of *AN-stability* for ODEs. Finally, we shall consider the general nonlinear system (1.2) with constant delay which leads to the concepts of *RN-stability* and *GRN-stability* as generalizations of the concept of *BN-stability* for ODEs.

As in Section 2.3, we assume that the functions involved in the various test equations are complex valued.

4.1 P-stability and GP-stability

Recall Barwell's definitions of P-stability and GP-stability (see [6]).

Definition 4.1 *The P-stability region of a numerical step-by-step method for DDEs is the set S_P of the pairs of complex numbers (α, β), $\alpha = h\lambda$, $\beta = h\mu$, such that the discrete numerical solution $\{y_n\}_{n\geq 0}$ of (2.4) satisfies $lim_{n\to\infty} y_n = 0$ for all constant delays τ and all initial functions $\varphi(t)$ and for any constant step size $h > 0$ under the constraint*

$$h = \tau/m, \tag{4.1}$$

where m is a positive integer.

Definition 4.2 *A numerical step-by-step method for DDEs is P-stable if*

$$S_P \supseteq \{(\alpha, \beta) \in \mathbb{C}^2 \mid \mathrm{Re}(\alpha) + |\beta| < 0\}.$$

In other words, a numerical method for DDEs is P-stable if it preserves the asymptotic stability properties of the solution $y(t)$ of (2.4) whenever (2.22) holds, under the constraint (4.1) on the step size.

Removing the constraint (4.1) leads to the following stronger concepts of stability.

Definition 4.3 *The GP-stability region of a numerical step-by-step method for DDEs is the set S_{GP} of the pairs of complex numbers (α, β), $\alpha = h\lambda$, $\beta = h\mu$, such that the discrete numerical solution $\{y_n\}_{n\geq 0}$ of (2.4) satisfies $lim_{n\to\infty} y_n = 0$ for all constant delays τ and all initial functions $\varphi(t)$ and for any constant step size $h > 0$.*

Definition 4.4 *A numerical step-by-step method for DDEs is GP-stable if*

$$S_{GP} \supseteq \{(\alpha, \beta) \in \mathbb{C}^2 \mid \mathrm{Re}(\alpha) + |\beta| < 0\}.$$

Observe that the constraint (4.1) defines, indeed, a constrained mesh. Therefore, the study of the P-stability properties of a method is relevant when we apply such a method in the constrained mesh strategy. On the other hand, when we remove the constraint (4.1) we allow the method to be used in a free mesh strategy, so that the study of the GP-stability properties is more appropriate in this case. However, in both cases, the step size is assumed to be constant, and hence the stability properties of the numerical solutions are not completely captured by such investigations.

It is clear that $S_{GP} \subseteq S_P$. Thus, a GP-stable method is P-stable, too. Moreover, a P-stable method for DDEs is A-stable for ODEs.

Although the investigation of P-stability and GP-stability regions of the numerical methods is not easy since the final product is always a difference equation of arbitrary order m depending on the ratio τ/h, the problem has been seriously reduced in complexity. For the case of LM methods, we quote the papers by Bickart [14] and by Watanabe and Roth [71]. In particular, the latter established a sort of equivalence between A-stability and GP-stability, in the sense that each A-stable LM formula for ODEs endowed with piecewise constant or linear Lagrange interpolation is GP-stable for DDEs. For the case we are mostly interested in, that of RK methods, Zennaro [80] found a procedure to compute the P-stability regions. Later on, in 't Hout and Spijker [42] found a more general method that allows the computation of GP-stability regions.

Let us begin with the results from [80] on P-stability.

The RK method for DDEs (3.11)–(3.12), applied to the test equation (2.4) with constant step size h satisfying the constraint (4.1), assumes the form

$$Y_{n+1}^i = y_n + h \sum_{j=1}^{s} a_{ij} (\lambda Y_{n+1}^j + \mu \eta(t_{n-m+1}^j)), \quad i = 1, \dots, s, \tag{4.2}$$

$$\eta(t_n + \theta h) = y_n + h\sum_{i=1}^{s} b_i(\theta)(\lambda Y_{n+1}^i + \mu\eta(t_{n-m+1}^i)). \tag{4.3}$$

After elimination of the stage values Y_{n+1}^i from (4.2) and computation of (4.3) for $\theta = c_1, \ldots, c_s, 1$, this pair of equations reduces to the following recurrence relation with constant coefficients for the sequence of $(s+1)$-dimensional vectors $\mathbf{H}_n = [\eta(t_n^1), \ldots, \eta(t_n^s), y_n]^T$:

$$\mathbf{H}_{n+1} = P(\alpha)\mathbf{H}_n + \beta Q(\alpha)\mathbf{H}_{n-m+1}, \tag{4.4}$$

where

$$P(\alpha) = \begin{bmatrix} \mathrm{O} & e + \alpha B(I-\alpha A)^{-1}e \\ 0\ldots 0 & 1 + \alpha b^T(I-\alpha A)^{-1}e \end{bmatrix},$$

$$Q(\alpha) = \begin{bmatrix} & 0 \\ B(I-\alpha A)^{-1} & \vdots \\ & 0 \\ b^T(I-\alpha A)^{-1} & 0 \end{bmatrix},$$

$b = [b_1, \ldots, b_s]^T$, $e = [1, \ldots, 1]^T$ is the unit s-vector, O is the s-dimensional zero matrix, I is the s-dimensional identity matrix and the matrices $A = [a_{ij}]_{i,j=1}^s$ and $B = [b_j(c_i)]_{i,j=1}^s$ are defined from the coefficients of the RK method.

The asymptotic behavior of the solutions of (4.4) is determined by the roots ζ of its characteristic equation

$$\det(\zeta^m I - \zeta^{m-1}P(\alpha) - \beta Q(\alpha)) = 0,$$

where, this time, I is the $(s+1)$-dimensional identity matrix (see Gohberg, Lancaster and Rodman [30]). In [80] the K-notation for the RK formula is used, and hence slightly different formulae are obtained. However, in a similar way it can be found that, instead of the roots ζ of the above characteristic equation, we can equivalently consider the solutions of the algebraic equation

$$\zeta = R_\alpha(\beta/\zeta^m), \tag{4.5}$$

where

$$R_\alpha(z) = 1 + (\alpha + z)b^T(I - \alpha A - zB)^{-1}e,$$

provided the very mild assumption on the RK method holds that the matrix $I - \alpha A - z^*B$ is singular if and only if z^* is a pole of $R_\alpha(z)$.

Now, for the rational function $R_\alpha(z)$, consider the curve $\Gamma_\alpha = \{z \in \mathbf{C} \mid |R_\alpha(z)| = 1\}$ in the complex plane and let $\sigma_\alpha = \inf_{z\in\Gamma_\alpha}|z|$ be the distance of the curve Γ_α from the origin of the complex plane. Moreover, let S_A denote the A-stability region of the underlying discrete RK method (3.3)–(3.4), that is $S_A = \{\alpha \in \mathbf{C} \mid |R_\alpha(0)| < 1\}$.

Lemma 4.5 *Let $R_\alpha(z)$ be nonconstant and such that $z = 0$ is not a pole. Consider the following three statements.*

(a) $\alpha \in S_A$ *and* $|\beta| < \sigma_\alpha$;

(b) *all the roots ζ of (4.5) are inside the unit circle for all $m \geq 1$;*

(c) $\alpha \in S_A$ *and* $|\beta| \leq \sigma_\alpha$.

Then (a) $\Longrightarrow$ (b) $\Longrightarrow$ (c).

If we define the set

$$\Sigma = \{(\alpha, \beta) \in \mathbf{C}^2 \mid \alpha \in S_A \quad \text{and} \quad |\beta| < \sigma_\alpha\},$$

by Lemma 4.5 we get the following characterization of the P-stability region.

Theorem 4.6 *The P-stability region S_P of the RK method for DDEs (3.11)–(3.12) is such that $\Sigma \subseteq S_P$. Moreover, under the mild assumption on the method that $\lim_{n\to\infty} \mathbf{H}_n = 0$ if and only if $\lim_{n\to\infty} y_n = 0$, it also holds that $S_P \subseteq \overline{\Sigma}$.*

We can conclude that, in general, the study of the P-stability properties of the RK methods for DDEs is reduced to a constrained minimum problem in the complex plane, namely to compute σ_α for $\alpha \in S_A$.

It is very easy to see that, if the RK method for DDEs (3.11)–(3.12) is natural, that is (3.17) holds, then $\sigma_\alpha = dist(\alpha, \partial S_A)$, where $dist(.,.)$ denotes the Euclidean distance in the complex plane and ∂S_A is the boundary of S_A. Moreover, if the underlying discrete RK method is A-stable, that is $S_A \supseteq \{\alpha \in \mathbf{C} \mid \text{Re}(\alpha) < 0\}$, then it clearly holds that $\sigma_\alpha \geq -\text{Re}(\alpha)$, and we have the following result.

Corollary 4.7 *A natural RK method for DDEs (3.11)–(3.12) is P-stable if the underlying discrete RK method (3.3)–(3.4) is A-stable.*

By virtue of Corollary 4.7, we can claim that the one-step collocation method at Gaussian points for DDEs is P-stable, since it is obviously natural and, as is well known, A-stable for ODEs. However, for a direct proof of this fact, see Zennaro [78].

Now consider the results in [42] about GP-stability.

For simplicity, we assume that $0 \leq c_1 \leq c_2 \leq \ldots \leq c_s \leq 1$. Since the constant step size h does not necessarily satisfy the constraint (4.1), there exists $\delta \in [0, 1)$ such that

$$h = \tau/(m - \delta), \tag{4.6}$$

where m is a positive integer. Therefore, if r is the integer such that $c_r + \delta \leq 1 < c_{r+1} + \delta$ with the convention $c_0 = 0$ and $c_{s+1} = 1$, the RK method for

DDEs (3.11)–(3.12), applied to the test equation (2.4), assumes the form

$$
\begin{aligned}
Y_{n+1}^{i} = y_n \quad &+ \quad h\sum_{j=1}^{r} a_{ij}(\lambda Y_{n+1}^{j} + \mu\eta(t_{n-m+1}^{j} + \delta h)) \\
&+ \quad h\sum_{j=r+1}^{s} a_{ij}(\lambda Y_{n+1}^{j} + \mu\eta(t_{n-m+2}^{j} + (\delta - 1)h)),
\end{aligned} \tag{4.7}
$$

for $i = 1, \ldots, s$, and

$$
\begin{aligned}
\eta(t_n + \theta h) = y_n \quad &+ \quad h\sum_{i=1}^{r} b_i(\theta)(\lambda Y_{n+1}^{i} + \mu\eta(t_{n-m+1}^{i} + \delta h)) \\
&+ \quad h\sum_{i=r+1}^{s} b_i(\theta)(\lambda Y_{n+1}^{i} + \mu\eta(t_{n-m+2}^{i} + (\delta - 1)h)).
\end{aligned} \tag{4.8}
$$

As for eqns (4.2)–(4.3), after elimination of the stage values Y_{n+1}^{i} from (4.7) and computation of (4.8) for $\theta = c_1 + \delta, \ldots, c_r + \delta, c_{r+1} + \delta - 1, \ldots, c_s + \delta - 1, 1$, we get the following equivalent recurrence relation with constant coefficients for the sequence of $(s+1)$-dimensional vectors $\mathbf{H}_n^{(\delta)} = [\eta(t_n^1 + \delta h)), \ldots, \eta(t_n^r + \delta h)), \eta(t_n^{r+1} + (\delta - 1)h), \ldots, \eta(t_n^s + (\delta - 1)h), y_n]^T$:

$$
\mathbf{H}_{n+1}^{(\delta)} = P^{(\delta)}(\alpha)\mathbf{H}_n^{(\delta)} + \beta T^{(\delta)}(\alpha)\mathbf{H}_{n-m+2}^{(\delta)} + \beta Q^{(\delta)}(\alpha)\mathbf{H}_{n-m+1}^{(\delta)}, \tag{4.9}
$$

where

$$
P^{(\delta)}(\alpha) = \begin{bmatrix} \mathrm{O} & e + \alpha B^{(\delta)}(I - \alpha A)^{-1}e \\ 0\ldots0 & 1 + \alpha b^T(I - \alpha A)^{-1}e \end{bmatrix},
$$

$$
Q^{(\delta)}(\alpha) = \begin{bmatrix} & 0 \\ B_1^{(\delta)} + \alpha B^{(\delta)}(I - \alpha A)^{-1}A_1^{(\delta)} & \vdots \\ & 0 \\ b_1^{(\delta)T} + \alpha b^T(I - \alpha A)^{-1}A_1^{(\delta)} & 0 \end{bmatrix},
$$

$$
T^{(\delta)}(\alpha) = \begin{bmatrix} & 0 \\ B_2^{(\delta)} + \alpha B^{(\delta)}(I - \alpha A)^{-1}A_2^{(\delta)} & \vdots \\ & 0 \\ b_2^{(\delta)T} + \alpha b^T(I - \alpha A)^{-1}A_2^{(\delta)} & 0 \end{bmatrix},
$$

$$
A_1^{(\delta)} = \begin{bmatrix} a_{11} & \ldots & a_{1r} & 0 & \ldots & 0 \\ \vdots & & \vdots & \vdots & & \vdots \\ a_{s1} & \ldots & a_{sr} & 0 & \ldots & 0 \end{bmatrix},
$$

$$A_2^{(\delta)} = \begin{bmatrix} 0 & \dots & 0 & a_{1,r+1} & \dots & a_{1s} \\ \vdots & & \vdots & \vdots & & \vdots \\ 0 & \dots & 0 & a_{s,r+1} & \dots & a_{ss} \end{bmatrix},$$

$$B^{(\delta)} = \begin{bmatrix} b_1(c_1+\delta) & \dots & b_s(c_1+\delta) \\ \vdots & & \vdots \\ b_1(c_r+\delta) & \dots & b_s(c_r+\delta) \\ b_1(c_{r+1}+\delta-1) & \dots & b_s(c_{r+1}+\delta-1) \\ \vdots & & \vdots \\ b_1(c_s+\delta-1) & \dots & b_s(c_s+\delta-1) \end{bmatrix},$$

$$B_1^{(\delta)} = \begin{bmatrix} b_1(c_1+\delta) & \dots & b_r(c_1+\delta) & 0 & \dots & 0 \\ \vdots & & \vdots & \vdots & & \vdots \\ b_1(c_r+\delta) & \dots & b_r(c_r+\delta) & 0 & \dots & 0 \\ b_1(c_{r+1}+\delta-1) & \dots & b_r(c_{r+1}+\delta-1) & 0 & \dots & 0 \\ \vdots & & \vdots & \vdots & & \vdots \\ b_1(c_s+\delta-1) & \dots & b_r(c_s+\delta-1) & 0 & \dots & 0 \end{bmatrix},$$

$$B_2^{(\delta)} = \begin{bmatrix} 0 & \dots & 0 & b_{r+1}(c_1+\delta) & \dots & b_s(c_1+\delta) \\ \vdots & & \vdots & \vdots & & \vdots \\ 0 & \dots & 0 & b_{r+1}(c_r+\delta) & \dots & b_s(c_r+\delta) \\ 0 & \dots & 0 & b_{r+1}(c_{r+1}+\delta-1) & \dots & b_s(c_{r+1}+\delta-1) \\ \vdots & & \vdots & \vdots & & \vdots \\ 0 & \dots & 0 & b_{r+1}(c_s+\delta-1) & \dots & b_s(c_s+\delta-1) \end{bmatrix},$$

$b_1^{(\delta)} = [b_1(1), \dots, b_r(1), 0, \dots, 0]^T$ and $b_2^{(\delta)} = [0, \dots, 0, b_{r+1}(1), \dots, b_s(1)]^T$.

Remark 4.8 *For $\delta = 0$, we have $\mathbf{H}_n^{(0)} = \mathbf{H}_n$, $P^{(0)}(\alpha) = P(\alpha)$, $Q^{(0)}(\alpha) = Q(\alpha)$, $T^{(0)}(\alpha) = \mathrm{O}$, the zero matrix, and (4.9) reduces to (4.4). In fact, for $\delta = 0$ the constraint (4.1) is satisfied.*

As for the recurrence (4.4), the asymptotic behavior of the solutions of (4.9) is determined by the roots ζ of its characteristic equation

$$\det(\zeta^m I - \zeta^{m-1}P^{(\delta)}(\alpha) - \beta\zeta T^{(\delta)}(\alpha) - \beta Q^{(\delta)}(\alpha)) = 0. \tag{4.10}$$

In [42] the K-notation for the RK formula is used, and hence slightly different formulae are obtained. However, the essence of the problem is that, for $\delta \neq 0$, the characteristic roots ζ are not the solutions of an algebraic equation similar to (4.5) any longer. Therefore, a more general result than Lemma 4.5 is needed. To this end, observe that the characteristic equation

(4.10) is a particular case of

$$\det(\zeta^m U(\zeta) + V(\zeta)) = 0, \tag{4.11}$$

where $U(\zeta)$ and $V(\zeta)$ are matrix polynomials.

Lemma 4.9 *Let $U(\zeta)$ and $V(\zeta)$ be matrix polynomials with $\det(U(\zeta))$ non identically zero, and let $m_0 \geq 0$ be the smallest integer such that $\deg(\det(\zeta^m U(\zeta) + V(\zeta))) = \deg(\det(\zeta^m U(\zeta)))$ for all integers $m \geq m_0$. Consider the following three statements.*

(a) *$U(\zeta)$ is invertible whenever $|\zeta| \geq 1$ and $\sup_{|\zeta|=1} \rho[V(\zeta)U(\zeta)^{-1}] < 1$;*
(b) *all the roots ζ of (4.11) are inside the unit circle for all $m \geq m_0$;*
(c) *$U(\zeta)$ is invertible whenever $|\zeta| \geq 1$ and $\sup_{|\zeta|=1} \rho[V(\zeta)U(\zeta)^{-1}] \leq 1$;*

where $\rho[\cdot]$ denotes the spectral radius of a matrix. Then (a) $\Longrightarrow$ (b) $\Longrightarrow$ (c).

If we define the sets

$$\begin{aligned} \Sigma^{(\delta)} \;\; = \;\; & \{(\alpha,\beta) \in \mathbf{C}^2 \mid \rho[P^{(\delta)}(\alpha)] < 1 \quad \text{and} \\ & |\beta| < \left(\sup_{|\zeta|=1} \rho[(\zeta T^{(\delta)}(\alpha) + Q^{(\delta)}(\alpha))(\zeta I - P^{(\delta)}(\alpha))^{-1}] \right)^{-1} \} \end{aligned}$$

and

$$\Sigma^* = \bigcap_{0 \leq \delta < 1} \Sigma^{(\delta)},$$

by Lemma 4.9 we get the following characterization of the GP-stability region.

Theorem 4.10 *The GP-stability region S_{GP} of the RK method for DDEs (3.11)–(3.12) is such that $\Sigma^* \subseteq S_{GP}$. Moreover, under the mild assumption on the method that $\lim_{t\to+\infty} \eta(t) = 0$ if and only if $\lim_{n\to\infty} y_n = 0$, it also holds that $S_{GP} \subseteq \overline{\Sigma^*}$.*

The most widely studied RK methods for DDEs have been the so called *θ-methods* with linear interpolation. They have the following form:

$$\begin{aligned} Y^1_{n+1} &= y_n, \\ Y^2_{n+1} &= y_n \;\; + \;\; h_{n+1}\left((1-\omega) f(t_n, Y^1_{n+1}, \eta(t_n - \tau))\right. \\ & \qquad\quad + \;\; \left.\omega f(t_{n+1}, Y^2_{n+1}, \eta(t_{n+1} - \tau))\right), \end{aligned}$$

$$\begin{aligned} \eta(t_n + \theta h_{n+1}) = y_n \;\; & + \;\; \theta h_{n+1}\left((1-\omega) f(t_n, Y^1_{n+1}, \eta(t_n - \tau))\right. \\ & + \;\; \left.\omega f(t_{n+1}, Y^2_{n+1}, \eta(t_{n+1} - \tau))\right), \end{aligned}$$

where ω is a parameter belonging to $[0, 1]$. The name θ-methods is due to the fact that, in the literature, our parameter ω is commonly denoted by θ.

Observe that, for $\omega = 0$, they reduce to the so called *forward Euler* method, whereas, for $\omega = 1$, they reduce to the so called *backward Euler* method, which are both 1-stage RK methods. The order of the θ-methods is always equal to 1 but for $\omega = \frac{1}{2}$, in which case the order is 2 and the method is also called *trapezoidal rule*.

The θ-methods are not as difficult to treat as a general RK method, so that Liu and Spijker [50] were able to compute their GP-stability regions by using a much less general form of Lemma 4.9. In particular, they showed that they are GP-stable if and only if $\frac{1}{2} \leq \omega \leq 1$.

As a negative result, by using the potential of Lemma 4.9, in 't Hout [38] proved that no one-step collocation method with abscissae in $[0, 1)$ can be GP-stable.

On the other hand, as a positive result, in 't Hout [39] defined a new type of interpolation procedure such that, if endowed with it, any A-stable RK method for ODEs gives rise to a GP-stable RK method for DDEs. However, we do not enter into details about this interpolation procedure because it is of multistep type and, therefore, it is not included in the class of methods that we consider in these notes.

4.2 PN-stability and GPN-stability

Recall Torelli's definitions of PN-stability and GPN-stability (see [66] and Bellen and Zennaro [11]).

Definition 4.11 *A numerical step-by-step method for DDEs is PN-stable if the discrete numerical solution $\{y_n\}_{n\geq 0}$ of any test equation (2.3) satisfying (2.19) is such that*

$$|y_n| \leq \max_{x \leq t_0} |\varphi(x)|, \quad n \geq 0, \tag{4.12}$$

for all constant delays τ and all initial functions $\varphi(t)$ and for any constant step size $h > 0$ under the constraint (4.1).

Definition 4.12 *A numerical step-by-step method for DDEs is GPN-stable if the discrete numerical solution $\{y_n\}_{n\geq 0}$ of any test equation (2.3) satisfying (2.19) is such that (4.12) holds for all constant delays τ and all initial functions $\varphi(t)$ and for any constant step size $h > 0$.*

Also for the test equation (2.3) we distinguish between a constrained and a free mesh, by imposing the condition (4.1) or not. Moreover, this time we do not consider the regions of PN-stability and GPN-stability (although they have been defined in [66]), but consider directly those methods which completely preserve the stability properties of the analytic solutions.

It is clear that a GPN-stable method is also PN-stable and that a PN-stable method for DDEs is AN-stable for ODEs.

PN-stability and GPN-stability are stronger concepts than P-stability and GP-stability because they are based on a more general test equation, to the same extent that AN-stability is a stronger stability concept than A-stability for ODEs. Anyway, there is also another difference due to the fact that PN-stability and GPN-stability are demands for contractivity of the numerical solution, whereas P-stability and GP-stability are demands for asymptotic stability (that is convergence to zero). Apart from the cases of equations and methods for which the coefficients belong to the boundary of the coefficient space, contractivity implies also asymptotic stability. The opposite implication is, of course, never true.

In order to cover also the case of contractivity of the numerical solution for the simplest test equation (2.4), in [11] the definitions of *P-contractivity* and *GP-contractivity* have been given, but we do not consider them here for the sake of brevity.

Contrary to the previous case in which A-stability of the underlying discrete method was sufficient to assure P-stability and GP-stability provided a suitable interpolation procedure is used, in our short survey of the main results concerning PN-stability and GPN-stability we shall see that such stability properties are not assured at all even if the underlying discrete method is AN-stable. In fact, a stronger concept of stability for ODE methods, based on the test equation (2.24) with forcing term, has to be involved.

We do not know of any results concerning LM methods. On the other hand, the theory concerning RK methods is rather developed (see Torelli [66, 67] and Bellen and Zennaro [11]).

Definition 4.13 *The RK method (3.3)–(3.4) is* $\mathrm{AN_f}$*-stable if the numerical solution* $\{y_n\}_{n\geq 0}$ *of (2.24) satisfies*

$$|y_{n+1}| \leq \max\{|y_n|, \max_{1\leq i\leq \nu} |g(t^i_{n+1})|\},$$

whenever $\mathrm{Re}(\lambda(t)) \leq 0$, $t \geq t_0$, *and for any mesh* Δ.

Definition 4.14 *The continuous RK method (3.3)–(3.6)–(3.5) is semi-*$\mathrm{AN_f}$*-stable if the numerical solution* $\eta(t)$ *of (2.24) satisfies*

$$|\eta(t_n + \theta h_{n+1})| \leq \max\{|y_n|, \max_{1\leq i\leq s} |g(t^i_{n+1})|\}, \tag{4.13}$$

for $\theta = c_1, \ldots, c_s, 1$, *whenever* $\mathrm{Re}(\lambda(t)) \leq 0$, $t \geq t_0$, *and for any mesh* Δ.

The method is $\mathrm{AN_f}$*-stable if (4.13) holds for all* $\theta \in [0,1]$.

Analogous definitions of A_f*-stability* are given with respect to the test equation with forcing term (2.26).

It is clear that requiring an RK method (3.3)–(3.4) to be $\mathrm{AN_f}$-stable is more than requiring that it be AN-stable. In fact, AN-stability is obtained as a particular case when the forcing term $g(t)$ is identically zero in the test equation (2.24).

The next theorem represents the link, almost an equivalence result, between the theory of PN-stability and GPN-stability and the theory of

AN$_f$-stability. The proof is easy and, once again, is based on the use of induction on the intervals $[\xi_{k-1}, \xi_k]$.

Theorem 4.15 *If the RK method for DDEs (3.11)–(3.12) is PN-stable, then the underlying discrete RK method (3.3)–(3.4) is AN$_f$-stable. Conversely, if the underlying continuous RK method (3.3)–(3.6)–(3.5) is (semi-) AN$_f$-stable, then the RK method for DDEs (3.11)–(3.12) is (PN-stable) GPN-stable.*

In order to translate the stability properties of RK method for DDEs (3.11)–(3.12) with respect to the test equation (2.3) in terms of the coefficients, we have to give necessary and sufficient conditions on the coefficients of the underlying continuous RK method for its AN$_f$-stability. The application of formulae (3.3)–(3.6)–(3.5) to the test equation with forcing term (2.24) yields

$$Y_{n+1}^i = y_n + h_{n+1} \sum_{j=1}^{s} a_{ij}(\lambda(t_{n+1}^j)Y_{n+1}^j + \mathrm{Re}(\lambda(t_{n+1}^j))g(t_{n+1}^j)), \quad i = 1, \dots, s,$$

$$\eta(t_n + \theta h_{n+1}) = y_n + h_{n+1} \sum_{i=1}^{s} b_i(\theta)(\lambda(t_{n+1}^i)Y_{n+1}^i + \mathrm{Re}(\lambda(t_{n+1}^i))g(t_{n+1}^i)).$$

Therefore, it is easy to prove the following results.

Proposition 4.16 *Assume that the RK method is nonconfluent, that is $c_i \neq c_j$ if $i \neq j$, and let $Z = \mathrm{diag}(z_1, \dots, z_s)$ denote a diagonal matrix with complex coefficients. Then we have:*

- *The discrete RK method (3.3)–(3.4) is AN$_f$-stable if and only if*
$$|1+b(\theta)^T(I-ZA)^{-1}Ze|+\|b(\theta)^T(I-ZA)^{-1}\mathrm{Re}(Z)\|_1 \leq 1 \quad \forall \mathrm{Re}(z_i) \leq 0, \tag{4.14}$$
for $\theta = 1$.
- *The continuous RK method (3.3)–(3.6)–(3.5) is semi-AN$_f$-stable if and only if (4.14) holds for $\theta = c_1, \dots, c_s, 1$.*
- *The continuous RK method (3.3)–(3.6)–(3.5) is AN$_f$-stable if and only if (4.14) holds for all $\theta \in [0, 1]$.*

With reference to (4.14), it is to be understood that $\|x\|_1 = \sum_{i=1}^{s} |x_i|$ for any s-dimensional row-vector x.

Sometimes it is more convenient to consider condition (4.14) in the following equivalent form

$$|1 + b(\theta)^T(I - ZA)^{-1}(Z + W)e| \leq 1 \quad \forall \mathrm{Re}(z_i) \leq -|w_i|, \tag{4.15}$$

where also $W = \mathrm{diag}(w_1, \dots, w_s)$ denotes a diagonal matrix with complex coefficients.

Checking condition (4.14) or, equivalently, (4.15) is not an easy task. Unfortunately, so far no equivalent easier conditions to handle have been found. This happens in spite of the fact that (4.14) and (4.15) are modifications of the condition for AN-stability of RK methods for ODEs (to see this, put W=0 in (4.15)) for which, as is well known, *algebraic stability* is an equivalent condition which is very easy to verify. As a consequence, very few $\mathrm{AN_f}$-stable methods have been found.

In the class of 1-stage continuous RK methods of order 1, the only one which is $\mathrm{AN_f}$-stable is the *backward Euler* method with linear interpolation, that is the θ-method for $\omega = 1$ presented in Section 4.1. It gives rise to the following GPN-stable RK method for DDEs

$$Y_{n+1}^1 = y_n + h_{n+1} f(t_{n+1}, Y_{n+1}^1, \eta(t_{n+1} - \tau)),$$
$$\eta(t_n + \theta h_{n+1}) = y_n + \theta h_{n+1} f(t_{n+1}, Y_{n+1}^1, \eta(t_{n+1} - \tau)).$$

In the class of 2-stage continuous RK methods of order 2, the only one which is $\mathrm{AN_f}$-stable is the *Lobatto III-C* method with linear interpolation. It gives rise to the following GPN-stable RK method for DDEs

$$Y_{n+1}^1 = y_n + \frac{1}{2} h_{n+1} \left(f(t_n, Y_{n+1}^1, \eta(t_n - \tau)) - f(t_{n+1}, Y_{n+1}^2, \eta(t_{n+1} - \tau)) \right),$$
$$Y_{n+1}^2 = y_n + \frac{1}{2} h_{n+1} \left(f(t_n, Y_{n+1}^1, \eta(t_n - \tau)) + f(t_{n+1}, Y_{n+1}^2, \eta(t_{n+1} - \tau)) \right),$$

$$\eta(t_n + \theta h_{n+1}) = y_n + \frac{1}{2} \theta h_{n+1} \left(f(t_n, Y_{n+1}^1, \eta(t_n - \tau)) \right.$$
$$\left. + f(t_{n+1}, Y_{n+1}^2, \eta(t_{n+1} - \tau)) \right).$$

As a negative result, Zennaro [82] proved that no 3-stage discrete RK methods of order 3 with abscissae in $[0, 1]$ exist which are $\mathrm{AN_f}$-stable. This makes us guess a very restrictive order barrier.

The first part of Theorem 4.15 already says that AN-stability of the underlying discrete RK method is not sufficient to guarantee PN-stability. But there is something more: in [67] it is shown how one-step collocation at 1 Gaussian point (also known as the one-step *midpoint rule*), which is AN-stable, gives rise to a numerical solution $\{y_n\}_{n\geq 0}$ which blows up as $n \to \infty$ for a particular test equation (2.3) with $\lambda(t) = -\mu(t) \leq 0$. In line with this counterexample, in 't Hout [37] proved the following general result.

Theorem 4.17 *Assume that the RK method for DDEs (3.11)–(3.12) is nonconfluent and that* 0 *and* 1 *are not both abscissae of the method. Moreover, assume that, when it is applied to any test equation (2.3) satisfying (2.19) with any constant delay* τ *and any initial function* $\varphi(t)$*, it produces a discrete numerical solution* $\{y_n\}_{n\geq 0}$ *which is bounded for any constant* *step size $h > 0$ under the constraint (4.1). Then the underlying discrete RK method (3.3)–(3.4) is AN_f-stable.*

This result is important, since it states that, for a wide class of RK methods for DDEs, AN_f-stability of the underlying discrete RK method is a necessary condition even just to assure the boundedness of the numerical solutions of the test equation (2.3) satisfying (2.19), no matter what interpolation procedure is used. A less general version of Theorem 4.17, valid for natural RK methods for DDEs (in particular, for collocation methods), is given in [40].

4.3 RN-stability and GRN-stability

Now we analyze the stability properties of the RK methods for DDEs with respect to the general nonlinear system (1.2) under condition (2.13). The theory is very similar to the one in Section 4.2. Therefore, we just give the definitions and the main results, which can be found, once again, in [66] and [11].

Definition 4.18 *A numerical step-by-step method for DDEs is RN-stable if the discrete numerical solutions $\{y_n\}_{n\geq 0}$ and $\{z_n\}_{n\geq 0}$ of any pair of systems (1.2) and (2.10) satisfying (2.13) are such that*

$$\|y_n - z_n\| \leq \max_{x \leq t_0} \|\varphi(x) - \psi(x)\|, \quad n \geq 0, \tag{4.16}$$

for all constant delays τ and all initial functions $\varphi(t)$ and $\psi(t)$ and for any constant step size $h > 0$ under the constraint (4.1).

Definition 4.19 *A numerical step-by-step method for DDEs is GRN-stable if the discrete numerical solutions $\{y_n\}_{n\geq 0}$ and $\{z_n\}_{n\geq 0}$ of any pair of systems (1.2) and (2.10) satisfying (2.13) are such that (4.16) holds for all constant delays τ and all initial functions $\varphi(t)$ and $\psi(t)$ and for any constant step size $h > 0$.*

It is clear that RN-stability and GRN-stability are stronger concepts than PN-stability and GPN-stability, respectively, since they are based on more general test equations.

Again, it turns out that the BN-stability of the underlying discrete method is not sufficient to guarantee the above stability properties, no matter what interpolation procedure is used. Therefore, in this case too, another stronger concept of stability for ODE methods, based on the test systems (2.15)–(2.16) with forcing terms, has to be involved.

Definition 4.20 *The RK method (3.3)–(3.4) is* $\mathrm{BN_f}$*-stable if, under conditions (2.17) and (2.18), the numerical solutions $\{y_n\}_{n\geq 0}$ and $\{z_n\}_{n\geq 0}$ of (2.15) and (2.16) satisfy*

$$\|y_{n+1} - z_{n+1}\| \leq \max\{\|y_n - z_n\|, \max_{1\leq i\leq \nu} (|r(t^i_{n+1})| \cdot \|u(t^i_{n+1}) - v(t^i_{n+1})\|)\}$$

for any mesh Δ.

Definition 4.21 *The continuous RK method (3.3)–(3.6)–(3.5) is semi-*$\mathrm{BN_f}$*-stable if, under conditions (2.17) and (2.18), the numerical solutions* $\eta(t)$ *and* $\gamma(t)$ *of (2.15) and (2.16) satisfy*

$$\begin{aligned} &\|\eta(t_n + \theta h_{n+1}) - \gamma(t_n + \theta h_{n+1})\| \\ \leq &\max\{\|y_n - z_n\|, \max_{1\leq i\leq s}(|r(t_{n+1}^i)| \cdot \|u(t_{n+1}^i) - v(t_{n+1}^i)\|)\} \end{aligned} \quad (4.17)$$

for $\theta = c_1, \ldots, c_s, 1$, *and for any mesh* Δ.

The method is $\mathrm{BN_f}$*-stable if (4.17) holds for all* $\theta \in [0, 1]$.

An analogous result to Theorem 4.15 holds.

Theorem 4.22 *If the RK method for DDEs (3.11)–(3.12) is RN-stable, then the underlying discrete RK method (3.3)–(3.4) is* BN_f*-stable. Conversely, if the underlying continuous RK method (3.3)–(3.6)–(3.5) is (semi-) BN_f-stable, then the RK method for DDEs (3.11)–(3.12) is (RN-stable) GRN-stable.*

It is well known that, for a nonconfluent RK method (3.3)–(3.4), the concepts of AN-stability and BN-stability are equivalent. On the other hand, so far no equivalence has been proved between $\mathrm{AN_f}$-stability and $\mathrm{BN_f}$-stability. Nevertheless, the only methods which are known to be $\mathrm{AN_f}$-stable are also known to be $\mathrm{BN_f}$-stable. Therefore we can conclude that the backward Euler method with linear interpolation and the Lobatto III-C method with linear interpolation are GRN-stable.

4.4 Further stability investigations

It could be possible to extend the stability analyses presented in Sections 4.2 and 4.3 to the same test equations with variable delay $\tau(t)$ and/or completely free meshes Δ. So, appearantly, stronger stability concepts would be introduced. On the other hand, it is easy to generalize Theorems 4.15 and 4.22 to get the following results.

Theorem 4.23 *If the underlying continuous RK method is* $\mathrm{AN_f}$*-stable, then the numerical solution* $\eta(t)$ *of any test equation (2.3) satisfying (2.19) is such that*

$$|\eta(t)| \leq \max_{x\leq t_0} |\varphi(x)|, \quad t \geq t_0,$$

for all variable delays $\tau(t)$, *all initial functions* $\varphi(t)$ *and all meshes* Δ.

Theorem 4.24 *If the underlying continuous RK method is* $\mathrm{BN_f}$*-stable, then the numerical solutions* $\eta(t)$ *and* $\gamma(t)$ *of any pair of systems (1.2) and (2.10) satisfying (2.13) are such that*

$$\|\eta(t) - \gamma(t)\| \leq \max_{x\leq t_0} \|\varphi(x) - \psi(x)\|, \quad t \geq t_0,$$

for all variable delays $\tau(t)$, all initial functions $\varphi(t)$ and $\psi(t)$ and all meshes Δ.

In view of these generalizations and of the necessary condition given by Theorem 4.17, we can conclude that, for a large class of RK methods, the only possible differences in the stability properties of the RK methods for DDEs with respect to the test equations considered in Sections 4.2 and 4.3 may be determined by the stability properties of the underlying interpolant.

Indeed, an interesting open problem is whether or not the interpolant is also subject to a necessary condition for boundedness of the numerical solution similar to the one expressed by Theorem 4.17 for the discrete underlying method.

Other important stability investigations have been done for numerical methods with respect to the test equation (2.27). For example, Lu [51] obtained results on θ-methods when the matrices L and M are diagonal and reverse diagonal, respectively, whereas in 't Hout [41] considered the general case where L and M are arbitrary matrices. He found that, if constant step size $h > 0$ is used, the θ-methods with $\frac{1}{2} \leq \omega \leq 1$ yield a numerical solution which is asymptotically stable whenever the sufficient condition (2.28) holds. To do this, he used again Lemma 4.9.

We conclude this section about stability by mentioning some investigations based on the pure delay test equations (2.7) and (2.8). Barwell [6] gave the definition of *Q-stability* and *GQ-stability* with respect to (2.8). This test equation was also considered by van der Houwen and Sommeijer [69] to analyze stability properties of LM methods. Finally, Torelli and Vermiglio [68] gave the definition of *QN_0-stability* with respect to (2.7) with real coefficients and found some conditions sufficient to guarantee this property for the continuous quadrature rule (3.18).

5 More general classes of delay differential equations

In this section we consider more general DDEs of the form (1.2) not subject to the restrictions on the delay τ imposed in Section 2. Moreover, we consider also equations which are not included in the formulation (1.2).

For each class of equations we say something about the theory and something about the numerical methods. However, we never go into much detail. In fact, our aim in this section is just to make the reader acquainted of some of the main problems connected with these more general DDEs and provide a list of references. Some issues in the subject are highlighted and discussed at a deeper level by Baker, Paul and Willé [4].

5.1 Equations with vanishing delay

We consider the DDE (1.2) where the delay τ is variable but not state dependent and may vanish for some $t \in [t_0, t_f]$. So the delay is subject to the sole condition $\tau(t) \geq 0$.

First of all we remark that also in this case the existence and uniqueness of solutions is assured, although the proof is not as trivial as under the hypotheses made in Section 2.

In order to give an idea of what may happen, we consider the following sample situation:

- the delayed argument $t - \tau(t)$ is a strictly increasing function for all $t \in [t_0, t_f]$;
- there exists a point $t^* \in (t_0, t_f)$ such that $\tau(t) > 0$ for all $t \neq t^*$ and $\tau(t^*) = 0$.

Under these assumptions, we are interested in the location of the discontinuity points ξ_k. It is clear that in the interval $[t_0, t^*)$ the discontinuity points are still generated inductively by the recursion (2.1). Anyway, although they still form an increasing sequence by virtue of the monotonicity of the delayed argument $t - \tau(t)$, they accumulate at the point t^* because the delay $\tau(t)$ vanishes at t^*, which is a fixed point of (2.1). On the other hand, to the right of t^* no other discontinuity points exist. In fact, the monotonicity of $t - \tau(t)$ and the fact that t^* is a fixed point of (2.1) again do not allow the limit discontinuity point t^* to propagate forward any further.

In such a situation, there is one main problem for the numerical treatment of (1.2). It consists of the fact that, as the integration proceeds towards t^*, the discontinuity points become closer and closer so that, at some point, the distance between two of them becomes smaller than the step size used by the numerical method. The consequence is that, if $[t_n, t_{n+1}]$ is the underlying step of integration, an approximation $\eta(t-\tau)$ of the delayed part $y(t-\tau)$ is already known only in a first part of the step. So, with reference to the RK method for DDEs (3.11)–(3.12), some of the quantities $\eta(t_{n+1}^j - \tau)$ might be unknown and should be computed together with the stage values Y_{n+1}^i. Thus, if we assume that $0 \leq c_1 \leq c_2 \leq \ldots \leq c_s \leq 1$ and if we define $c_0 = 0$ and the integer r, $0 \leq r \leq s-1$, such that $t_{n+1}^r - \tau \leq t_n < t_{n+1}^{r+1} - \tau$, then (3.11)–(3.12) could be modified into

$$Y_{n+1}^i = y_n + h_{n+1} \sum_{j=1}^{r} a_{ij} f(t_{n+1}^j, Y_{n+1}^j, \eta(t_{n+1}^j - \tau)) \qquad (5.1)$$

$$+ h_{n+1} \sum_{j=r+1}^{s} a_{ij} f(t_{n+1}^j, Y_{n+1}^j, Z_{n+1}^j), \quad i = 1, \ldots, s,$$

$$Z_{n+1}^i = y_n + h_{n+1} \sum_{j=1}^{r} b_j(\theta_{n+1}^i) f(t_{n+1}^j, Y_{n+1}^j, \eta(t_{n+1}^j - \tau)) \qquad (5.2)$$

$$+ \quad h_{n+1} \sum_{j=r+1}^{s} b_j(\theta_{n+1}^i) f(t_{n+1}^j, Y_{n+1}^j, Z_{n+1}^j),\ i = r+1, \ldots, s,$$

$$\eta(t_n + \theta h_{n+1}) = y_n \quad + \quad h_{n+1} \sum_{i=1}^{r} b_i(\theta) f(t_{n+1}^i, Y_{n+1}^i, \eta(t_{n+1}^i - \tau)) \qquad (5.3)$$

$$+ \quad h_{n+1} \sum_{i=r+1}^{s} b_i(\theta) f(t_{n+1}^i, Y_{n+1}^i, Z_{n+1}^i), \quad 0 \le \theta \le 1,$$

where $\theta_{n+1}^i = \frac{t_{n+1}^i - \tau - t_n}{h_{n+1}}$, $i = r+1, \ldots, s$, and the unknown quantities $Z_{n+1}^i = \eta(t_{n+1}^i - \tau)$ play the role of additional stage values, related to the delayed term. It is clear that, even if the underlying RK method (3.3) is explicit, the resulting method (5.1)–(5.2)–(5.3) is implicit in the additional stage values Z_{n+1}^i. However, the number of stages is only apparently increased. In fact, by using the K-notation, which is obtained by setting $K_{n+1}^i = f(t_{n+1}^i, Y_{n+1}^i, \eta(t_{n+1}^i - \tau))$ for $= 1, \ldots, r$, and $K_{n+1}^i = f(t_{n+1}^i, Y_{n+1}^i, Z_{n+1}^i)$ for $i = r+1, \ldots, s$, we easily get a system in the K_{n+1}^i's alone.

A minor problem is that one or more (or even infinitely many) discontinuity points ξ_k lie inside the underlying step $[t_n, t_{n+1}]$. This fact may cause a break down in the order p of accuracy of the method as long as the index k of the smallest of them satisfies the inequality $k < p$. To overcome this, we recall that the solution $y(t)$ becomes one order smoother at each discontinuity point. It is clear that, at some point, the step $[t_n, t_{n+1}]$ contains all the rest of the sequence of the discontinuity points, including the limit t^*. In particular, if $t_{n+1} = t^*$, in the next step $[t_{n+1}, t_{n+2}]$ we have to apply formulae (5.1)–(5.2)–(5.3) with $r = 0$. Afterwards, the integration may proceed without any problem following the usual strategy explained in Section 3.3.

In some cases the delay τ vanishes already at t_0, and afterwards it is strictly positive. Thus no discontinuity points are spread ahead and the solution $y(t)$ is determined just by the initial value at t_0. For a theoretical analysis of DDEs of this type, see, for example, Iserles [44] and Feldstein, Iserles and Levin [28]. From the numerical point of view, in this situation we have problems only in the first step of integration.

Example 5.1 A particular important example of DDEs with vanishing delay is

$$\begin{cases} y'(t) = \lambda y(t) + \mu y(\alpha t), & t \ge 0, \\ y(0) = y_0, \end{cases} \qquad (5.4)$$

where λ and μ are complex numbers and $0 < \alpha < 1$. Some authors call (5.4) the *pantograph equation.*

A detailed theoretical study of equation (5.4) has been done by Kato and McLeod [47], whereas stability properties of LM methods for its numerical solution have been investigated, for example, by Bakke and Jackiewicz [5].

Now consider the more general case in which the retarded argument $t - \tau(t)$ is not monotone.

Again, we are interested in the propagation of the discontinuity points ξ_k. The difference from the monotone case consists of the possible non-uniqueness of the solution to equation (2.1). That means that each discontinuity point may generate, in turn, more than one such point on its right hand side. Moreover, if the delay vanishes for some t^*, then we have also the phenomenon of the accumulation of discontinuity points. It is clear that, in such a situation, the a priori control of the location of the discontinuity points becomes very difficult, although still possible.

From the numerical point of view, it is quite clear that a constrained mesh strategy is almost impossible for all kinds of DDEs with vanishing delay, and that the underlying continuous method must have uniform order p or, at least, $p-1$, according to the step size selection mechanism.

5.2 Neutral delay differential equations

Now we consider DDEs of the form

$$\begin{cases} y'(t) = f(t, y'(t-\tau), y(t), y(t-\tau)), & t \geq t_0, \\ y(t) = \varphi(t), \quad t \leq t_0, \end{cases} \tag{5.5}$$

which are called *neutral* delay differential equations (NDDEs).

For NDDEs too, we distinguish between the case in which the two hypotheses of positivity and monotonicity of the delay made in Section 2 hold and the case in which they do not. If the hypotheses hold, then it is easily seen that, again, existence and uniqueness of the solution are guaranteed if the right hand side function $f(t, w, y, x)$ is sufficiently smooth. Moreover, it is clear that the problem of the location of the discontinuity points is also unchanged. Nevertheless, the presence of the neutral term $y'(t-\tau)$ prevents the smoothing of the solution at the discontinuity points ξ_k as the index k increases.

From the numerical point of view, one has to take care of the approximation of the derivative of the delayed term. This may be done by using the derivative of the interpolant used for the approximation of the solution or, alternatively, another type of approximation (see, for example, Jackiewicz [46]). So, formulae (3.11)–(3.12) modify to become the following *RK method for NDDEs*:

$$Y_{n+1}^i = y_n + h_{n+1} \sum_{j=1}^{s} a_{ij} f(t_{n+1}^j, \psi(t_{n+1}^j - \tau), Y_{n+1}^j, \eta(t_{n+1}^j - \tau)), \; i = 1, \ldots, s,$$

$$\eta(t_n + \theta h_{n+1}) = y_n + h_{n+1} \sum_{i=1}^{s} b_i(\theta) f(t_{n+1}^i, \psi(t_{n+1}^i - \tau), Y_{n+1}^i, \eta(t_{n+1}^i - \tau)),$$

where $\psi(t-\tau)$ is an approximation to $y'(t-\tau)$, possibly being $\psi(t) = \eta'(t)$. Of course, for $t \leq t_0$ we define $\eta(t) = \varphi(t)$ and $\psi(t) = \varphi'(t)$.

Once again, for a free mesh strategy, it is important that the approximation $\psi(t)$ to the derivative of the solution $y'(t)$ also be of uniform order p or, at least, $p-1$. For this reason we may create problems by using the derivative of the interpolant $\eta'(t)$ because, whenever we differentiate the interpolant, we loose one order of accuracy. On the other hand, if we use RK methods in a constrained mesh strategy with a NCE $\eta(t)$ of degree d as the underlying interpolant and its derivative $\eta'(t)$ for the approximation of the derivative of the solution, then, by virtue of Theorem 3.4, we have again a result analogous to Corollary 3.6, that is (3.15) and (3.16) hold.

Particular attention has been given to the analysis of the stability of numerical methods with respect to the following neutral test equation with constant delay, which is a generalization of (2.27)

$$\begin{cases} y'(t) = Ky'(t-\tau) + Ly(t) + My(t-\tau), & t \geq t_0, \\ y(t) = \varphi(t), & t \leq t_0, \end{cases} \tag{5.6}$$

where K, L and M are constant $m \times m$-matrices. Brayton and Willoughby [16] considered the case where all the matrices K, L and M are real and symmetric, whereas, more generally, Tian, Kuang and Xiang [65] considered the case of arbitrary complex matrices K, L and M. In both cases a sufficient condition has been given on the matrices in order that the equation be asymptotically stable. For $K = 0$, the condition derived in [65] reduces to the condition (2.28) derived in [41] for the asymptotic stability of (2.27). As far as numerical methods are concerned, in both papers the authors considered the stability of the θ-methods and, as in [41], in both cases it turns out that the θ-methods with $\frac{1}{2} \leq \omega \leq 1$ yield a numerical solution which is asymptotically stable whenever the sufficient condition for the asymptotic stability of the test equation (5.6) holds, provided that constant step size $h > 0$ is used. Finally, we mention the results by Bellen, Jackiewicz and Zennaro [9], who consider the test equation (5.6) where K, L and M are complex numbers. They introduced a generalization of the definition of P-stability and, following the line in [80], they made a similar complete analysis of the stability properties of the RK methods for NDDEs.

If the delay does not satisfy the hypotheses of positivity and monotonicity, then we have the same problems mentioned in the previous Section 5.1 for the location of the discontinuity points. In addition, since there is no smoothing of the solution at the discontinuity points ξ_k as the index k increases, the eventual accumulation of discontinuity points is a serious problem for the maintenance of the accuracy order of the numerical methods. Moreover, the existence and uniqueness of the solution is no longer such a trivial matter and, in general, may not be guaranteed.

In the last few years a deep theoretical analysis of NDDEs with vanishing delay has been done (see, for example, Iserles [43] and Iserles and Terjéki [45]).

Example 5.2 Particular emphasis has been given to the following equation, called the generalized pantograph equation, which is a generalization of the pantograph equation (5.4):

$$\begin{cases} y'(t) = Ky'(\alpha t) + Ly(t) + My(\alpha t), & t \geq 0, \\ y(0) = y_0, \end{cases} \tag{5.7}$$

where K, L and M are complex matrices and $0 < \alpha < 1$.

Numerical methods for the generalized pantograph equation (5.7) with particular stress on their stability properties have been studied by Buhmann and Iserles [18, 19] and Buhmann, Iserles and Nørsett [20].

5.3 Equations with state dependent delays

Finally, we briefly consider the most difficult case of DDEs (1.2), that is when the delay has the form $\tau(t, y(t))$. In such a situation, it is not possible to locate the discontinuity points a priori without any knowledge of the solution. For this problem, a good reference is Neves and Feldstein [55].

Of course, the impossibility of detecting the discontinuity points a priori makes the numerical solution of (1.2) rather a complicated task, because it becomes very hard to be able to include the discontinuity points in the mesh Δ, which is important for accuracy requirements.

Consider in a little more detail the case of strictly positive delay $\tau(t, y) \geq \tau_0 > 0$. It is convenient to choose the step size $h_{n+1} \leq \tau_0$, so that the delayed term is already known. A first adaptation of the RK formulae (3.11)–(3.12) is

$$Y_{n+1}^i = y_n + h_{n+1} \sum_{j=1}^{s} a_{ij} f(t_{n+1}^j, Y_{n+1}^j, \eta(t_{n+1}^j - \tau(t_{n+1}^j, Y_{n+1}^j))), \; i = 1, \ldots, s,$$

$$\eta(t_n + \theta h_{n+1}) = y_n + h_{n+1} \sum_{i=1}^{s} b_i(\theta) f(t_{n+1}^i, Y_{n+1}^i, \eta(t_{n+1}^i - \tau(t_{n+1}^i, Y_{n+1}^i))).$$

Along with the above RK formulae, in order to approximate the discontinuity points, we can use a rootfinding routine to solve in sequence the equations

$$\xi_k - \tau(\xi_k, \eta(\xi_k)) = \xi_{k-1}, \quad k \geq 1, \tag{5.8}$$

making use of the already computed continuous approximation $\eta(t)$ to the solution $y(t)$. However, besides the possible non-uniqueness of the solution (in fact, in general the delayed argument $t - \tau(t, \eta(t))$ is not monotone), there is another delicate problem to face: whenever we have solved equation (5.8) and hence found the approximation to the next discontinuity point ξ_k, we have necessarily already passed such point ξ_k with our numerical integration. Therefore, unless it has been accidentally included in the mesh Δ, a loss of accuracy has possibly occurred, which might have compromised the reliability of the discontinuity location procedure itself. In order to overcome this difficulty, more sophisticated strategies are necessary.

In general, the delay $\tau(t,y)$ is also allowed to vanish, and therefore the situation is much more complicated.

For a more detailed discussion about the numerical detection and computation of the discontinuity points, see Willé [73] and Willé and Baker [74, 75, 77].

5.4 Equations with several delays

Sometimes in applications we may find DDEs or NDDEs where the right hand side depends on more than one retarded argument, that is equations such as

$$\begin{cases} y'(t) = f(t, y(t), y(t-\tau_1), \ldots, y(t-\tau_r)), & t \geq t_0, \\ y(t) = \varphi(t), \quad t \leq t_0, \end{cases}$$

or

$$\begin{cases} y'(t) = f(t, y'(t-\tau_1), .., y'(t-\tau_r), y(t), y(t-\tau_1), .., y(t-\tau_r)), \; t \geq t_0, \\ y(t) = \varphi(t), \quad t \leq t_0, \end{cases}$$

respectively.

Although there exist some papers where the authors stress the presence of more than one delay, there are no particular additional difficulties with respect to (1.2) and (5.5). This is true as long as all the delays τ_i, $i = 1, \ldots, r$, are of the same type, in which case existence and uniqueness theorems, as well as stability results, modify easily to the more general situation.

The main complication lies in the possibly more chaotic proliferation of the discontinuity points. For example, even if all the delays satisfy the positivity and monotonicity conditions imposed in Section 2, each discontinuity point generates, in turn, another r such points. Thus, if we denote by ξ_{k-1}^j a discontinuity point of *level* $k-1$, then we find the corresponding discontinuity points of level k by solving the r equations

$$\xi_k^{(j-1)r+i} - \tau_i(\xi_k^{(j-1)r+i}) = \xi_{k-1}^j, \quad i = 1, \ldots, r.$$

The only point of level 0 is $\xi_0^1 = t_0$. It may happen that two or more discontinuity points, possibly of different levels, coincide.

From the numerical point of view, the constrained mesh strategy may become impracticable even if the delays are all constant. This is the case when the ratio between two of them is not a rational number. However, in the implementation of the numerical methods, in most cases the difference with respect to the one-delay case is technical rather than conceptual.

Bibliography

1. Al-Mutib, A.N. (1984). One-step implicit methods for solving delay differential equations. *Intern. J. Computer Math.*, **16**, 157–168.
2. Arndt, H. (1984). Numerical solution of retarded initial value problems: local and global error and step size control. *Numer. Math.*, **43**, 343–360.
3. Baker, C.T.H., Butcher, J.C. and Paul, C.A.H. (1992). *Experience of STRIDE applied to delay differential equations.* NA Report 208, Dept. of Mathematics, University of Manchester.
4. Baker, C.T.H., Paul, C.A.H. and Willé, D.R. (1994). Issues in the numerical solution of evolutionary delay differential equations. In preparation
5. Bakke, V.L. and Jackiewicz, Z. (1986). Stability analysis of linear multistep methods for delay differential equations. *Internat. J. Math. Math. Sci.*, **9**, 447–458.
6. Barwell, V.K. (1975). Special stability problems for functional differential equations. *BIT*, **15**, 130–135.
7. Bellen, A. (1984). One-step collocation for delay differential equations. *J. Comput. Appl. Math.*. **10**, 275–283.
8. Bellen, A. (1985). Constrained mesh methods for functional differential equations. In *Delay Equations, Approximation and Application. ISNM*, **74**, 52–70.
9. Bellen, A., Jackiewicz, Z. and Zennaro, M. (1988). Stability analysis of one-step methods for neutral delay-differential equations. *Numer. Math.*, **52**, 605–619.
10. Bellen, A. and Zennaro, M. (1985). Numerical solution of delay differential equations by uniform corrections to an implicit Runge–Kutta method. *Numer. Math.*, **47**, 301–316.
11. Bellen, A. and Zennaro, M. (1992). Strong contractivity properties of numerical methods for ordinary and delay differential equations. *Appl. Numer. Math.*, **9**, 321–346.
12. Bellman, R. (1961). On the computational solution of differential-difference equations. *J. Math. Anal. Appl.*, **2**, 108–110.
13. Bellman, R. and Cook, K.L. (1965). On the computational solution of a class of functional differential equations. *J. Math. Anal. Appl.*, **12**, 495–500.
14. Bickart, T.A. (1982). P-stable and P$[\alpha, \beta]$-stable integration / interpolation methods in the solution of retarded differential- difference

equations. *BIT*, **22**, 464–476.

15. Bock, H.G. and Schlöder, J. (1981). Numerical solution of retarded differential equations with state dependent time lags. *ZAMM*, **61**, 269–271.
16. Brayton, R.K. and Willoughby, R.A. (1967). On the numerical integration of a symmetric system of difference-differential equations of neutral type. *J. Math. Anal. Appl.*, **18**, 182–189.
17. Buchacker, U. and Filippi, S. (1989). Step size control for delay differential equations using a pair of Runge–Kutta formulae. *J. Comput. Appl. Math.*, **26**, 339–343.
18. Buhmann, M.D. and Iserles, A. (1992). On the dynamics of a discretized neutral equation. *IMA J. Numer. Anal.*, **12**, 339–363.
19. Buhmann, M.D. and Iserles, A. (1993). Stability of the discretized pantograph differential equation. *Math. Comp.* **60**, 575–589.
20. Buhmann, M.D., Iserles, A. and Nørsett, S.P. (1993). Runge–Kutta methods for neutral differential equations. In *Contributions in Numerical Mathematics*, edited by R.P. Agarwal. World Scientific Series in Applicable Analysis, World Scientific, Singapore.
21. Butcher, J.C. (1987). *The numerical analysis of ordinary differential equations.* Wiley, London.
22. Butcher, J.C. (1992). The adaptation of STRIDE to delay differential equations. *Appl. Numer. Math.* **9**, 415–425.
23. Cryer, C.W. (1972). Numerical methods for functional differential equations. In *Delay and functional differential equations and their applications*, edited by K. Schmitt. Academic Press, New York.
24. Dekker, K. and Verwer, J.G. (1984). *Stability of Runge–Kutta methods for stiff nonlinear differential equations.* North Holland, Amsterdam.
25. Driver, R.D. (1977). *Ordinary and delay differential equations.* Springer-Verlag, Berlin.
26. El'sgol'ts, L.E. and Norkin S.B. (1973). *Introduction to the theory and application of differential equations with deviating arguments.* Academic Press, New York.
27. Feldstein, A. (1964) *Discretization methods for retarded ordinary differential equation.* Ph.D. Thesis, Department of Mathematics, UCLA, Los Angeles.
28. Feldstein, A., Iserles, A. and Levin, D. (1991). *Embedding of delay equations into an infinite-dimensional ODE system.* Report NA21, Dept. of Applied Mathematics and Theoretical Physics, University of Cambridge.
29. Feldstein, A. and Neves, K.W. (1984). High order methods for state-dependent delay differential equations with nonsmooth solutions.

SIAM J. Numer. Anal., **21**, 844–863.

30. Gohberg, I., Lancaster, P. and Rodman, L. (1982). *Matrix polynomials.* Academic Press, New York.
31. Hairer, E., Nørsett, S.P. and Wanner, G. (1993). *Solving ordinary differential equations I, Nonstiff problems.* Springer-Verlag, Berlin.
32. Hairer, E. and Wanner, G. (1993). *Solving ordinary differential equations II, Stiff and differential algebraic problems.* Springer-Verlag, New York.
33. Hale, J.K. (1977). *Theory of functional differential equations.* Springer, New York.
34. Hale, J.K. (1986). Homoclinic orbits and chaos in delay equations. In *Proceedings of the Ninth Dundee Conference on Ordinary and Partial Differential Equations*, edited by B.D. Sleeman and R.J. Jarvis. John Wiley and Sons, New York.
35. Hayashi, H. (1994). Numerical solution for delay differential equations. Algorithm and numerical results. In preparation.
36. Higham, D. (1993). Error control for initial value problems with discontinuities and delays. *Appl. Numer. Math.*, **12**, 315–330.
37. in 't Hout, K.J. (1989). Unpublished handwritten notes.
38. in 't Hout, K.J. (1992). The stability of a class of Runge–Kutta methods for delay differential equations. *Appl. Numer. Math.*, **9**, 347–355.
39. in 't Hout, K.J. (1992). A new interpolation procedure for adapting Runge–Kutta methods to delay differential equations. *BIT*, **32**, 634–649.
40. in 't Hout, K.J. (1992). *Runge–Kutta methods in the numerical solution of delay differential equations.* Ph.D. Thesis, The University of Leiden.
41. in 't Hout, K.J. (1993). *The stability of θ-methods for systems of delay differential equations.* Report Series n. 282, Dept. of Mathematics and Statistics, The University of Auckland.
42. in 't Hout, K.J. and Spijker, M.N. (1991). Stability analysis of numerical methods for delay differential equations. *Numer. Math.*, **59**, 807–814.
43. Iserles, A. (1993). On the generalized pantograph functional- differential equation. *Europ. J. Appl. Math.*, **4**, 1–38.
44. Iserles, A. (1992). *On nonlinear delay differential equations.* Report NA13, Dept. of Applied Mathematics and Theoretical Physics, University of Cambridge.
45. Iserles, A. and Terjéki, J. (1992). *Stability and asymptotic stability of functional-differential equations.* Report NA1, Dept. of Applied Mathematics and Theoretical Physics, University of Cambridge.

46. Jackiewicz, Z. (1984). One-step methods of any order for neutral functional differential equations. *SIAM J. Numer. Anal.*, **21**, 486–511.
47. Kato, T. and McLeod, J.B. (1971). The functional-differential equation $y'(x) = ay(\lambda x) + by(x)$. *Bull. Amer. Math. Soc.*, **77** 891–937.
48. Kolmanovskii, V. and Myshkis, A. (1992). *Applied theory of functional differential equations.* Kluwer Academic Publishers, Dordrecht.
49. Lambert, J.D. (1991). *Numerical methods for ordinary differential equations.* Wiley, Chichester.
50. Liu, M.Z. and Spijker, M.N. (1990). The stability of the θ-methods in the numerical solution of delay differential equations. *IMA J. Numer. Anal.*, **10**, 31–48.
51. Lu, L. (1991). Numerical stability of the θ-methods for systems of differential equations with several delay terms. *J. Comput. Appl. Math.*, **34**, 291–304.
52. Meinardus, G. and Nürnberger, G. (1985). Approximation theory and numerical methods for delay differential equations. *ISNM*, **74**, 13–40.
53. Neves, K.W. (1975). Automatic integration of functional differential equations: an approach. *ACM Trans. Math. Software*, **1**, 357–368.
54. Neves, K.W. (1975). Algorithm 497 — automatic integration of functional differential equations. *ACM Trans. Math. Software*, **1**, 369–371.
55. Neves, K.W. and Feldstein, A. (1976). Characterization of jump discontinuities for state dependent delay differential equations. *J. Math. Anal. Appl.*, **56**, 689–707.
56. Neves, K.W. and Thompson, S. (1992). Software for the numerical solution of systems of functional differential equations with state dependent delays. *Appl. Numer. Math.*, **9**, 385–401.
57. Nørsett, S.P. and Wanner, G. (1979). The real-pole sandwich for rational approximation and oscillation equations. *BIT*, **19**, 79–94.
58. Oberle, H.J. and Pesh, H.J. (1981). Numerical treatment of delay differential equations by Hermite interpolation. *Numer. Math.*, **37**, 235–255.
59. Oppelstrupp, J. (1978). The RKFHB4 method for delay-differential equations. *Lecture Notes in Mathematics*, **631**, 133–146.
60. Paul, C.A.H. (1992). Developing a delay differential equation solver. *Appl. Numer. Math.*, **9**, 403–414.
61. Paul, C.A.H. (1994). *A test set of functional differential equations.* NA Report 243, Dept. of Mathematics, University of Manchester.
62. Stetter, H.J. (1965). Numerische Lösung von Differentialgleichungen mit nacheilendem Argument. *ZAMM*, **45**, 79–80.
63. Tavernini, L. (1971). One-step methods for the numerical solution of Volterra functional differential equations. *SIAM J. Numer. Anal.*, **4**,

786–795.

64. Tavernini, L. (1978). The approximate solution of Volterra differential systems with state dependent time lags. *SIAM J. Numer. Anal.*, **15**, 1039–1052.
65. Tian, H., Kuang, J. and Xiang, J. (1993). On the stability of one-parameter methods for systems of neutral differential equations. Report Series n. 1, Dept. of Mathematics, Shanghai Normal University.
66. Torelli, L. (1989). Stability of numerical methods for delay differential equations. *J. Comput. Appl. Math.*, **25**, 15–26.
67. Torelli, L. (1991). A sufficient condition for GPN-stability for delay differential equations. *Numer. Math.*, **59**, 311–320.
68. Torelli, L. and Vermiglio, R. (1993). On the stability of continuous quadrature rules for differential equations with several constant delays. *IMA J. Numer. Anal.*, **13**, 291–302.
69. van der Houwen, P.J. and Sommeijer, B.P. (1984). Stability in linear multistep methods for pure delay equations. *J. Comput. Appl. Math.*, **10**, 55–63.
70. Vermiglio, R. (1985). A one-step subregion method for delay differential equations. *Calcolo*, **22**, 429–455.
71. Watanabe, D.S. and Roth, M.G. (1985). The stability of difference formulas for delay differential equations. *SIAM J. Numer. Anal.*, **22**, 132–145.
72. Weiner, R. and Strehmel, K. (1988). A type insensitive code for delay differential equations basing on adaptive and explicit Runge–Kutta interpolation methods. *Computing*, **40** 255–265.
73. Willé, D.R. (1989). *The numerical solution of delay-differential equations.* Ph. D. Thesis, University of Manchester.
74. Willé, D.R. and Baker, C.T.H. (1994). Step size control and continuity consistency for state-dependent delay differential equations. To appear in *J. Comput. Appl. Math.*.
75. Willé, D.R. and Baker, C.T.H. (1992). The tracking of derivative discontinuities in systems of delay differential equations. *Appl. Numer. Math.*, **9**, 209–222.
76. Willé, D.R. and Baker, C.T.H. (1992). DELSOL — a numerical code for the solution of systems of delay differential equations. *Appl. Numer. Math.* **9**, 223–234.
77. Willé, D.R. and Baker, C.T.H. (1994). *Some issues in the detection and location of derivative discontinuities in delay differential equations.* NA Report 238, Dept. of Mathematics, University of Manchester.
78. Zennaro, M. (1985). On the P-stability of one-step collocation for delay differential equations. *ISNM*, **74**, 334–343.

79. Zennaro, M. (1986). Natural Continuous Extensions of Runge–Kutta methods. *Math. Comp.*, **46**, 119–133.
80. Zennaro, M. (1986). P-stability properties of Runge–Kutta methods for delay differential equations. *Numer. Math.*, **49**, 305–318.
81. Zennaro, M. (1988). Natural Runge–Kutta and projection methods. *Numer. Math.*, **53**, 423–438.
82. Zennaro, M. (1993). Contractivity of Runge–Kutta methods with respect to forcing terms. *Appl. Numer. Math.*, **10**, 321–345.

The manufacturer's authorised representative in the EU for product safety is Oxford University Press España S.A. of El Parque Empresarial San Fernando de Henares, Avenida de Castilla, 2 - 28830 Madrid (www.oup.es/en or product.safety@oup.com). OUP España S.A. also acts as importer into Spain of products made by the manufacturer.
Printed and bound by CPI Group (UK) Ltd, Croydon, CR0 4YY
06/07/2026
02157629-0001